Inflamed Invisible

Sonics Series

Atau Tanaka, editor

Sonic Agency: Sound and Emergent Forms of Resistance, Brandon LaBelle

Meta Gesture Music: Embodied Interaction, New Instruments and Sonic Experience, Various Artists (CD and online)

Inflamed Invisible: Collected Writings on Art and Sound, 1976–2018, David Toop

Goldsmiths Press's Sonics series considers sound as media and as material – as physical phenomenon, social vector, or source of musical affect. The series maps the diversity of thinking across the sonic landscape, from sound studies to musical performance, from sound art to the sociology of music, from historical soundscapes to digital musicology. Its publications encompass books and extensions to traditional formats that might include audio, digital, online and interactive formats. We seek to publish leading figures as well as emerging voices, by commission or by proposal.

Inflamed Invisible

Collected Writings on Art and Sound, 1976–2018

David Toop

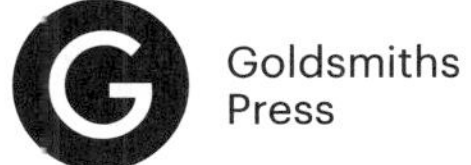

Goldsmiths
Press

Copyright © 2019 Goldsmiths Press
First published in 2019 by Goldsmiths Press
Goldsmiths, University of London, New Cross
London SE14 6NW

This paperback edition published in 2021 by Goldsmiths Press

Printed and bound by Versa Press, USA
Distribution by the MIT Press
Cambridge, Massachusetts, and London, England

Copyright © 2019 David Toop

A CIP record for this book is available from the British Library

ISBN 978-1-913380-62-5 (pbk)
ISBN 978-1-912685-16-5 (hbk)
ISBN 978-1-912685-17-2 (ebk)

www.gold.ac.uk/goldsmiths-press

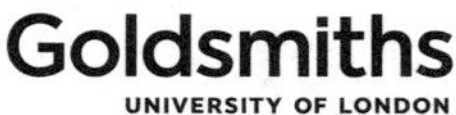

Contents

Note to the Reader

This is a book about music and sound. We wanted to bring the print text to sonic life. We have compiled a series of web links to take you to recordings of the music, musicians, and artists David Toop describes, as well as to artists' websites. These tracks and links are listed at the end of the book.

We have placed bar codes in the margins, so you can listen to the music written about as you read. These codes can be scanned by a smart-phone camera. On some phones, the built-in camera app will automatically recognize a code. On other phones, you would need to download a QR code reader app.

We have endeavored to find as much of the music online as possible whether it has been commercially released or not. Many of the links take you to the Discogs database. There, scrolling down, you will find links to videos and audio on YouTube. Other links take you to the artist's gallery website or personal site.

For the music that is commercially available, we have compiled an Inflamed Invisible playlist on the Spotify music streaming service. The playlist is accessible here:
https://open.spotify.com/user/atau/playlist/6ksANkcLBAuBVsSIgEKClc
Individual tracks from this playlist are seen as Spotify codes at the bottom of the page. To scan the codes and listen to the tracks, please download and use the Spotify app on your phone. Select the magnifying glass icon to search, click again on the search box, then select the camera icon and scan the code.

Wishing you a sonic reading and listening,

Atau Tanaka
Sonics series editor

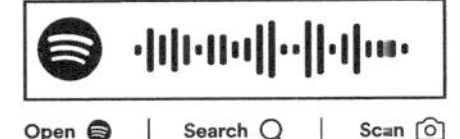

Introduction

Inflamed Invisible collects together in one volume a selection of my essays, reviews, interviews, profiles, lectures, blogs and experimental texts on the subject of sound and art – the ways in which these two categories of art making have been entangled both historically and in the present when this intensely close relationship is better known as a relatively new category called sound art.

What obsessed me in the early 1970s was a possibility that music might no longer be bounded by the formalities of an audience: the clapping, the booing, the drinking, the short attention span, the demand for instant gratification. Thinking more expansively about sound and listening as foundational practices in themselves led music into unknown and thrilling territory: stretched time, stretched fences and wilderness spaces, under the water of swimming pools, ruined buildings, video monitors, bus journeys, singing sculptures, tactile electronics, weather, meditations, vibration and the interior resonance of objects, autonomous sonic devices, interspecies communications, sine waves, explosions, improvisations, instructional texts, silent actions and the strange rituals of performance art.

The situation was complicated further by musicians seeking a foothold within the art world along with energetic hybridisations of previously self-contained art practices: poetry becoming electronic music, for example, or film becoming practice-as-theory.

When I began writing about art and sound outside my own practice there was no existing category explicitly named "sound art," only (still youthful) ancestors such as Alvin Lucier, Max Neuhaus, Atsuko Tanaka or Annea Lockwood whose work specialised in the behaviour and conditions of sounding phenomena. Art was in a state of flux with many emergent styles asserting themselves. It was clear that sound and listening practices were implicated in certain examples of video art, experimental film and theatre, performance art, text scores, land art, sculpture, kinetic art and site-specific installations of diverse forms. Even the most cursory glance at documented work by (among many others) Nam June Paik, Joan Jonas, Robert Rauschenberg, Marina Abramović, Joseph Beuys, Alison Knowles, John Latham, Yoko Ono, Michael Snow, Charlotte Moorman, Mieko Shiomi or Merce Cunningham shows up early examples.

This transition from music into less certain territory was so gradual as to be imperceptible. Speaking personally I felt an accumulation of stimuli, all of which set out challenges to orthodox categories of practice. They included viewings of Tony Conrad's *Flicker* at the Electric Cinema in Portobello Road, John Latham's *Speak* during the Hornsey College of Art sit-in of 1968, Michael Snow's *Wavelength* at the Kentish Town incarnation of the London Film-Makers' Co-op and Dick Fontaine's film of John Cage and Rahsaan Roland Kirk, *Sound??*, shown on British television in 1966. At Watford School of Art in 1968 I sat in a small music room and heard Christian Wolff perform his composition, *Stones*; on record I heard *The Glass World of Anna Lockwood* and at a Bond Street gallery watched one of Jean Tinguely's noisy sculptures flailing itself to death. There were many of these experiences, a high proportion encountered in the company of the artists with whom my own musical and conceptual experiments were entangled in the early 1970s – Marie Yates, Paul Burwell, Carlyle Reedy, Bob Cobbing, Hugh Davies and others – all of them adding to a sense that sound, silence and listening were in a slow process of becoming detached from familiar associations with the world of music.

Only gradually did an expanding group of artists find themselves operating under a rubric known as sound art, operating somewhat within the long shadows of John Cage and David Tudor. Naturally, the majority protested against such categorisation but the characteristic behaviour of sound – invasive, immersive, lacking in objects and generally unsuited to reverberant, optically optimised gallery and museum spaces – ensured that their difference would always be conspicuous.

Curating *Sonic Boom: The Art of Sound* at the Hayward Gallery, London, in 2000 was a turning point for me. This exhibition, along with others that preceded it, particularly *Sonambiente Festival für Hören und*

Sehen, staged by Akademie Der Künst, Berlin in 1996, focused a new public attention on to this previously unknown field. The question was frequently raised: why sound art, why now? The sensation of moving into a new era was palpable at that time, given the rise of digital communications, what became the World Wide Web and what was known by the anachronistic term of the Information Superhighway. Some shift in the sensory balance was evident, through which seeing could no longer be considered an isolated arbiter of truth or reality. Games, virtual reality and the Internet all promised immersion into a parallel reality that would supersede pictorial distance.

Since then, many other exhibitions have featured sound and a wealth of books has appeared, largely aimed at an academic market. Predominantly these are theoretical and philosophical texts dealing with particulars of the sound art discourse and its internal conflicts. In general, they keep a cool scholarly distance from practitioners who make work. What interests me more at this point is an examination of this making, along with the context lives and ideas of those active or ancestral artists who constitute the scene, albeit a scattered and heterogeneous one. As a reflection of the contexts within which I practice, this book maps my own struggles with the conceptualisation of sound work over a period of nearly fifty years. The question – then as now – was how to think through the originality and unfamiliarity of this activity from my perspective as a practitioner and writer: how to write it without drawing it back into the domain of music; at the same time acknowledging that much of it grew from the vitality and hybridity of twentieth-century musics of all kinds, even as it moved toward the world of art galleries, museums and site-specific installations. In part, the answer was to focus on practitioners, if only because the stories, ideas and motivations of individual artists are as compelling as the more theoretical and abstracted implications of their works.

Though not organised chronologically, the thought process of *Inflamed Invisible* begins in 1976 with quizzical reflections on video works by the late Stuart Marshall, a pioneer of UK audio and video art, later queer cinema and AIDS activism before his death in 1993. Born within two days of each other, we first met, shared a work table and became friends as students at Hornsey College of Art in 1967. Like many other artists in this collection, Stuart is not a standard reference within canons of sound art yet his evolving practice exemplified the complex origins of this crossing point of sensory, intellectual, philosophical preoccupations through which objects, thoughts, questions of identity and the air itself come alive as the inflamed invisible.

There are emphases and assertions in my own writing with which I would now quarrel, omissions I regret, terminologies I reject (not least the term "sound art"), all of them markers of this personal struggle so not to be effaced. Lines are not taken for a walk. They are smudged, erased, broken, backtracked, dissipated in foams, watery pools and cloudy lostness. As a description of the writing this also mirrors the course of music in its relation to the sensorium since humans first sprayed red ochre on an outstretched hand or painted the outline of animals and supernatural beings onto the rock of reverberant cave spaces, marvelling that the non-human nothingness of these spaces was an active, living being, to be understood along with the flesh and skin beings that moved within firelight, darkness and resonance, the inflamed invisible.

For encouraging me to collect these essays together, offering patient guidance and reflection, steering the book through turbulence and contributing immensely to the editing process I am grateful to my editor at Goldsmiths Press, Atau Tanaka. Without him I would never have stayed on course. I am also grateful to the many editors and curators who commissioned essays and the artists who were prepared to take time to discuss their work.

1

archiving clouds of sounding dust

Sonic Boom

Pauline Oliveros

Sonic Boom

Exhibition catalogue essay, Hayward Gallery, London, 2000

Sound art is not a new invention. "Composers and painters alike," wrote
Karin v. Maur in *The Sound of Painting* (1999), "have frequently gleaned
new ideas from an approximation to, or borrowings from, procedures
used in the sibling art. This reciprocal relationship runs like a continuous
thread through the entire century. Music stood behind the birth of pictorial abstraction and the revolutionary unrest in the arts that, in the years
before World War I, pervaded the great art centres from Paris to Moscow
and Prague, from London to Rome and, finally, New York."

So sound art – sound combined with visual art practices is not a novelty. Its relevance seems to grow as the material world fades to the immaterial, fluid condition of music. Despite our design specifications as fully
articulated graspers and shapers, we humans are busily constructing an
environment that marginalises our own corporeal presence. Our fingers
no longer grip; they click and drag. For better or worse, the 21st century
promises to be an aetherial landscape of images, sounds and disembodied
voices, all connected by invisible networks and accessed through increasingly transparent interfaces.

Being a denizen of the aether, music has reacted favourably to this
remarkable situation. The possibility of downloading music as digital files
from the Internet is proving to be as popular as porn and many musicians
now create music with computer software that replicates state-of-the-
art recording studios. Yet music is as ancient as human culture. A social
activity responsible for some of the biggest gatherings of humanity in history, music is matter and aether all at once. The same piece of music that

persuades one person to jump up and dance can provoke migraine, grief, sexual arousal or doctoral dissertations in others. Music marks death, marriage and other rites of passage, yet in the digital age musical production and participation are melting into the virtual sphere, leaving only nostalgic echoes of life when it was fleshy, physical and acoustically imperfect.

In this context, the medium of sound is both fascinating and problematic as a communicative tool. Modern city dwellers are immersed in audio to the extent that music is becoming just one filament in a web of electronic signals and machine noise. This absorption of music into the sonic environment (and the sonic environment into music) was foreseen by a number of significant, if wildly contrasting, 20th-century composers and musical inventors: Erik Satie, the Italian Futurists, Duke Ellington, John Cage, Karlheinz Stockhausen, space age bachelor pad king Juan Garcia Esquivel and the inventor of Muzak, George Owen Squier.

Notes sent by Jean Cocteau to Erik Satie as Satie prepared his musical score for Cocteau's ballet of 1917, *Parade*, included references to elevators, steamship apparatus, dynamos, airplanes, the telegraph operator from Los Angeles who marries the detective in the end, gramophones, palatial cinemas and the Brooklyn bridge. As well as signalling this new intoxication with machines and electric media by incorporating the sounds of a siren, lottery wheel, typewriter and pistol shots, Satie composed a score that integrated popular music forms such as French music hall, American ragtime and exotica parodies into his own spare and inimitable style. In an antecedent of today's vortex of quotation and digital sampling, Satie even paraphrased an Irving Berlin composition, "That Mysterious Rag."

Composing *Socrate* in 1918, Satie devised the idea of *musique d'ameublement* or "furniture music." "Two years later, on 8 March 1920," wrote Nancy Perloff in *Art and the Everyday*, "Satie and Milhaud introduced the idea of background music to the public when they organized a concert at the Galerie Barbazanges in which 'furnishing music,' made up of popular refrains, was played by a small band during the intermissions." Their experiment in relegating music to the outer fringes of social activity was not entirely successful, since Satie had to rush around the room, instructing the audience to stop listening.

This emerging and heady sense of convergence – music colliding with the noise of life, art music borrowing from popular songs and jazz, the

plastic arts confronting the performing arts as mixed-media – was hugely important to the evolutionary rush of sound work in the 20th century.

Deep Silence, Speed, Noise

Sound is the stuff of music. While the organising principles of music have been disputed, shattered and redefined throughout the past 100 years, sound remains at the core. At the cusp of the 20th century, Claude Debussy anticipated the future. As a student in 1883, he played "groaning" chords on the piano, evoking the noise of buses as they drove through Parisian streets. Six years later he heard a Javanese gamelan ensemble perform at the Paris Exposition of 1889, one of a series of celebrations of French colonial conquest, and was captivated by the rich vibrating fusion of Indonesian melody and percussion. Then in 1913, only a few years away from death, he questioned the domestication of sound and its history, "this magic that any one can bring from a disk at his will," arguing instead that "[t]he century of aeroplanes has a right to a music of its own."

With World War I imminent, the Italian Futurists grasped the electrical, mechanical and frankly destructive character of European life with a fervour that both swept aside and built on Debussy's hopes and fears. Speed was celebrated, along with the movement of electricity and the sounds of modern war. Luigi Russolo, a painter, wrote his *Art of Noises* manifesto in the same year as Debussy's plea for a truly modernist music. "The Italian Futurists could soon be heard berating Italy as the land where museums and ruins were spreading across the cultural landscape like a crop of tombstones," wrote Douglas Kahn in *Noise, Water, Meat* (1999), "and were leading [the European avant-gardes] forward with Marinetti's revelation in The Founding and Manifesto of Futurism that the roaring car is more beautiful than the Victory of Samothrace."

Rooted in magic, machines that can play music independently of humans invoke that most modern of fears. Like HAL in Stanley Kubrick's *2001: A Space Odyssey*, they ask a chilling but logical question: "Humans, are they really necessary?" Futurist theorist, poet, activist and pugnacious foghorn Filippo Marinetti posed a similar threat in an essay called "Multiplied Man and the Reign of the Machine." "We look for the creation of a non-human type," he wrote, "in which moral suffering, goodness of heart, affection and love, those sole corrosive poisons of inexhaustible vital energy, sole interruptors of our powerful bodily electricity, will be abolished ... This nonhuman and mechanical being, constructed for an omnipresent velocity, will be naturally cruel, omniscient and combative."

The noise machines or intonarumori invented by Luigi Russolo and constructed with Ugo Piatti just before the outbreak of World War 1, embodied a kind of cruelty. Russolo had written with clinical ecstasy about the noises of war in his *Art of Noises*. The vocal agonies of dying horses and injured men were silently present, after all, in Uccello's *The Rout of San Romano*; Russolo was interested only in modern war, and particularly its technology. So he wrote about the glissandi of falling shells, the harsh repeating ejaculations of machine guns, the tomcat screams of shrapnel. "A man who comes from a noisy modern city," wrote Russolo, "who knows all the noises of the street, of the railway stations, and of the vastly different factories will still find something up there at the front to amaze him. He will still find noises in which he can feel a new and unexpected emotion." Great boxes amplified by monstrous horns, Russolo's noise machines growled, hummed, whizzed and crackled in acoustic imitation of this chaotic new soundscape unleashed by the industrial revolution. Audiences for intonarumori performances in Modena, Lilan, Genoa, Paris, Prague and London failed to justify Russolo's optimism, though they confirmed Marinetti's dream of combative non-human types. "I have the impression of having introduced cows and bulls to their first locomotive," wrote Marinetti, contemptuous of the public derision that the *Art of Noises* provoked.

The aesthetic barbarism of the intonarumori was overshadowed in 1914 by the barbarity of total war. A tour was cancelled, the noise instruments were lost and Russolo enlisted in the Italian army. Now ghosts at the millennial feast, the intonarumori stand mute, an intangible beginning for the 20th-century, now 21st-century fascination with noise, industry and the operations of nonhuman mechanical and electrical beings.

Spirit Voices on Demand

"The men of the Middle Ages were so mechanically minded they could believe that angels were in charge of the mechanisms of the universe: a fourteenth-century Provençal manuscript depicts two winged angels operating the revolving machine of the sky."
Jean Gimpel, *The Medieval Machine* (1976)

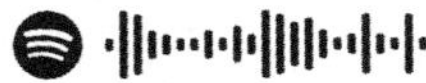

The impulse to create sounding machines goes back much further than Russolo and Piatti. "This imaginal relationship between man and machine was a long time coming," wrote Erik Davis in *TechGnosis: Myth, Magic and Mysticism In the Age of Information* (1998). "The ground was laid by the mechanistic cosmologists of ancient Greece, and it seized the imagination when tinkerers like [Alexandrian inventor] Heron started building those fanciful protorobots we call automata – mechanical gods, dolls and birds that fascinated ancient and medieval folks as much as they fascinate kids at Disneyland today."

In 1650, the Renaissance Jesuit polymath Athanasius Kircher published 1,500 copies of a treatise on music and acoustics called *Musurgia Universalis*. Taking over the role of the angels, Kircher had invented an eccentric collection of mechanical devices that generated, amplified and ordered sound. Floating in a Renaissance netherworld of science and mysticism, Kircher's designs for sound machines included solar powered singing statues, Aeolian harps powered by the wind, a hydraulic organ that seemed to sound through automata representing Pan and Echo, and spiral tubes that projected sound out of the mouths of statues or eavesdropped on conversations in adjacent rooms. He also built an elementary computer, described by Joscelyn Godwin, researcher in esoteric sound, as a "musarithmetic ark" or box of sliders on which the patterns are written, that serves as a composing machine."

The ventriloquism of speaking statues and articulated masks, used by the priesthood to conjure spirit voices on demand, was a formative stage in the history of automata. In their turn, as Erik Davis suggests, automata were the forerunners of robots, replicants and recording. In 18th-century Japan, where an optimistic belief in the robotic future of classic sci-fi still survives, Dutchmen were entertained by karakuri performances staged by live musicians and mechanical dolls. An illustration from a guide to Osaka published in 1798 shows the Takeda theatre, where a Kabuki style percussion ensemble accompanied a mechanical cockerel banging a large drum.

These mechanical inventions played an important role in technological evolution. "Just as the European automata of men like Vaucanson anticipated the machines of the Industrial Revolution," wrote Mary Hillier in *Automata & Mechanical Toys* (1976), "the Japanese performance of karakuri was an awakening of automation." According to Hillier, this example of human-machine interfacing led to improvements in the making of medicines and sugar with the use of treadmill machines.

Similarly, the development of virtual veality has been traced to another mechanical musical instrument, the player piano. In *Virtual Worlds* (1992), Benjamin Wooley's exploration of VR simulation, a genealogy is mapped: notions of computer simulated reality, formulated in the late 1960s by computer graphics pioneer Ivan Sutherland, were inspired by the Link Trainer flight simulator. In turn, the source of inspiration for the Link Trainer was the Pianola. Having been born into a family business of mechanical musical instruments makers, Edwin Link used the pneumatic mechanism of player pianos as a basis for his invention of the first flight trainer in 1930.

A technology that allowed music to be perfectly and repeatedly reproduced until the mechanism broke, mechanical music also anticipated the age of phonography. Playfully sinister creations such as Alexandre Theroude's violin-playing monkey, designed in 1862, became refined and miniaturised for public consumption. A brisk luxury trade in musical boxes, clocks adorned with mechanical singing birds, even musical pictures enhanced by chiming bells, only declined with World War I as other forms of recorded sound became more widely available. This evolutionary obsolescence inevitably becomes a sign of mutated history within the work of composers and performers who create with machines, whether Conlan Nancarrow's or James Tenney's compositions for player pianos, Karlheinz Stockhausen's Zodiac piece for music boxes or the extraordinary diversity of post-John Cage, post-Grandmaster Flash turntablism that transforms the record deck, its needle and its vinyl records into a phonographic instrument of the past and the future.

An Ocean of Sound

"A gasoline-driven generator in the entrance hall was soon pounding away, its power plugged into the mains. Even this small step immediately brought the building alive ... However, in the tape recorders, stereo systems and telephone answering machines, Halloway at last found the noise he needed to break the silence of the city."
J. G. Ballard, *The Ultimate City* (1976)

While 20th-century robots and androids continued their inexorable march towards the goal of artificial intelligence, speaking machines such as the

wireless, the phonograph, the telephone and cinema added new imagery and strange speculations to the interface between human life and invisible mysteries. In *TechGnosis*, Erik Davis describes an incident in 1924, when Mars passed closer than usual to Earth. Activity from many transmitters was temporarily suspended so that "the Martians" would have clear air through which to send messages. "Radio hackers were treated to a symphony of freak signals," he wrote, "Scientists today would describe the bulk of these sounds as sferics – a wide range of amazing radio noises stirred up by the millions of lightning bolts that crackle through the atmosphere every day. Skeptics would chalk up the rest to the human imagination and its boundless ability to project meaningful patterns into the random static of the universe. But this argument, however true in its own terms, distorts the larger technocultural loop: New technologies of perception and communication open up new spaces, and these spaces are always mapped, on one level or another, through the imagination."

The music of the past 100 years has been characterised by a feeling of immersion. Musical boundaries have spread until they are no longer clear. Music has become a field, a landscape, an environment, a scent, an ocean. Media such as radio, television and cinema, or more recently, the Internet and the mobile phone, have fostered an image of a boundless ocean of signals. "Children, parents and grandparents gathered by the Grebe, Radiola, or Aeriola set in the radio room," wrote Gene Fowler and Bill Crawford in *Border Radio* (1987), "and marvelled at the sounds they heard transported mysteriously from faraway lands ... Listeners who bought radio sets were sometimes disappointed, though. Shrieks, grunts, groans, and cross talk ruled the airwaves, which were described by some as a hertzian bedlam."

Humans float in this ocean, this bedlam, their existence and identity represented by icons, cursors, passwords, credit card and PIN numbers, avatars, disembodied voices. In the cinema, we sit in darkness, transported by an intricate fusion of sound and image; playing electronic games we are bombarded by sound effects, hyperactive music and the interactive drama that unfolds or collapses as we frantically thumb the controller button; or mobile phones, voices and information travel with us wherever we go; or the Internet we move through endless layers of data flow, often unaware of a geographical source, lost to the idea that we are anywhere other than indescribable space.

Musicians and sound artists have made significant contributions to the exploration and mapping of this indescribable, entirely unfamiliar space that now envelops humanity. Premiered in 1952, John Cage's *4' 33"* required a musician to present a timed performance on an instrument without making a sound. Half a century later, the piece can still stimulate startling results, even though audiences are more aware of the expectations such work demands. For the premiere, an audience sophisticated in its understanding of contemporary music discovered new frontiers of disorientation. "A local artist finally stood up," wrote David Revill in *Roaring Silence* (1992), his biography of Cage, "and suggested with languid vehemence, 'Good people of Woodstock, let's drive these people out of town.'" David Tudor, who performed the piece on piano that night, described it as "one of the most intense listening experiences you can have." Beyond restlessness, a feeling of being cheated or conned, lies an acute awareness of the immediate sonic environment and its atmosphere. Beyond that lies a plateau of memory and feeling that may have been unexplored within individuals for decades.

Cage's pivotal composition was inspired by Zen Buddhism, by his experience in the Harvard University anechoic chamber where he heard the sounds of his own body and so concluded that silence did not exist, and by the all-black and all-white paintings of Robert Rauschenberg. Night plants, Rauschenberg called the black paintings. "Far from echoing Malevich's famous White On White," wrote Calvin Tomkins in *Ahead of the Game* (1962), "Rauschenberg's white paintings have been seen as the purest possible statement of the idea that life (that is to say environment) can enter directly into art; they have also been seen as definitive proof of the impossibility of creating a void (one might have thought science had already proved this, but art takes nothing for granted)." Tomkins quotes Rauschenberg's observation that the white paintings were hypersensitive rather than passive. "One could look at them," he told an interviewer in 1963, "and almost see how many people were in the room by the shadows cast, or what time of day it was."

From such ideas it was a short step to creating sound installations that made their own music. Perhaps music was no longer an issue. As Michael Nyman wrote in *Experimental Music* (1974), "Cage's piece is hindered by being set in a concert hall, by containing no specific directive for

the audience, and by leaving what is heard completely to chance." Many sound artists who followed Cage were impatient with such hindrances. Composers or musicians were absented from the frame; in some cases, the frame was broken and thrown away. Sounds might generate themselves, through sound sculptures or kinetic machines such as those made by Takis, Pol Bury, Harry Bertoia, Jean Tinguely, Len Lye, Tsai Wen-Ying and the Baschet Brothers. The music of sound sculptures could become a cartography of emergent phenomena: continuous, diffuse, immersive, a conglomerate of inner rhythms that was endlessly engaging, an enactment of a process that seemed to hover on the threshold of nature and culture.

In work that was more conceptual than material, the sounds were foregrounded in the perception of the auditor who discovered an art work taking place, sometimes without being aware that they were the recipient of art. Sound, after all, can surround us on all sides – in the air, through the ground and within our bodies – yet we can still marginalise its presence. In the late 1960s Max Neuhaus composed a piece called *Listen*. An audience expecting a lecture was put on a bus, their hands were stamped with the word "listen" and they were driven to distinctive sound environments such as power stations and the subway. Context, as John Latham has said, was half the work. As the identity or imprint of the artist faded into the background, so the experience of the work took over.

Sound without boundaries culminated in a period of intense, often extreme activity: the conceptual and performance work of the Fluxus group: the American free jazz movement that embraced such diverse talents as Ornette Coleman, Albert Ayler, Cecil Taylor and Sun Ra: minimalists and systems composers such as La Monte Young, Terry Riley, Steve Reich, Philip Glass and Charlemagne Palestine; a thriving scene of European-based improvisation groups that included AMM, MEV and the Spontaneous Music Ensemble.

All of this work proposed new relationships between creators and listeners, between structure and performance, between event and venue between form and time, between the sound object and its environment (and between musicians and their financial masters). John Cage's example had encouraged many musicians to abandon rigid compositional systems and pursue indeterminate or open methods. These new initiatives linked to art movements such as happenings, land art, conceptual art, kinetic sculpture and underground film, even overlapping with psychedelic rock. The Fluxus movement proposed musical events that questioned all definitions of music, using settings that relocated art into unfamiliar, ridiculous and even impossible environments. Pianos were fed with hay, guitars were dragged along the streets of New York, pianos were demolished. As a speculative exercise in absurdity that reversed received expectations of artists as ambassadors of high culture, Walter De Maria's *Art Yard* (1960, New York) portrayed an imaginary scene in which composers such as La Monte Young dug a hole in the ground in front of spectators.

The influence of this type of work, along with the audio ecology researches of R. Murray Schafer and the Vancouver based World Soundscape Project, contributed to the growth of a loosely defined movement now known as Sound Art or Audio Art. Detaching itself from the organising principles and performance conventions of music, Audio Art explored issues of spatial and environmental articulation, the social and psychosomatic implications of sound or the physics of sound using media that included sound sculptures, performance and site-specific installations.

Until the Piano Vanishes

"He liked the happy-looking row of electrical meters and the fact that they ticked off in 3/ 2 time, claves time, that the multiple row of pipes with their valves whistled, water whirring through them. He liked the crunching noises when faucets were turned on, the conga-drum pounding of the washroom dryer: the thunder of the coal-bin walls."
Oscar Hijuelos, *The Mambo Kings Play Songs of Love* (1989)

The key was listening. In his 1977 manifesto of soundscape ecology, *The Tuning of the World*, R. Murray Schafer argued for the urgent need for a coherent method of auditing the sound environment. "The soundscape of the world is changing," he wrote. "Modern man is beginning to inhabit a world with an acoustic environment radically different from any he has hitherto known. These new sounds, which differ in quality and intensity from those of the past, have alerted many researchers to the dangers of an indiscriminate and imperialistic spread of more and larger sounds into every corner of man's life." Researchers in acoustics, psychoacoustics, structural analysis of language and music, noise abatement and other sonic studies were united in asking two questions, Schafer claimed: "What is the relationship between man and the sounds of his environment and what happens when those sounds change?"

Conceptual art, land art, ecology and the aftermath of Fluxus performance were pervasive influences on sound works during the 1970s. Many of these works seemed to be spiritual descendants of both Athanasius Kircher's magic-science inventions and Yoko Ono's whimsical conceptual pieces. Annea Lockwood's *Piano Transplant – Pacific Ocean Number 5*, composed in 1972, gave the following instructions: "Materials: a concert grand piano, a heavy ship's anchor chain. Bolt the chain to the piano's back leg with strong bolts. Set the piano in the surf at the low tide line at Sunset Beach near Santa Cruz, California. Chain the anchor to the piano leg. Open the piano lid. Leave the piano there until it vanishes."

Frustrated with the confines of the concert hall and the educated expectations of new music's small audience, sound art aspired to a closer engagement with the environment and the auditor. Either directly or tangentially, the results were a critique of musical behaviour that was suffocated by tired conventions, even within the so-called avant-garde Techniques, technology, performance, musical structure and context were all called into question. Just as many painters and sculptors no longer felt locked to a specific medium, sound artists used the musical reference as a starting point rather than a defining category. So Laurie Anderson performed on her Viophonograph, a turntable mounted on a violin and played by a needle in a bow, or played violin while standing on a melting block of ice; Bill Fontana proposed a project that amplified the singing tones produced by traffic crossing the Brooklyn Bridge, then sent a mixed version via satellite to other parts of the world; Alan Lamb recorded the aeolian humming of telegraph wires in Australia and David Dunn immersed himself in the emergent systems of bioacoustics.

R. Murray Schafer had contextualised artistic and scientific approaches to sound within a wider framework of global ecology and social imperatives. At the same time, sound artists were moving out of concert halls and galleries into city streets, office buildings, harbours, wilderness, even into the furthest reaches of the earth's atmosphere.

Sound artist Felix Hess has described his work as "my way to find out more about the intelligence of our senses." In this sense, sound art can be a fanciful form of science as well as an art. The work of Alvin Lucier, in particular, researches acoustical, biological and psychological phenomena, transforming his physical explorations into pieces that are haunting and strange. "What interests me is sound moving from its source out into space," Alvin Lucier told Michael Parsons in *Resonance* magazine, "in other words what the three-dimensional quality is. Because sound waves, once they're actually produced, they have to go somewhere, and what they do as they're going interests me a lot."

The spirit of this approach has its roots in science spanning the years from Debussy to the Futurists. "This branch of physics has received renewed attention from research workers during the past decade," wrote E. G Richardson in the 1927 preface to *Sound: A Physical Text Book*, "stimulated no doubt in part by the European War and by the development of broadcasting." Musical instruments such as the piano – embodiments of the aesthetic values of European art music – were threatened by challenges from the electrical world of the radio, the phonograph or early 20th-century electronic musical instrument inventions such as the Theremin. E. G. Richardson's text book updated the work of late 19th-century physicists such as Hermann Helmholtz and John Tyndall and predecessors such as E. F. Chladni, scientists whose researches have been echoed in the music of Lucier, Edgard Varèse and Harry Partch. Tyndall, for example, summarised many experiments in *Sound*, first published in 1898: bowing long monochords, optical illustration of acoustical beat frequencies, the action of fog, hail and snow on sound, echoes from flames, vibrations in metal plates, an analysis of sirens and the "clang of piano wires."

Although they were conducted with scientific rigour, the aetherial nature of sound imbued these experiments with an air of mystery. Smelling faintly of the alchemist's laboratory, they were less torrid versions of Raymond Roussel's literary creations. Staged for one week at the Parisian Théâtre Fémina in 1911, Roussel's *Impressions d'Afrique* featured among its scenes the trained earthworm whose undulations in a mica trough dripped mercurial water onto the strings of a zither to produce complex melodies. Roussel's fantastic inventions lay in an interzone between exotic vaudeville, anthropological surrealism, voyeuristic travel narratives and future audio art. A fictive art that was improbable yet tantalisingly possible, the living sound sculptures of *Impressions d'Afrique* touched on sensitive

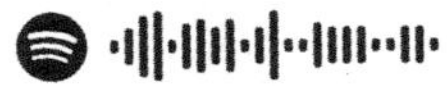

areas of cruelty, dream, perverted science, alien systems and an atavistic social subversion.

Glitch, Bug Noise, Trance Spaces

In the 1990s, these strands – acoustic science, the art of listening, theatres of the impossible, sentient machines, the phenomenology of perception, the articulation of space – converged with concerns pursued by younger musicians and artists who held a certain distance from the worlds of high art and mainstream music. In Japan, for example, the determinedly non-musical sound processes of Minoru Sato and Atsushi Tominaga documented the peripheral bug noise and fugitive crackle of loudspeakers saturated by steam or disconnecting electrodes planted in vibrating window frames. "When we reflect on the condition that most sound works have been requisitioned by music," Minoru Sato wrote in his catalogue essay for the *Sonic Perception 1996* exhibition in Kawasaki City Museum, "we are forced to think that the perception/consciousness of the aspect of sound as a phenomenon has not been valued." Another Japanese composer, Mamoru Fujieda, echoed those sentiments with his claim that "The common notion that any art form using sound as its material is in itself music has begun to lose validity."

A new development in sound art during the past 10 years has emerged out of a club context. With its ancestry in disco, funk and the techno-pop of Europe and Japan along one line of the family tree, Stockhausen and Pierre Schaeffer on the other, this work evolved from popular dance genres such as house, techno, electro, hip-hop, acid house, industrial and ambient. In 1975, Brian Eno had described ambient music as an environmental tint, a background that could be listened to at various levels of attention or simply ignored. Like Stéphane Mallarmé, who devised a perfume that would use scent to express the essence of his poem, *L'Après-Midi d'un Faune* (just as Debussy had expressed its essence through music), Eno envisaged music as perfume. In the wake of disco, clubs in the 1980s featuring electronic dance music catalysed related feelings amongst its followers, though the results were a dramatic contrast to Eno's meditative sound and video installations. Since DJs were playing the records in dance clubs, layering them, transforming them or blending them into one seamless flow motion

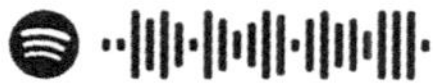

the authorship of individual tracks began to lose importance. Sound in a club was a powerful articulation of space, an extreme expression of musical duration and sonic physicality. Though at the forefront of perception due to its high volume and relentless beats, music was no longer the focal point of socialisation. The idea of sitting in parallel rows, staring at musicians on a distant stage, still endured. The potency of this ritual, however, was fading to a nostalgic memory.

Technology aligned with the nature of clubbing to raise difficult contradictions for musicians still called upon to perform. How could a live performance be meaningful in any traditional sense if the sound had been created through complex sample montage, laborious mathematical calculations and mouse clicks on a laptop computer? Fully immersed in the characteristics of the digital age, recording artists such as Oval, Bernard Günter, Pole, Christian Fennesz, Terre Thaemlitz and Sachiko M are the archaeologists of digitisation and its glitches, their music described by Rob Young in *The Wire* as "an urban environmental music – the cybernetics of everyday life – that reflects the depletion of 'natural' rhythms in the city experience, and in the striated plateaux of the virtual domain." This post-techno universe of clicks, scratches and audible silences drew upon a century of musical innovation yet posited something entirely new: "as if the Ambient soundfields on the Cage-Eno axis," wrote Young, "have been zoomed in on until we are swimming amid the magnified atoms of sound."

Many of the old divisions between so-called "high" and "low" arts have been blurred by the relationship of sound art and experimental electronica to their more danceable cousins. A hip-hop turntable virtuoso like DJ Disk is capable of working with improvising guitarist Derek Bailey while still retaining his credibility in the underground hip-hop scene. A fascination with the qualities of the record deck – an icon of phonography, a noisemaker, a signifier of phonographic memory, a mechanical device that lends itself to performance, a tool for real-time montage and transformation – has captivated both sound artists and hip-hop DJs, with both categories of turntable artist overlapping in their obsessions with vinyl and stylus. Similarly, digital samplers and scanners entrap ghostly traces of our fragmented, mediated present through their capture and mutation of snapshots from sound archives and invisible communications.

None of us know where media arts will go, since their fate is bound up with the uncertain and overheated future of electronic communications. We can only guess. One aspect of sound art that is compelling, at this stage in its history, is the way that dramatic contrasts in working practice and materials can still link to common historical sources. Mechanical or organic, electronic or acoustic, delicate or brutal, hi-tech or ramshackle, solid or intangible, complex or simple, all of the works are linked at a profound level of sonic disturbance.

References

Bailey, Derek, *Improvisation*, The British Library, London 1992.

Benthall, Jonathan, *Science and Technology In Art Today*, Thames and Hudson, London 1972.

Cage, John, *Silence*, Marion Boyars, London 1978.

Davis, Erik, *Techgnosis:* Myth, Magic and Mysticism In the Age of Information, Serpent's Tail, London 1999.

Godwin, Joscelyn, *Athanasius Kircher*, Thames & Hudson, London 1979.

Helmholz, Hermann, *On the Sensations of Tone*, Dover, New York 1954.

Hillier, Mary, *Automata & Mechanical Toys*, Bloomsbury Books, London 1976.

Kahn, Douglas, *Noise, Water, Meat*, MIT Press, Cambridge 1999.

Maur, Karin v., *The Sound of Painting*, Prestel, Munich 1999.

Nyman, Michael, *Experimental Music: Cage and Beyond*, Cambridge University Press, Cambridge 1999.

Revill, David, *The Roaring Silence*, Bloomsbury, London 1992.

Russolo, Luigi, *The Art of Noises*, Pendragon Press, New York 1986.

Schafer, R. Murray, *The Tuning of the World*, Alfred A. Knopf, New York 1977.

Shimoda, Nobuhisa, *2010: The Centrifugal Force of Space, Sound & People: A Document on Xebec*, TOA Corporation, Japan 1993.

Tomkins, Calvin, *Ahead of the Game*, Penguin, Weidenfeld and Nicolson, London 1965.

Toop, David, *Exotica*, Serpent's Tail, London 1999.

Toop, David, *Ocean of Sound*, Serpent's Tail, London 1995.

Toop, David, *Rap Attack 3*, Serpent's Tail, London 2000.

Sonic Meditations: Pauline Oliveros

interviewed 6 January 1995

DT: There's a few subjects I want to talk to you about: the Deep Listening concept, obviously your work in reverberant spaces and natural resonating chambers, the relationship of sound to health, your Deep Listening training sessions and maybe a little bit of your background in electronic music.

PO: In 1988 I think it was, that when we went into that big cistern and recorded and then we discovered that we had material for a CD. When I was planning to write the liner notes I was trying to come to some conclusion about what it was we were actually doing in there and these two words came together – deep listening – because we had a very challenging space to create music in, when you have forty-five seconds of reverberation coming back at you. The sound is so well mirrored, so to speak, that it's hard to tell direct sound from the reflected sound, and that's very challenging, so it puts you in the deep listening space. You're hearing the past of sound that you made and you're continuing it, possibly, so you're right in the present and you're anticipating the future sound that's coming to you from the past so it puts you in the simultaneity of sound, which is quite wonderful. But it's challenging to maintain it and stay concentrated. So that was what I thought we were doing, and listening to one another as well. So then the space itself becomes a very active partner in the creation of music but how you listen to it is how it gets really shaped. From there, once I had made that discovery of putting those two words together – deep listening – it felt like a very good way, a nice logo even, to describe or to make a connection to the work that I've been doing for all these years and continue to do.

DT: From what I've researched into the mythical significance of sound in pre-industrial societies you always come across the thing that resonant spaces had an awesome presence, echoes and so on …

PO: Otherworldly. Of course cathedrals are constructed for that purpose to have sound which has supernatural presence and so resonant spaces are what we get in the cathedrals. In other places of religious significance it was a communion with other worlds. That's what it feels like, I guess.

DT: Was that a factor for you?

PO: In the cistern? Well the main thing was that both Stuart Dempster and myself had been very interested in reverberant spaces over a lifetime. Both of us being horn players. I play the French horn as well as the accordion and Stuart is playing the trombone, didgeridoo and so on. Actually, I've written quite a lengthy article on what I call "Virtual and Acoustic Space as a Dynamic Parameter of Music." I described in the article how as a young performer I was aware of the kinds of rooms I was playing my instrument in and I noticed of course that in a very dry space that the sounds and tone didn't sound so good and in a more reverberant space I began to get a more full, round, rich tone. I didn't know really what that was about at the time but through the years I was learning about reverberation, reflections and resonance and so on, so a resonant space can be very enhancing for your tone quality but it also modifies it tremendously. The cistern is a very extreme example of that. The sounds of the instruments are transformed by standing waves and all the reflections but that's the fun of it. I think that's what was interesting to us.

DT: Maybe you could tell me about the Deep Listening training sessions, which presumably come from these insights?

PO: Yes, well every year I lead a retreat in New Mexico, in the Sangre de Cristo mountains, it's the fifth year coming where people spend a week together doing these different things which I call Deep Listening or we try to find out what Deep Listening is because it opens out, you know, it continues. There's a lot of different aspects to listening. It implies listening below the surface and also listening inwardly. So in these sessions with people I'm trying to provide guidelines so that different forms of attention can be practiced, so that people come away with maybe a fuller sense of the variety of experience they can have by listening and not tuning out.

DT: Do you think there's a relationship between body vibrations and acoustic resonance of this kind?

PO: Oh absolutely, and in these sessions we do in the retreats we try to get into the more subtle vibrations of the body, close to the cellular level as possible and even the vibrations of the energy systems of the body. Then also working on sounding and sounding to increase the sensitivity to the more subtle vibrations of the body.

DT: When you say sounding?

PO: Sounding vocally. Yes, we always do these sessions without instruments because it's better to get to the real fundamentals by vocalising.

DT: Is there a relationship to shamanic methods?

PO: Yes, I'm very interested in that and we do journeying almost every night in the retreat, journeying in different ways: journeying for oneself, journeying for a partner, journeying for the group or journeying on a theme. Listening for me, I want to connect in the broadest sense – it's not only listening to sound but also vibrations, understanding that we are vibrations. We're made of it.

DT: This seems to me one of the fundamental directions in which a certain area of music has gone in this century.

PO: Yes, I think it has done. It's working its way back to a spiritual development. Sound is a leading energy in developing that. Giving people space to do that. So I think it's possible for people to get together and create a resonance that can help and amplify that spirituality, and without it having to be full of trappings or whatever, it *is*. It's an experience and it can accumulate.

DT: It's very interesting but if I compare the notes you wrote for your *I of IV* piece (1966) on the Odyssey album (*New Sounds In Electronic Music*, Odyssey, 1967), which are very technical, and now you've become involved in something that is very intuitive, to do with improvisation and so on ...

PO: Of course, *I of IV* was very intuitive as well except I had set a system very carefully so I could take a free flight.

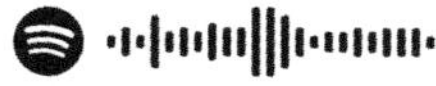

DT: Yes, if I listen to it I can make connections between that and what you're doing now with the Deep Listening albums but the language is so different.

PO: They're definitely connected. I was a young person in a field in which everybody was hiding in technical considerations – hiding their feelings, hiding their motivations and everything else. It still goes on – the techno-jocks that are out there. It's so easy to do that – technocracy.

DT: But that was also a very masculine arena, wasn't it?

PO: Yes, of course, and I had to prove myself, that I could be a jock too.

DT: When did improvisation become important to you?

PO: In the fifties. Terry Riley and I and Loren Rush got together because Terry had a film score he had to do in a very short time. We went to the studio at Radio KPFA and improvised a five minute soundtrack for Terry, together. We enjoyed our improvisation so much that we decided to meet often and improvise. Nobody was doing that at the time – it was considered a very unsavoury activity. So we would go and record our improvisations and talk about it. We discovered that if we kept our mouths shut and just improvised we did better than if we talked and then improvised or tried to make some scheme. That was good fun. Terry was playing piano, I was playing my horn actually, Loren was playing string bass and koto. Recordings of that exist that are in my archive at the University of California.

DT: Do you ever listen to them?

PO: I haven't listened to them in God knows how long, maybe twenty or thirty years. It might be really embarrassing. [laughs]

DT: There is a whole new young audience out there that is hungry for this stuff.

PO: Yes, they really want to know. It's just now there are lots of young musicologists who are writing about my Sonic Meditations and Deep Listening pieces. They're writing really interesting stuff – to me – because my work was so marginalised and not well understood, but they understand it, they really do, and it makes me feel good. It's happening after twenty-five years.

DT: You used the phrase Sonic Meditations. Is that something that you apply to the Deep Listening improvisations?

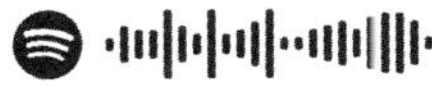

PO: Sonic Meditation refers to a body of work that I did, starting around 1970. I wrote a number of meditations. I continue to this day to write them. I have a large number but the first collection was called Sonic Meditations. They're instructions. The very first one was teach yourself to fly. The instruction goes something like this: observe one's breath and try to remain an observer of the breath, when you feel in tune with the breath cycle breath can become audible but without trying to place your voice. Just to let the air vibrate the vocal chords naturally and then to try to increase the intensity of that until it comes to the point when it's time to decrease and then goes back to the breath. So that was the first meditation. It's appeared in a variety of forms with different people. Like Jill Purse, for example [laughs] – I went to one of her workshops and she read that meditation practically word for word. I was astounded. [laughs]. They continue to work. They have gotten around but as I say, at the time I started to compose those it was a very strange activity, in the context I was in then, at the University of California.

DT: Going back to the idea of performing in reverberant space, do you think that the ritualistic sense that imposes itself from performing in such a strange or awesome environment is a big factor.

PO: I don't feel in awe exactly, myself, but I can tell you this. Stuart Dempster has a tape called *On the Boards*, which is a very wonderful thing that he's working with an audience and doing things with them. There was somebody who was listening to the *Deep Listening* CD and then to *On the Boards* and really preferred *On the Boards* because the *Deep Listening* CD did seem awesome because there was no sense of audience presence. I think that's a very interesting observation. *On the Boards* you can feel and hear the audience.

DT: That is very interesting, particularly in this whole idea of the virtual, that oscillation between the social.

PO: Right, so I think when you subtract the social or audience relationship you get that feeling, where is this? How am I to enter this space?

DT: That could almost be a definition of the effect of the cathedral and its significance in Christianity.

PO: I think so. In Christianity, in Islam, in Buddhist temples and Hindu temples. We have a traditional Tibetan Buddhist monastery here, up in Woodstock. It's a brilliant place, a wondrous place. You certainly do feel that.

DT: Do you think the attraction of shamanism …

PO: It's the in-thing now [laughs]. You can be a weekend shaman, take a course. What we have is a sort of spiritual supermarket going on where all these different tastes of different traditions are being presented in different ways. It's not necessarily bad. It depends on the motivation. There's a value to getting some understanding of different practices and traditions but I think that shamanic journeying practice is very powerful and can be very rewarding and rich in terms of gaining access to the inner world and getting valuable messages from the inner teachers that are there.

DT: That awesome aspect of the Buddhist monastery or the Christian cathedral, that distancing effect, the remoteness of it in a sense is maybe what leads people to something like shamanism.

PO: Yes, it's a non-ordinary reality. You can't be hanging out in reverberant spaces in the office. Nobody would be able to figure out what's going on. We have a number of them. Grand Central Station is a tremendous reverberant space. Railroad stations were built like cathedrals Big pyramids!

2

spirits eruptions inflammations

Áine O'Dwyer

Stephen Cripps

Daphne Oram and Lily Greenham

Victor Grippo and Tania Chen

Joachim Koester

sleep music, the underworld: on hearing Áine O'Dwyer's *Gallarais* and the subsequent conversations between November 2016 and January 2017

Published on the cover of *Gallarais* LP, MIE Music, 2017

Twenty-six letters were sent, not far to travel, from Proust, hypersensitive writer to his upstairs neighbour, Marie Williams – "I was rather troubled by noise ... I was trying to sleep off an attack. But at 8am the tapping on the parquet was so distinct that the Veronal didn't work and I woke with the attack still raging." Marie Williams played harp, though it was not the harp that dragged Proust from the fumes of a bedroom armed against asthma attacks, cork-lined to smother noise. Hammerings disturbed him, the banging of carpet fitters, crates nailed shut. Similar disturbances creep into his books periodically: the "disagreeable" noise of a newly installed heating boiler, "a sort of spasmodic hiccup," which nonetheless served on every re-hearing to revive images of misty landscapes caught in his memory and "studied" during the first morning the heater began its work. Proust spoke of an imperceptible breath, "like the wind breathing into the stem of a reed," mingling with the subdued song of his dying grandmother's breathing, "swift and light ... gliding like a skater towards the delicious fluid," the human sighs released at the approach of death, suggesting pain or pleasure "in those who can no longer feel."

"Who's there?" cries out the old man, stark terror pulled awake at the faintest of noises from pitch black vicinity of an unseen doorway. Harsh

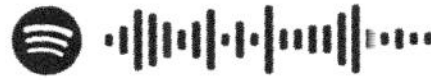

scrape, the tin fastening of a lantern, phantom heartbeat, subdued breath, asphyxiated breath.

The interior is dimply lit, utterly dark even. There is one voice or two, whispers of shaping breath thrown into far obscure and occult recesses of the space as if spirits on the wing whose feathers shriek and keen. They are swans with near-human heads, carrying the lightness of souls, moving between dry land of the living, subterranean rivers of the dead. Sleep music they make, its murmurs written by the method of "passive writing," a transcribing of tongues unknown to all but the most open of listeners. "To look at [animals] from an underworld perspective," writes James Hillman in *The Dream and the Underworld*, "means to regard them as carriers of souls … there to help us see in the dark. To find out who they are and what they are doing there in the dream, we must first of all watch the image and pay less attention to our own reactions to it."

The space was a cave, a tunnel, a room without windows. A skull without eyes, ears, nostrils, mouth, though as Beckett had noticed, the soul turns in this cage as in a lantern, silence "beating against the walls and being beaten back by them'; the space was a chapel, upturned boat, perhaps the curragh that carried Maildun and his crew to the Isle of Weeping, the Isle of Speaking Birds, the Palace of Solitude.

"Who is there? Is anybody there?" A roaring encroaches on the silence and voices ascend out of water, wails of grief, in Dante's world "no one there to make those sounds." Voices flared in pitch black, spouting flame; buried in deep water, they rise up. Within a house described by Mary Butts, in *Ashe of Rings*, the bronze note of a clock rings, "like a body falling bound into deep water." "Can you feel," says the guardian of rings to the woman, "now time is made of sound and we listen to it, and are outside it? Have you thought what it is to be outside time?"

The body descended into the tunnel, never to return as itself.

Stephen Cripps: Exploding Practice Inevitable

Published in *Stephen Cripps: Performing Machines* exhibition catalogue, Museum Tinguely, Basel, 2017

Handmade flyer, black and white, a cheap A4 photocopied collage whose text is a mixture of typewriter, stencil, and handwriting. Four images: at the top, adapted from what we can only assume to be publicity material for a charismatic church, promising the subject for Friday, 25 August at 8pm. London Musicians Collective, 42 Gloucester Avenue, London NW1, at a cost of £1.00 (claimants 60 pence) to be EVERLASTING PUNISHMENT. Below that, a film still. This shows four characters in superhero costumes, taken from a clearly risible Hong Kong kung fu spin-off called *Supermen Against the Orient* (1973). Listed below this, names of the four main performers: Steve Cripps (clarinet and arc welder), Terry Day (percussion, alto saxophone, 'all sorts'), Paul Burwell (percussion, junk, fan, "all sorts"), Marie Leahy (with Steve Beresford, Pam Tait, Ann Brooks, and Paul Burwell), those five performing (as indicated by adjacent drawing), a piece called "Human Percussive Staircase" on the metal stairs leading to the London Musicians Collective (LMC) and the London Filmmakers' Co-op. The year of the event is not given – though a short article by Marie Leahy describing the stair piece is included in *MUSICS* 21 (March 1979), suggesting a probable date of 1978.

From my written description the poster is surely confusing but even to see it at the time, stuck to the window or door of a local bookshop or record shop in North London, would have been mystifying to all but the most committed followers. What, for example, were "all sorts"? Such posters (this one made by Paul Burwell) were one of the only

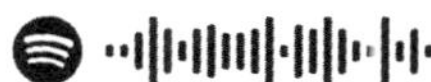

available channels to publicise evenings like this, events flourishing in the interstices of shifting, marginal arts that shared some common aims. The self-appointed task of the A4 flyer was not to explain or sell a predetermined outcome to a virginal audience. It was simply to notify the cognoscenti of a date, time, and place at which unexpected happenings would occur.

That task of finding coherence, patterns, affiliations, and histories within the fluid art practices of the 1970s falls to the twenty-first century archivist or curator. Caution should be exercised, however. The fiercely independent nature of such performances can be surmised not only from the wilful obscurity of their "publicity" but also from small clues buried within all aspects of the layout. At the foot of this poster lies a note (all lower case typewriter) almost erased by the frame limitations of the photocopier on which it was duplicated: "this performance subsidised by the performers, their friends and members of the london musicians collective." In other words there was no state funding or assistance directly awarded to the performers; whatever modest payment they took home at the end of the evening would come from the collection of door takings. If some organisational assistance was required, the performers, their friends, and colleagues provided it. The implicit defiance of this footnote has more or less evaporated through intervening decades, yet its significance as a statement of purpose is as revealing of the times as any other signifier decipherable from the poster as object, image, or text.

The ostensible subject of this essay is Stephen Cripps, but my aim is to contextualise his practice within a milieu of the time when he was most active, from 1975 until the year of his death in 1982. I have written about Cripps before, notably the essay included in a monograph devoted to his life and work: *Stephen Cripps: Pyrotechnic Sculptor*.[1] At that time, memories were still relatively fresh. Since then a number of key figures have died or disappeared, recollections have grown foggy; it falls upon those who survive and were present to emphasise once more the resistance (or obliviousness) of practitioners such as Cripps to established categories of art practice. Were they sculptors, performance artists, musicians, or something entirely new? At the time it hardly mattered.

What did matter was the context that encouraged and facilitated their work, allowing it to maintain a fluidity of practice during its formative phase. For Cripps the milieu and community of experimental improvised music, its collectively produced magazine, *MUSICS*, and its base in the premises of the London Musicians Collective provided spaces in which to create a discourse and presence within the London scene without necessarily committing to its ethos or day-to-day politics. The emphasis of these "spaces" inclined more to the performative aspect of his practice, if only because their history was founded in the performance of music. This assumption was unsettled, however, both by internal discourse subverting conventions of music performance and by the welcoming of precariously placed artists like Cripps into the scene. As Paul Burwell wrote, almost as an afterthought voiced out loud, in his interview with Cripps (published in *MUSICS* magazine, July 1976):

During our conversation it became apparent that sound was just one element in his perceived world. Misunderstandings developed. Questions like, what instruments do you play, got answered by lists of things like "slot machines, turning on the TV, driving a car ..." I thought that it would be unlikely for him to do any of these things just to hear the sound (voicing my unconscious definition of a musical instrument as something that you use just for its sound) but for him sound is just part of other activities. "The sound would be part of the journey I was making (of driving a car)." Later, he grasped the idea of what I was asking, and said "In the actual classical interpretation of the term instrument ... straight instruments, I began by rapping with my knuckles on things, dragging sticks along railings ... things I have forgotten about – but I'm getting back into them. These things developed into hitting things. Gongs, temple blocks. I also use a harmonium." (Later it came up that he used to play kit drums, but like Lester Young and Babs Gonzalez before him, had given them up in favour of something more portable "a harmonica would fit into my pocket").[2]

This passage is fascinating for a number of reasons. Voluntarily it exposes a misjudgement, a slow comprehension that something new and unfamiliar is unfolding, yet by creating an initially dissonant picture by imposing musicological assumptions onto an emergent practice it allows a more complicated reading of Cripps: the way in which the aimless sound-making of childhood – dragging sticks along railings – progressed through hitting resonant objects more purposefully, then playing a drum kit, finally by creating machines or explosives that would recreate the randomness of boyish experimentation while amplifying its effects to a level that invoked simultaneously a childish pleasure in bangs, a teenage love of amplified music, and a rather more conflicted adult preoccupation with war. Burwell arrives at this realisation not as an innocent. In the early 1970s he was making his own instruments, experimenting with percussion, strings, and winds in unconventional shapes but also building drums to be played by rain (see, for example, his contribution to *New/Rediscovered Musical Instruments*, published in 1974).[3]

The underlying feeling of the interview is that Burwell is recognising something of his own trajectory in Cripps, along with a perception that music (in a magazine called *MUSICS*) was moving (had for some time been moving) into a territory beyond the grasp of even the most expansive definitions of music, into an interstitial practice that we began at that time to call sound work. As Burwell wrote as the introductory line to his article – "Radical Structure: 1" – for the November/December 1976 issue of *Studio International*:

At present there is a large amount of music and sound work based wholly or in part on improvisation. As with all art forms, the principles that inform it are constantly being analysed or reassessed, making the task of describing them and outlining the salient points of the work difficult or inexact.[4]

In my own contribution to the same issue, a companion piece to Burwell's entitled "Radical Structure: 2," I applied John Latham's theory of event structure to the problem of music as entertainment and spectacle:

As the perfect time-based medium, music has been betrayed and stands as an example of the seemingly impossible feat of transmuting process into object. The consequences are evident in the spectacle of music as a high-entropy consumable … Despite the pressure exerted upon sound-workers by a space-based, object-oriented ideology there are nevertheless pockets of resistance.[5]

Exemplar of all these difficulties, Cripps provides a clear illustration of what I describe as organology without bodies. This reverses the Deleuzian formulation of bodies without organs, derived from Antonin Artaud's use of the term in his 1947 radio play, *Pour en finir avec le jugement de Dieu*. For Artaud, the body without organs was a human being delivered from automatic responses, restored to true freedom by leaving matter (the meat of the body) behind. Organology without bodies, then, is the study or usage of sound producing instruments that lack clear physical boundaries.

The "instrument" can be a cluster of loosely connected elements – like the clarinet and simultaneously "played" arc welder presented by Cripps at the LMC's *EVERLASTING PUNISHMENT* event – or may "borrow" parts in the way that the Japanese gardening concept of *shakkei* borrows distant scenery to incorporate it within the design of a garden. So in the *MUSICS* interview with Burwell there is a description of an instrument:

Two cassette recorders, one playing a tape of Battersea Dogs Home, the other a tape of jets taxiing at London Airport, are fixed to a revolving arm, with extension speakers on the ends, which revolve past a microphone connected to an amplifier and speakers.

Visceral, sensational sounds attracted Cripps but he was fascinated by their structure and dynamic range. In many cases they were a starting point – materials and processes which catalysed, triggered a reaction, or could be manipulated in various ways. As he says in the *MUSICS* interview (speaking of a piece called "Cymbal Machine"): "It was more like a piece to see what happens." For the cassette piece described above, exhibited at the Fitzrovia Cultural Centre in Whitfield Street, London, 1976, the revolving arm rotating the cassette players moved two dissimilar sounds through space, while the microphone fragmented these sounds by amplifying whichever section was closest. For listeners the impact might have associations of panic and hysteria; for Cripps they were vivid sounds, worthy of appreciation but neglected because of their invasive, disturbing nature.

The technical nature of his materials, unusual as they were, was both centrally important and a means to an end. In a conversation recorded between Paul Burwell and myself, discussing Cripps as I prepared to write the essay of 1992, I asked Burwell to talk about these materials in detail (Cripps died in 1982, so Burwell was to some extent the inheritor of his legacy and well informed about his methods).[6] His response is transcribed below, almost in full, for its value as an insight into both method and materials, not to mention the dark humour of the activity:

"We talked earlier about some of the specific rigs for shows and also some of the equipment he had in his studio. You've been to some of his shows so you remember some of the things he made … most of the things he used were theatrical pyrotechnics and what is called shop goods, which is stuff you buy on November the 5th, basically. He introduced me to the guy he bought stuff off … He had a little office just off Covent Garden where he kept all his bloody explosives [laughter]. It was in an office block. He opened the filing cabinet and pulled out little bombs, basically. Theatrical pyrotechnics which fall into various categories: coloured smokes, which have different durations; maroons, which are bangs, basically, and they come in different strengths or grades – small, large, giant, and thermonuclear. They're detonated by two different methods. One is by ignition and one is by electric ignition, i.e. using a detonator or fuse, you know, burning fuse which is about three different kinds – blue touch paper like you get on November the 5th type things. There's pipe match, which is the oldest, as the name implies medieval. I've seen people make it in buckets in Mexico. But it's faster than anything other than electronic detonation. Flash pots, which is basically magnesium, which gives you a flash and a burst of heat. And some chemicals, actually. He occasionally used a mixture of potassium permanganate and glycerol which is liquid paraffin, basically, and that mixed together spontaneously combusting. It burns with a bright blue flame. I've still got some of those chemicals of his. I keep them in different rooms cause I don't want them getting to know each other [laughter]. That's essentially what he used. He used very little actually – a very limited palette in terms of pyrotechnics. It's kind of what he did with them: the exploding chocolate cake, which is a fairly self-explanatory piece [laughter]. He went through a food phase of blowing up bags of cornflakes. Flour is quite good because flour, if you're lucky, will ignite. And also lycopodium powder, which magicians use to make a flash. It's very, very fine pollen. They use it for Chladni plates, for the vibration. It's very, very fine and I guess not so sensitive to moisture as talcum powder … His interest was with what the things did – like strapping a maroon on the back of a gong so that the percussive force would sound the gong. So he wasn't doing it so much for the bang as for the resonance afterwards. Why he didn't think of hitting the things instead of giving us all split eardrums is anybody's guess [laughter] but there are reasons obviously. It was very different from someone hitting them. The fact that he used these things in very confined spaces – too confined spaces [laughter]. I got nervous a couple of times in performances – not if I was in the performances but if I was in the audience, thinking, my God, this is going to resonate hard in the room, but if you're in there and your adrenalin's up and you're performing, you're stepping over the things virtually and not noticing. Fool! Blue touch paper, he used. He made some quite nice things. You could buy blue touch paper in half imperial sheets, which meant you could do things with it. Like you could stick it together and get a really large figure; and he used to draw with gunpowder, you know, glue and gunpowder, so you could make a very ephemeral drawing. In the video we watched recently there's a whole line of blue touch paper burning in a darkened room. He just set fire to the bottom edge. You know blue touch paper just burns red. It doesn't burst in flame, it just glows. It's impregnated paper – this ever-changing red line, moving slowly up the space. Things like that ... He used tape recordings sometimes, and instruments."

Towards the end of this section of the conversation I remarked that the blue touch paper drawings made me think of Zen calligraphy. "His drawing style had a feeling of that about it," Burwell replied, "taking a pen and going at the paper, gestural drawing." As I commented at the time, this recalled the *Destruction In Art Symposium* (DIAS), convened in London in September 1966, organised by a committee including Bob Cobbing and Gustav Metzger and featuring artists such as Metzger, John Latham, Wolf Vostell, and Yoko Ono. I was also reminded of a story told to me in that same year by one of my art masters – Harman Sumray – at grammar school. When I tried to provoke him by talking about Metzger's performative paintings of acid thrown at stretched nylon Sumray countered by telling me that he and Metzger were students together at the Sir John Cass Institute in East London between 1945–8. During life drawing classes Metzger would cover the paper with charcoal marks, the intensity of his marks gradually tearing through the paper onto the board below.

The underlying politics of DIAS were complex. For Metzger there was the shadow of World War II. As orthodox Jews, almost all of his family were killed in Poland during the war. Metzger, then a young teenager, made a last-minute escape from Germany to England in January 1939, assisted by the Refugee Children's Movement. Many of Metzger's subsequent ideas linked the destructive tendencies of post-war art to his experience of the holocaust, the growth of capitalism, environmental destruction, and nuclear threat during the cold war.

For his contribution to DIAS, Raphael Montañez Ortiz destroyed a piano, not the first artist to murder this object so charged with symbolism nor surely the last, but he locates his action in a moment when "many people were searching for a non-European aesthetic framework and many Native American and African American people were searching for an identity outside of the existing colonial ones." For Ortiz it was a matter of bridging a gap between his Catholic background and Yaqui ancestry "that took in a shamanic sense of reconciliation through sacrificial processes."[7]

In addition to DIAS Cripps could count among his forebears Jean Tinguely and Nam June Paik: Paik for elaborate exhibitions of sounding and moving objects (such as *Exposition of Music. Electronic Television*, Galerie Parnass, Wuppertal, 1963), Tinguely for his machines and moving sculptures, the destructive, noisy tendencies of large-scale creations such as "Homage to New York" (1960) and his own description of this notorious event: "It's a machine that makes a spectacle, it's a sculpture, it's a picture, makes a picture, makes sounds, it's a componist [sic], it's a poet, it's a declaration, this machine is a situation."[8] What emerged in the aftermath of all these events of the 1960s was a new kind of art, uncontainable by terms such as "happening," nor did it fit in existing categories of studio and gallery art, cinemas, theatres, musical concert halls, and clubs. It was closer to life and politics, overlapping any or all art forms, challenging orthodox conditions and durations, demanding new settings.

By 1975–6, Cripps was already creating events in venues that positioned themselves as part of (or outliers to) the art world: the Serpentine Gallery, Artists For Democracy (AFD), Acme Gallery, and others, though their stories were far from straightforward. AFD, for example, was founded in 1974 by a group of artists and critics in response to the military coup in Chile.

Two organisations shared the same semi-derelict building in Gloucester Avenue, Camden Town, North London, in the 1970s–1980s. If you turned left after the entrance at the top of the steps you would find yourself in the home of structural film, experimental and expanded cinema: the London Film-Makers' Co-op; if you turned right you would enter London's main venue for improvised and experimental music performance: the London Musicians Collective (LMC).

A photograph taken at the LMC in 1981 shows Cripps in a trio performance with Paul Burwell and Anne Bean. Another photograph, taken at the LMC in 1980, shows Cripps and five other performers – Burwell, Bean, Max Eastley, Sylvia Hallett, and Sarah Hopkins – standing on chairs, blowing up balloons connected to whistles. As with the poster described at the beginning of this essay, the title of the piece – "How to Explain Music to a Dead Critic" – reveals a hostility to music's infrastructure in the 1970s What all of these partially documented events reveal is a context in which Cripps could collaborate, experiment, and separate himself from existing institutions. Though he has come to be posthumously celebrated as a dramatic exponent of explosive sculptures and self-destructive machine art this informal and inclusive setting of improvised music, *MUSICS* magazine, and the London Musicians Collective was crucial to the development of his practice.

Notes

1 David Toop, "Aftershock," in *Stephen Cripps: Pyrotechnic Sculptor*, ACME, 1992.
2 Paul Burwell, "Stephen Cripps," *MUSICS*, no. 8, July 1976, p. 21.
3 David Toop (ed.), *New/Rediscovered Musical Instruments*, Quartz/Mirliton, London, 1974.
4 Paul Burwell, "Radical Structure 1," *Studio International*, November/December 1976 London, p. 319.
5 David Toop, "*Radical Structure 2*," *Studio International* (ibid), p. 324.
6 Paul Burwell died in 2007 His performance group – Bow Gamelan Ensemble – formed in 1983 with Anne Bean and Richard Wilson, was to some extent an extrapolation from the work of Stephen Cripps.
7 Raphael Montañez Ortiz, quoted in "Art Attack at Tate Britain," *The Guardian*, 28 September 2013.
8 Jean Tinguely, Breaking It Up At the Museum, Pennebaker Hedegus Films, 1960.

The Future Is Not Always Bright: Daphne Oram and Lily Greenham

Published in *The Wire*, March 2007 (issue 277) as a review of two CDs: Daphne Oram, *Oramics*, and Lily Greenham, *Lingual Music*, released by Paradigm in 2007

Two sonic artists, both different in almost all respects other than being neglected, almost forgotten, both women, and both based in England; neither of them living, and both now reconstructed through the fragments of sound that have remained as traces of two complex lives. As with all archive projects hedged by incomplete sources, there is a sense of privileged gazing, the battered scrapbook lying open, the photo album partially revealed, a glimpse through to the hidden person; and pathos, those few moments when access to the inner sanctum was granted: a temporary opening of microphones in a BBC radio studio; a concert in some official palace of the high arts.

Inevitably, the simultaneous release of these retrospectives acts as a reproach to the flawed utopianism of post-war music. Within the male technocracy of electronic music and masculine, even combative world of sound poetry, women were considered rare exotics; their presence and difference highlighting the pathetic subjectivity of aesthetic choices that a male majority battled among themselves to dignify as Theory and Law.

A composer and inventor of the Oramics "drawn sound" system, Daphne Oram is currently the better known of the two, if only because the kind of early electronic music in which she specialised is now fetishised. Her major work, *Four Aspects*, composed in 1960 and described by Hugh Davies as an uncanny anticipation of Brian Eno's *Discreet Music*, was a genuine glimpse into one version of the future. Another futurism, the

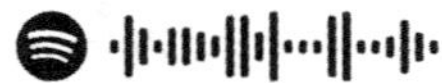

1960s techno-paradise, has become insufferably cute and kitsch, as illustrated by the current use of Raymond Scott's *Baltimore Gas and Electric Co* ("395") for a TV commercial by the beleaguered online bank, Egg. Before his unexpected death in 2005, Hugh Davies had plans to catalogue the Oram archive, of which he was custodian, and prepare material for release. He had noted the CD issue in 2000 of Scott's advertising jingles, film collaborations and musique concrète experiments from the 1950s-60s and believed Oram's largely unknown work to be a British equivalent.

Having worked within the BBC, first as a balance engineer during World War II, then as a founder member of the Radiophonic Workshop in 1958, Oram had to supplement the financing of her studio by making short electronic pieces for radio and television commercials. Themes of machine futurism – leisure through robotics, labour-saving devices and miracle substances – surface only too easily in her jingles for Atlas Copco power tools, Lego, English Electric washing machines, Nestea instant tea and Schweppes Kia-Ora. These were recorded between 1962 and 1966, which suggests that sonic experimenters of my generation were almost certainly affected by them at an impressionable age (is our vintage of experimental music just another side-effect of media manipulation, re. Vance Packard's Hidden Persuaders?).

Predictably, these vignettes are charming but rather one-dimensional. Her music for dramas of a more philosophical nature – plays by Professor Fred Hoyle and Arthur Adamov – is necessarily episodic but evident within the grain and fracture, ominous, melancholy, dystopian, of these distorted micro-compositions is the sense of a composer who never found resources or support to extend her potential. No, the future is not always bright.

Lily Greenham lived a very different story. Born in Vienna, she studied painting in Paris. Returning to Vienna for further study she encountered experimental writers such as Gerhard Rühm, and her performance of a text piece by Rühm and Konrad Bayer is the earliest unreleased tape included in this collection. A permanent move to London came in 1972  and though insistent on her independence from all classification she was a familiar, dynamic, wryly humorous and often humanising figure on the sound poetry scene, then centred at the Poetry Society in Earl's Court King of that court was Bob Cobbing, and the first of these CDs opens with her reading his *ABC In Sound*, the date 1968, the listeners apparently teenage and highly amused by the whole thing. Perhaps the setting was a school. Her commitment to audience feedback and abilities as a natural communicator are evident, though the discomfiting aspect of sound poetry – robust, historically logical yet still faintly ridiculous – anticipates its declining prospects within the performing arts.

Bob Cobbing described her as a performer, not a poet. The distinction seems contrary to his own rejection of art categories and is not borne out by the material collected for this release. Her interest in process art, or "programmed art" as she described it in 1995, is consistently pursued through experiments with visual geometric patterns, magic figures, grids, and what she described as "lingual music." In collaborations with Paddy Kingsland and flautist/saxophonist Bob Downes and his Open Music group, she accumulated dense patterns of sound derived from speech: narrated texts, multi-lingual phrases, single words (a keyword, she called this) or sounds extracted from elements of speech. Repetition, either electronically generated or vocally performed in real time, gives the best of these pieces a hypnotic quality, a slow build to overloaded density during which meaning erodes yet lingers as traces in language fog.

Painstakingly curated and nicely presented, these CDs burst with ideas, energy, ventures into territory only barely explored, moments of absolute clarity and inspiration. Obscurity clouds the issue just as much as tape deterioration or incomplete documentation, and though an understanding of 20th-century audio culture is partial without access to work by innovative, important if marginal figures such as Oram and Greenham, a feeling persists: these are sonic remnants of frustrated ambitions, lives which drifted out of earshot and return in a compressed, somehow indecipherable form.

The Sweat of Toads: Victor Grippo and Tania Chen

Blog post, David Toop, *a sinister resonance*, July 21, 2017

"I was in search of something – a small detail which I remembered with special intensity as part of my vision."
George Eliot: *The Lifted Veil* (1859)

The man whispers in Spanish as he pisses, sniffs, sighs, washes his hands, all sounds of the higher frequency. He is breathing in reverberance. He goes. Upstairs in Malba, the Buenos Aires Museum of Latin American Art, Victor Grippo's *Vida-Muerte-Resurrection*, ten lead vessels – cylinders, square and rectangular boxes, cones – face each other like an architect's model of a western frontier town set in the future. Beans moistened with drops of water spill out from this sombre architecture, their germination wreaking havoc among grey sobriety. 

The alchemy and hermetic symbolism of materials is central to Grippo's work, a radio drawing electrical energy from a potato, his writing – An Observation "In Vitro": "It lived in the intestine of a toad. It was carefully extracted with a pipette and placed more carefully still on a glass slide, isolated, solitary and mobile in a drop of water. The refringent cilia ..." and so on. "For Grippo," wrote Guy Brett, "it is an article of faith that instruments of work and works of art have a common starting point ... the irony of an inchoate lump of unstable matter forming a 'homage to constructors' could be read as an acid comment on the perversion of construction and order by fascist regimes."

Writing on Lotto's *Toilet of Venus* in *Pissing Figures 1280–2014*, Jean-Claude Lebensztejn surmised that "Lotto seems to have been

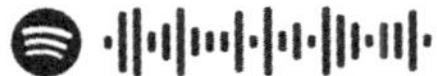

keenly interested in alchemy, where urine plays an important role, and in the illustration of hermetic symbols, of which the wooden covers in the basilica's choir provide so many stupefying examples." C. G. Jung's understanding was that base materials such as urine were instruments of a kind of folly. In *Alchemical Studies* he favoured examples of commentators ridiculing the "frivolous triflers," the literalists who worked with urine, salts, metals. His sympathies lay with those who passed beyond the torturing of arcane materials into a contemplative symbolism of the psyche, in which lead, for example, was "identical with the subjective state of depression."

This dismissal of materials, to rid the world of objects in favour of pure spirit, is a denial. I watch keenly for the way in which materials and objects are tortured in the pursuit of that illusion of pure spirit we call music. Tania Chen at Café Oto on the fifteenth June 2017 drinks coffee to dispel jetlag, creates feedback with small walkie-talkies, plays back voicemail messages both private and banal, shows brief video clips from her travels: rooms, corridors, aircraft interiors in which flight attendants wrestle with food trolleys. The corridors have a disquieting aspect, if only because there is a weight of cinematic evidence to show that corridors are dangerous places. She speaks about her dog Lychee, allowed in the cabin of the plane because Lychee offers emotional support in a world that has the potential to be as grey as lead.

The set up is like a living room, a table and chair, an open laptop, an awkward passage between table and piano. Tania moves to the piano, leans into it, her body sinister in the way it hovers over the keyboard, seducing it into softness before suddenly shooting out quick, stabbing motions of immense force, shocks that unlock the violence latent within every piano. Without drama she speaks, phrases plucked out of life's banalities, poignant emails about family, friends and forgotten birthdays, the sacrifices of certain choices and what must be forsaken to make something this fragmented, raw and compelling. Lychee the companion dog has sad eyes, we might say, knowing nothing of a dog's sadness. "Soft and fluffy," she says, repeating, "soft and fluffy." But the mood is as sticky as the sharp rasp of Sellotape pulled from a roll as if ripping the dressing off an open wound. There is nothing slick here. We see all the video clips laid out as a thumbnail world of atomized clips, observe the uncertain process of choosing which one, see the dislocation of time, the disconnections from earthing and familiarity, the fracturing of emotional ties, represented by these disparate materials as they acknowledge travel itself as a material (just another complex of refringent cilia that form the so-called "instrument," it could be said) with which a musician works.

Joachim Koester: The Art of Memory and Forgetting

Published in exhibition catalogue produced by Bergen Kunsthall and Camden Arts Centre, 2019

In the early 1970s I would occasionally stay with my then partner – Marie Yates – in a dilapidated rural barn belonging to John and Barbara Latham. From there we would walk onto the wild hills of Dartmoor and make artworks together, her with impermanent sculptures of sticks and muslin, me with the equally impermanent material of sound. Situated on the edge of the moor, the barn was once a water mill. The water wheel was gone but in one corner of the ground floor the river was diverted to run through the building. On the upper floor a double bed was positioned over this racing stream and its reverberant chamber. Drifting into a hypnagogic state at night I would hear indecipherable voices below, murmuring from within the turbulent white noise of the water, pulling me down into the dark flow of its riverrun dreams.

How ancient is guided meditation? As old as we care to imagine. In themselves, the enticement of the question and the vagueness of the answer are a form of guided meditation. They invite a journey into the unknown, undertaken with imagination as the primary technique. Just to hear quiet voices buried within implicitly vast chambers of white noise is to create virtual terrain and space, to walk and explore, return and remember, all without moving a physical muscle.

Think of the chthonic resonance of a deep cave, marks spat onto its walls with a pigment made from charcoal or oxides of iron mixed with animal fat, sometimes blown from the mouth or spat through a bone tube. Imagine the sound of this expulsion, then the sound of flutes made from vulture bone wings reverberating in the vaulted darkness. Then we think of stories, narrative, fragile remembering. Soon we are in monasteries, perplexed by the problem of how to remember. How to remember our stories, ancestors, places, words, events and beliefs? In *The Art of Memory*, historian Frances A. Yates quotes the story of a singer – Simonides of Ceos – who chanted a lyric poem honouring his host but made the error of also praising Castor and Pollux. Scopas, vain and devious host, refused to pay for the section not dedicated to him. Castor and Pollux can pay for that, he said. Shortly after, Simonides was called out of the hall to meet two young men. There was nobody in sight and while he looked for them the hall collapsed, crushing the cheapskate Scopas and all his guests to a pulp that rendered them unrecognisable. "And this experience suggested to the poet [Cicero] the principles of the art of memory of which he is said to be the inventor," wrote Yates. "Noting that it was through his memory of the places at which the guests had been sitting that he had been able to identify the bodies, he realized that orderly arrangement is essential for good memory."[1]

How simple it is that in a Joachim Koester exhibition, visitors lie down, wearing headphones, and listen to what at first seem to be meditation tapes. Despite knowing these are digital files I say "tape" because guided meditation and relaxation was central to the New Age audio techniques that emerged in the 1970s, and that economy grew and distributed itself as a separate network through the medium of cassette tapes. At Camden Arts Centre in 2017 a lot of people would fall asleep while listening through the headphones, as if able to slough off normal behaviour and vulnerabilities in public spaces. "[This was] strangely moving for me to observe," says Koester. "I think that the inclusion of different attentive modes of engagement, from sleep, inner space travel, to a more common interaction with the art works, in the same space holds a potential – the exhibition venue becomes a habitat for different modes of consciousness, communication and of being."[2]

The memory system of antiquity describe by Yates was, as she put it, "... ordered and neo-classical in form."[3] The technique was to create rooms in the mind, an imagined theatre populated by images and objects, all of which stimulated the recall of memory – a sermon, say, or a scholarly text – for the person who "moved" amongst them. Others discerned more occult, less orthodox possibilities for mnemonic techniques. Of the sixteenth-century philosopher, wanderer and condemned heretic Giordano Bruno she wrote: "Here was a man who would stop at nothing, who would use every magical procedure however dangerous and forbidden, to achieve that organization of the psyche from above, through contact with the cosmic powers"

"When it comes to the meditations," Koester says, "I've been very inspired by Yates' *Art of Memory* and her descriptions of Bruno, which can be seen as a sort of hacker of the memory technology. But I've also been interested in the older monastic techniques, before Bruno, where meditations were used as rhetorical devices to generate words and meanings or imagine past events."[4] This brings to mind the remarkable example of a particular Vedic chant of India, the Sāmaveda. Already charged with magical force, the Sāmaveda contains chants that are considered too dangerous to be heard by uninitiated people. The text of one of these – Rathamtara – is hidden by syllabication, each syllable replaced by another. "While the chanters recite these meaningless syllables," writes Professor J. F. Staal in his notes to John Levy's recordings of Vedic chant, "they should in their mind concentrate on the real syllables of the underlying text; in this way the chant will reach the gods for whom it is intended, and no one else."[5]

Guided meditation, whether in the yoga class or the disembodied voice that leads a headphoned listener through sound to an unknown destination, where else would it come from, other than the trope of the shamanic journey? In this account of a shamanic séance in Greenland, breath activates the drum which moves autonomously. Life has to be blown into the skin and frame for it to become inspired: "The drum, which was lying a bit to the left of Maratse, moved, roaring, hopping on the handle towards him. It sprang up on his back and 'sparkled' for a long time against his naked shoulders. Then it jumped down and moved back to its starting point. A second time it moved rocking towards Maratse – and flew up on his back. His vision was blown into layers, and the knowledge of all the world gave his spirit-possessed vision wide horizons to contemplate. As the drum for the third time had been up on the back of Maratse and had jumped down again, the lamps were extinguished. The drum roared and thundered, possessed by enormous force."[6]

In our era, journeying for self-discovery may have become the emptiest cliché of them all, yet without such fraught journeys into the depths

of the unconscious there would be no literature, no drama, no cinema The wanderer dies and is reborn. Think of a classic western from 1950 – Henry King's *The Gunfighter*. Prelude: stirring music; a man (Gregory Peck, though seen from long shot at this point) rides a horse across vast terrain, deserts, mountains, deep canyons, scrub, the creature noises of dusk, finally human settlement. Then, like the beginning of a joke, the man walks into a bar. Minutes later, two shots: another man is dead on the barroom floor and so the gunfighter's journey moves inexorably toward its summation in death, the end an explosion of gunfire that renews the same cycle all over again. Only the bodies change.

Ancestral lines run through here; not straight but convoluted, tangled lines that vanish and reappear. Threaded through shamanic forays into spirit worlds, their presence as oral guidance to strange haunted territories persists in more familiar cultural manifestations: *The Epic of Gilgamesh Beowulf*, the *Knýtingla* and *Skjöldunga* sagas, *The Egyptian Book of the Dead*, the Tibetan *Bardo Thodol*, Dogon funerary rites that used music to stir the souls of grain, the dead and the living, Japanese Noh theatre and *The Tale of the Heike*. Exemplars of self-discovery through journeys, Homer's epic poems – the *Iliad* and *Odyssey* – eventually bring us to James Joyce *Ulysses* and the dream text of *Finnegans Wake*, and what we call, for the sake of convenience "stream of consciousness" or its close cousin, "automatic writing," a gateway to the unconscious associated with the early modernists and pre-moderns of literature – Joyce, Virginia Woolf, Dorothy Richardson, Mary Butts, Marcel Proust and, in some passages, Jane Austen.

But more covert, perhaps closer to the surrealists and their automatism, were examples of spontaneous speech, writing and painting found in the occult revival and spiritualist boom of the late Victorian era, whether the spirit drawings of Georgiana Houghton and her contemporaries or the so-called "passive writing" practiced by nineteenth-century spiritualists, in which messages from the spirit world were transcribed, apparently without any conscious intervention from the writer. "Women adapted particularly quickly to the knack of 'mind passivity,' somewhat akin to the meditative process of ridding the mind of all thoughts," writes Alex Owen in her study of women, power and spiritualism, *The Darkened Room*, "and readily complied with spirit instructions to receive "prayerfully and passively" whatever might come."[7]

Though the tendency of moderns is to think of ancient sagas and epics as written literature, in many cases their origins lay in oral poetry, often read or sung in a confluence of entrancing voice and the hypnotic entrainment of instrumental sound from a stringed instrument such as a kithara, phorminx, kora, khun tovshur or biwa. So it is in *The Tale of the Heike* from Japan, a collection of oral stories compiled in the fourteenth century or earlier and traditionally sung by *biwa hoshi*, itinerant monks who played the plangently expressive stringed instrument, the biwa. So too it was in *The Nine Songs*, Chinese shamanistic poetry from the third century B.C. According to Arthur Waley's translations, encounters between the wu shamans and their spirit guides were enacted to the sound of sonic costume – "My girdle-gems tinkle with a ch'iu-ch'iang"[8] – and ritual music of drums, reed organs, flutes and stringed zither. In song VIII, *The River God*, the female shaman wanders down the Nine Rivers with Ho-po, the River God, later, after they part, she climbs the mythical mountain of K'un-lun and in a state of agitation ima-gines the underwater life of her supernatural companion:

> In his fish-scale house, dragon-scale hall,
> Portico of purple-shell, in his red palace,
> What is the Spirit doing, down in the water?[9]

Music is intrinsic to these journeys through mythical landscapes. A deep chasm of time separates Homer from Freud but is it such a leap from the Homeric tales of musical enchantment – Calypso, Circe and the Sirens all enmeshing listeners in the charm of song – to Franz Anton Mesmer's eighteenth-century experiments with animal magnetism, hypnosis and music (using piano and glass 'armonica to guide the moods of his patients) and from there to Sigmund Freud in 1892, abandoning hypnotism for what he called a "concentration technique"? Freud's original method of free association was to ask the patient, lying down with closed eyes, to concen-trate on a particular symptom and try to recall memories that could lead to a better understanding of its causes. "Freud was still given to urging, pressing and questioning, which he felt to be hard but necessary work," wrote his biographer, Ernest Jones. "On one historic occasion, however, the patient Frl Elisabeth [von R.] reproved him for interrupting her flow of thought by his questions. He took the hint, and thus made another step toward free association."[10]

"Freud," Alex Owen writes, "was intrigued by psychical research and pronounced himself reluctant to dismiss 'prophetic dreams, telepathic experiences, manifestations of supernatural forces and the like.' Despite later characterisations of his work and ideas, Freud himself remained highly sensitive to pictures, 'real dreams,' actual psychic or occult experiences that were superficially, at least, outside the province of established psychoanalysis."[11] Elizabeth von R.'s intervention is symbolic of the precarious nature of Freud's enterprise, oscillating between empirical science and the mellifluous flow of a voice outpouring all the hidden things that subvert and complicate human hopes and motivations.

Though all of these techniques originate in a hazy realm of intuition, atmosphere and spirit, there are obvious attractions to the idea that some causal trigger exists that can be identified and exploited. Jean Rouch, the French filmmaker and anthropologist, described this as the "strange mechanism," a switch flicked by music to arouse a trance state. Practitioners working in the field of brain wave training promote theories of subliminals – binaural beats and isochronic tones buried within so-called relaxing music – that allegedly alter consciousness without any significant effort on the part of the listener. Binaural beats, for example, are sine waves pitched at slightly different frequencies – ie. 100 Hz and 107 HZ. Heard through headphones, one tone on the left side, the other on the right side, they produce a third "beating" or phantom tone.

"Discovered in 1839 by German physicist Heinrich Wilhelm Dove" write the authors of *Brain Power*, "binaural beats were just a curious anomaly until 1973, when Gerald Oster, MD, published his landmark article entitled "Auditory Beats In the Brain." Working at Mount Sinai Hospital in New York City, Dr. Oster found that binaural beats evoked change in the electrical activity of the listener's brain. This tendency for brain waves to resonate sympathetically is known as the frequency following response."[12] Many such recordings can be found on YouTube, often claiming specific tuning of Beta and Gamma, Alpha or Theta waves, according to the desired outcome. In this sense they could be considered as a legal, quasi-medical equivalent of contemporary trends such as LSD microdosing and the offspring of radical experiments in total body immersion and psychedelics undertaken by John Lilly in the 1960s (as documented in *The Scientist* and *The Centre of the Cyclone*).

In *Music and Trance*, Gilbert Rouget is sceptical of physical effects, inclining instead to the moral action of the music. "Ultimately, then," he writes, "one might search in vain for the reasons why the role of music in triggering the 'strange mechanism' has so often been viewed as physiological in nature and, consequently ... as more or less comparable in its action to the use of a drug ... Although it is perfectly permissible to say, metaphorically, that music is a drug, in the present context, which is not that of metaphor, it simply contributes to general confusion."[13] In other words, the onset of trance is a culturally learned or acquired response to a situation, rather than a causal reaction to a particular pattern of drumming.

There is no question that music has a profound impact on the body. Through the intimacy of close listening we allow change within ourselves. If I listen to Javanese gamelan, for example, am I entranced by the beating tones of the gongs – characterised by analysts as "... signal behaviour that resembles the primary and secondary mistuned beats of its second and third harmonic frequencies ..."[14] or am I led by the hand through an exploration of time in which the world seems to untether itself from clocks, slow down and move in cycles rather than arrows? I would suggest all of these effects are entwined, complementary and cumulative.

When I listen to the sound works for headphones created by Joachim Koester and Stefan A. Pedersen I follow the example of those gallery visitors who precede me: eventually I fall asleep. "Breath out and let go," says the voice, as it might in so many other relaxation tapes, but immediately I am led by listening to an unexpected place, not a forest clearing opening out to a lake blanketed by lotus leaves but an urban contemporary art museum and its department of eagles. The reference is to Musée d'Art Moderne, Département des Aigles, an installation created for some months in 1968 by Marcel Broodthaers in his home. Described by Broodthaers as a response to the political and social upheavals that occurred in 1968, the museum interrogated the idea of a museum and its potential. This is where I have been led, to a dissident memory palace of another age. Gradually I relax, allowing my active, intellectual mind to seep away into the soft low tones, humming like a ship's engines, and washing white noise, like a distant sea, giving myself up to the unfolding story in much the same way that Joseph Conrad's unnamed company falls under a spell of silence and deepening gloom to listen to Marlow's tale of *Heart of Darkness*. "We live in the flicker," said Marlow. "But darkness was here yesterday."[15]

Just before I fall asleep, flickering in and out of the hypnagogic state that tells me I am falling, I am conscious of being led to a place of potential, not a closed destination but a space in which many things could be possible. Still conscious at this point that my body is prone and vulnerable in a room within a physical museum I am led into the conceptual domain of a virtual museum, an invisible place filled with empty crates, postcards, equipment, a ladder, props within a theatre of memory but also the museum as a shell, waiting for works. I exit the building, descending stairs (as if deep into my reservoir of ideas and forms) into a garden. Shadows and flashes of light flicker across the scene of a near-buried greenhouse, inviting me to project into them my own imaginings. "And as you are contemplating this changing sight." says the guide, "allow your mind to travel. Let your thoughts ramble. Let any image, scene or word appear. And you may recall the museum you have just visited, or you may see patterns, structures, models or maps, or something completely different. Just let your mind drift. Embrace all you encounter. You'll spend some time here and then you will return." For six or seven minutes, there I stay in the sound world, a visitor to Plato's cave, perhaps. My guide is suddenly absent. Half-conscious that I am immersed in the imaginative and generative process itself I am free to roam, to allow my own works to populate this museum garden or to sleep until the sound of trickling water brings me back.

As Joachim Koester has said, his practice is to some degree a recovery of lost opportunities, lost or abandoned knowledge but it also indicates openings to innovative discovery. These are actions, techniques, objects and phenomena that generate new ways of thinking about natural science, cosmology, social structures, the body and its potential at a time when new ways of thinking are urgently needed. Then there are the forces unleashed by these techniques, strange, also unpredictable, not altogether benign. Maybe the question for us is how these two strands are entangled.

Notes

1   Frances A. Yates, *The Art of Memory*, Penguin Books, 1969, p. 17.
2   Joachim Koester, personal communication, 2018.
3   Frances A. Yates, op. cit, p. 208.

4 Joachim Koester, personal communication, 2018.

5 J. F. Staal, *The Four Vedas*, Asch Mankind Series LPs, 1969, p. 9.

6 Merete Demant Jakobsen, *Shamanism*, Berghahn Books, 1999, p. 123.

7 Alex Owen, *The Darkened Room*, Virago Press, 1989, p. 213.

8 Arthur Waley, *The Nine Songs: A Study of Shamanism in Ancient China*, George Allen and Unwin, 1956, p. 23.

9 Arthur Waley, ibid., p. 47.

10 Ernest Jones, *The Life and Work of Sigmund Freud*, Basic Books Inc., 1961, p. 158.

11 Alex Owen, op. cit, p. 237.

12 Michael J. Gelb and Kelly Howell, *Brain Power*, New World Library, 2012, p. 167.

13 Gilbert Rouget, *Music and Trance*, University of Chicago Press, 1985, p. 183.

14 Matias H. Budhiantho and Gunawan Dewantoro, *"Javanese Gong Wave Signals,"* *Proceedings of Meetings of Acoustics*, 20, 035003 (2013), p. 1.

15 Joseph Conrad, *Heart of Darkness*, Penguin Books 1973, p. 8.

3

hysteria machines ghosts tremors

Camille Norment

James Richards

Gilbert & George

Seymour Wright

Realm of the Senses, for Camille Norment

Published in *Rapture 1,* Camille Norment, Office for Contemporary Art Norway, 2015

To comprehend the romantic sublime of the 18th century we only have to look at the landscape paintings that expressed this view of nature, then compare them with our own twenty-first century touristic perceptions of the same scene. Picturesque mountains were depicted then as vast, looming peaks; lakes as eternal seas. At the birth of Lake District tourism in the northwest of England, professional men ventured into the poverty, social neglect and sublime scenery of the Lakes with a sense of trepidation. Travelling in 1772, a Salisbury clergyman named William Gilpin paid to experience one of the diversions made available to these adventurers: cannon fired from a boat to set off dramatic echoes from surrounding woodland and mountains. "Such a variety of awful sounds," he wrote, "mixing, and commixing, and at the same moment heard from all sides, have a wonderful effect on the mind; as if the very foundations of every rock on the lake were giving way; and the whole scene from some strange convulsion, were falling into general ruin."[1]

By instinctively reaching for the trope of ruination as a way to describe the effects of sound vibrations, Gilpin inveigled the act of listening (harmless in itself) into other dark corners of the highly charged eighteenth-century imagination, notably the art of ruins as brought into being by Piranesi's inventions of vast, gloomy and labyrinthine prisons, and the gothic fiction of oppressive castles and fainting heroines satirised by Jane Austen. There is an air of hysteria within these accounts of Lake District echoes. First, these respectable men would calm their nerves

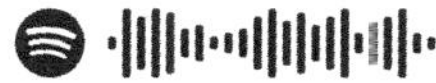

by listening to the echoes of flutes and French horns as they took their lunch (a precursor of ambient music); then they would jolt themselves into a state of high terror with a more sinister, martial version of the same effect. It was as if they first bathed in an invisible architecture of soothing vibrations, new age style, then blasted at the same unseen structure to reduce it to rubble. As Norman Nicholson, a writer on Lake District lore, has observed, the latter strategy has all the paternalistic signs of the pulpit or courtroom: "They sailed into the middle of the lake, fired off the guns of their own ego, and waited, patiently yet excitedly, to hear the echoes return to them. The world itself did not matter – what concerned them was the sound of their own voices."

In a post-Freudian world the pleasure they took in surveying the sublime also reveals a fascination with the uncanny. Sound and listening are ghostly in their susceptibility to the unseen, unverifiable, ambiguous and fleeting, yet vibration can be felt as a bodily sensation. Those who consign sound to the bottom of a hierarchy of sensory importance must always be wary of those apocryphal ram's horns and massed voices that flattened Jericho's walls. Sound may seem to be literally nothing but its material affect could be seen by the eyes. In 1680, Robert Hooke used flour, a glass plate and a violin bow to demonstrate nodal patterns; then in 1787 Ernst Chladni published his theories of sound, including the famous Chladni figures produced by bowing the side of a metal plate, sand arranging itself into symmetrical patterns as the plate reached resonance.

Sound vibrations were demonstrably a physical phenomena but the eighteenth-century understanding of listening was a site of contesting theories. "The idea of 'sympathetic vibration' between music and literally vibrating nerves," James Kennaway writes in his book, *Bad Vibrations*, "was one model of the impact of sound that proved highly influential. It was especially suited to the rhetoric of sensibility since it combined nerves with older ideas of harmony and sympathy."[2] Nerves were imagined as musical strings, capable of being in tune or wound too tightly but generally susceptible to the refinements of music. By 1800 music became implicated in medical theories that cautioned against excessive stimulation. Nervousness became central to a conception of what Kennaway calls "cultural hygiene," in which the body, particularly the female body, was at risk from music's strange quasi-erotic power.

Two sound producing instruments in particular – the aeolian harp and glass 'armonica – acquired dangerous reputations for their peculiar volatility in this realm of the senses. In 1802, Samuel Taylor Coleridge, considered one of the Lake poets for his residence in Keswick at that time, wrote *Dejection: An Ode*. The poem laments its narrator's inability to write and the loss of any connection with nature. Downcast, he observes the storm, hears its effect on the strings of an aeolian harp – "Mad Lutanist" – as a scream of agony by torture, hopeful that those sounds produced by the wind's "dull sobbing draft, that moans and rakes upon the strings" might startle his own dull pain back into life. The parallels with eighteenth-century electrotherapy theories and the twentieth-century use and abuse of Electroconvulsive Therapy as a treatment for serious depression are too tempting to ignore. Nearly a century after Coleridge, in Robert Louis Stevenson's short story – *The Beach of Falesa* – the aeolian harp had become a subterfuge to induce fear and simulate supernatural voices. "It's my belief a superstition grows up in a place like the different kind of weeds," writes the narrator, "and as I stood there, and listened to that wailing, I twittered in my shoes."[3]

As for the glass 'armonica: "The instrument's novelty, feminine character and unusual sound made it a key battleground in discourses about women's nerves," writes James Kennaway. Women who specialised in the instrument retired from playing, allegedly after suffering its cumulative effects, even died from conditions diagnosed at the time as a direct consequence of the instrument's uniquely disturbing characteristics. There is no question that the glass 'armonica has an uncanny, unearthly timbre, its lack of attack and pure tones intensified by the audible friction of wetted fingers against glass. Sounds seems to emerge from nowhere, suspended spirits floating then dissipating in the air like icy breath. In 1786 a musician named Karl Leopold Röllig claimed that the instrument could "make women faint; send a dog into convulsions, make a sleeping girl wake screaming through a chord of the diminished seventh, and even cause the death of one very young." This perception of an unnaturally close connection to the body's life force led to strange usages reminiscent of Edgar Allen Poe's *Tales of Mystery and Imagination*. In a Weimer mortuary, for example, corpses were attached by strings to a glass harmonica as a safety measure to ensure that nobody had been buried alive.

More notorious was the use of glass harmonica as a curative instrument by Franz Anton Mesmer. A German physician from Swabia, Mesmer developed a theory of animal magnetism, a form of sympathetic vibration founded in the hypothesis of an imperceptible vital fluid, transmitted from one body to the next to create animal electricity. In his 18th-century group healing sessions he would lay on hands for long periods, stare into the eyes of patients, make "passes" with his hands across their field of vision and conclude by playing the glass harmonica. As with George Gurdjieff's salons in occupied Paris in the 1940s, sessions in which his followers gathered to listen to their leader play harmonium after dinner, the music seems to have been improvised. Gluck heard Mesmer improvising in 1779 and was sufficiently impressed to advise him that improvisation was a better path for him than notated music. Improvisation, in which the audience is drawn into their own subjective construction of an invisible architecture as the music emerges and unfolds, fulfils the conditions of what I called in my book, *Sinister Resonance*, the "mediumship of the listener," an intensive listening close to hyperacusis. In this state of near-trance, the relative formlessness of such music and its unknown trajectory can seem miraculous.

Themes of spontaneity, susceptibility and entrancement link all of these narratives. I. M. Lewis's book, *Ecstatic Religion*, published in 1971, was unusual in taking a sociological approach to the anthropology of spirit possession and shamanism. For Lewis, many ecstatic cults are expressions of marginalisation. "... we shall see," he wrote, "how a widespread form of possession, which is regarded initially as an illness, is in many cases virtually restricted to women ... For all their concern with disease and its treatment, such women's possession cults are, I argue, thinly disguised protest movements directed against the opposite sex."[4] Merete Demant Jakobsen reinforces this thesis in her book, *Shamanism: Traditional and Contemporary Approaches to the Mastery of Spirits and Healing*, with the example of a Greenlandic *angakkoq*, a female shaman named Kaakaaq who was humiliated as an orphan: "After experiencing her father's attack of madness, she becomes more aware of her own potentials as an *angakkok*, helped by a grandmother. It is pointed out that she looks like a boy. Again, the angakkok apprentice is orphaned and therefore marginalised in society."[5] The illness that precipitated the calling, the becoming-shaman (a new life that released the marginalised person from one form of social subjection only to pitch them into another), was commonly known as "Arctic hysteria," a psychological and physical crisis that would give shaman and patient-to-be a shared experience of sickness.

Central to the "performance" of the *angakkok* was sound, a medium that could manipulate the perception of space to envelop listeners in a dramatic embodiment of the spirit world, an architecture of the supernatural. Even in the 1930s it was possible for an East Greenlander descendant of a famous shaman to give a performance: "... with lamps extinguished and for two hours filled the air with the most remarkable sounds, knocking, drumming, singing, screaming, wheezing, howling, hissing, and distorted speech ..." though when the lights came on the audience laughed at this antique display.

Interpretations of spirit sounds, verbal auditory hallucinations and similar auditory phenomena associated with illness are interpreted according to the beliefs and knowledge of their day. Ringing bells, ticking, whistling and knocking have been identified as signs of musicogenic epilepsy, degenerative hearing and alcoholic hallucinosis, yet they correspond closely to the sonic repertoire of possession and shamanism. Shortly before my mother died she experienced hallucinations: the floor was covered with water and she heard children's voices singing. These were symptoms of an infection; when a doctor asked her if she was troubled by what she saw and heard she told him that she wasn't bothered by them as long as she knew they were not real. This question of what is real (in the world) or unreal (in the mind) is particularly acute with auditory hallucinations. The mind already teems with voices; the world teems with sound. Who can say which are real or not?

If sound itself is ambiguous and unsettling then sounds that possess an extra-human or unearthly origin will exacerbate these attributes. In the 1950s, when the BBC was beginning to experiment with electronic sound one committee cautioned that musicians and engineers should be exposed to electronic sound effects only for a limited time. According to Louis Niebur in his book, *Special Sound: The Creation and Legacy of the BBC Radiophonic Workshop*, this directive warned that they would succumb to "mental instability,"[6] as if exposed to radiation or toxic chemicals. Music made from the conjunction of electricity, invisible but lethal, and sound, invisible, impermanent yet deeply affecting, can be an object of suspicion.

Yet music in itself, as a source of both pleasure and pain, is also an object of suspicion: painful because music stirs deep emotions, feeds on nostalgia, seduces, invades silence, reminds us in its transience that life is short and all things pass, all of these qualities identical to those that give so much pleasure.

Notes

1 William Gilpin, *Observations on Several Parts of England, particularly the Mountains and Lakes of Cumberland and Westmoreland relative chiefly to Picturesque Beauty made in the year 1772*, London, 1808, p. 61.

2 James Kennaway, *Bad Vibrations: The History of the Idea of Music as a Cause of Disease*, Ashgate, 2012, pp. 44–45.

3 Robert Louis Stevenson, *The Complete Stories of Robert Louis Stevenson*, The Modern Library, 2002, p. 559.

4 I. M. Lewis, *Ecstatic Religion*, Penguin Books, 1971, pp. 30–31.

5 Merete Demant Jakobsen, *Shamanism: Traditional and Contemporary Approaches to the Mastery of Spirits and Healing*, Berghahn Books, 1999, p. 128, p. 85.

6 Louis Niebur, *Special Sound: The Creation and Legacy of the BBC Radiophonic Workshop*, Oxford University Press, 2010, pp. 36–37.

A Piercing Silence: James Richards

Published in James Richards: To Replace a Minute's Silence With a Minute's Applause, Whitechapel Gallery, London, 2015

Sound is invisible stuff. Those who have expertise in its properties and potentialities also have a tendency to lack a full understanding of the worlds of tactile matter, visible surfaces, the volume of sound. Sound is a thing and no-thing, like air, money, time or love, complex to infinitude as one of the ungraspable phantoms of life. All these metaphors we use to bring into being the property of sound and the sensation of its hearing: a honeyed voice, a rough voice, a piercing scream; the taste of viscosity, a hand passes over splintered wood, a needle punctures the skin. Think of sound – that high sound of hearing and air – pouring into the volume of a space, translucent block of air like colourless jelly flecked and warped with every passing noise event and its trail of decomposing matter, something like a stiff liquid or intangible runny paste through which the body passes without resistance yet it enfolds and penetrates the body with the insistence of abyssal pressure and the clotted emotions of memories as active entities, in flight like birds, insubstantial as papery moths.

Samuel Beckett wrote of this, in *Malone Dies*: "The noises, too, cries, steps, doors, murmurs, cease for whole days, their days. Then that silence of which, knowing what I know, I shall merely say that there is nothing, how shall I merely say, nothing negative about it. And softly my little space begins to throb again. You may say it is all in my head and that these eight, no six, these six planes that enclose me are of solid bone. But thence to conclude the head is mine, no, never. A kind of air circulates, I must have said so, and when all goes still I hear it beating against the walls and being beaten back by them. And then somewhere in midspace other waves, other onslaughts, gather and break, whence I suppose the faint sound of aerial surf that is my silence."

Within this skull that is the originary space from which we move out into the spaces of the world there is a teeming of inaudible, unverifiable sound that is indistinguishable from the self. A head in the darkness of a room, barely possessed of body, face jutting forward alert and listening to the air circulating, the monumentality of time articulated by these periodic vibrations we call sound. Of the heads of Francis Bacon: Gilles Deleuze spoke of them as an answer to the question posed by the history of art – "how can one make invisible forces visible?" – writing: "The extraordinary agitation of these heads is derived from a movement that the series would supposedly reconstitute, but rather from the forces of pressure, dilation, contraction, flattening, and elongation that are exerted on the immobile head."

The immobile head looms, thrusts, seated in a darkness that may be absence of light or consciousness, convulsed by the rhythms of being, described by Henri Lefebvre as an integration of things: "... this wall, this table, these trees – in a dramatic becoming, in an ensemble full of meaning, transforming them no longer into diverse things, but into presences." Then there is a merging of presences audible and inaudible, asking which is alive, which sounds are in the realm of the real, which are born airless from the inner hearing of the listener. "If normal voice-hearing leads to the point that all experience is internal," writes Daniel B. Smith in *Muses, Madmen and Prophets*, "hallucinated voice-hearing begins there. Hallucinated voice-hearing conflates source and destination. All preliminary steps are bypassed. There is no breath, no manipulation of air, no movement of bones or cochlea, not even a stimulation of the auditory nerve. With voice-hearing the brain, working alone in its watery chamber, creates a voice out of nothing but its own duplicitous silence."

Can we ever be sure that sound exists (as a manipulation of air)? Rooms themselves have presence, resonate strangely in hypnotic currents. In 1907, Victor Segalen, friend of Claude Debussy, wrote a small novel about this – *Dans un Monde Sonore* – in which the world of things becomes audible, the sense organs become interchangeably confused, sound can be crystal, light or an indeterminate uncanny resonance that renders speech indecipherable in much the same way that Alvin Lucier's *I Am Sitting In a Room* gradually foregrounds resonance, obliterating writing and the sense of speech. Spaces breathe quietly to themselves, singing and whistling as if thinking in some ultra-dimension inaccessible to human perception. Make an audio recording of a room and the living force of it is immediately evident, not only its resonant frequency, its size, emptiness or fullness and reverberant characteristics but the sound through which all of these auditory properties are translated into what we call atmosphere. This is the subliminal sublime, always acting upon the way we behave in any given place. It is part of the language of cinema. "I'm real fascinated by presences," David Lynch has said, "What you call 'room tone.' It's the sound that you hear when there's silence, in between words and sentences. It's a tricky thing because in this seemingly kind of quiet sound, some feelings can be brought in, and a certain kind of picture of a bigger world can be made. And all those things are important to make that world."

Just as Segalen's protagonists become gradually deaf to normal auditory conditions by their hypersensitive turn to the world of resonance, so Beckett's narrator in *Malone Dies* speaks of "hearing things confusedly" all those noises of the world that were once to easy to differentiate, he now hears as a "single voice ... one vast continuous buzzing ... unbridled gibberish." What he was once able to decompose into its component strands reverts into a composition of the world's sound. The human no longer plays an active part in its own theatre.

Within this domain of ambiguity and transference lies the potentiality for hearing what is seen; for seeing, tasting and touching what is heard. In the mid-seventeenth-century Netherlands, the theatre of sound and listening in painting was dramatically articulated by *An Eavesdropper with a Woman Scolding* (1655) by the Dutch artist Nicolaes Maes. As a young man, Maes painted a series of six works in which the act of listening was represented by eavesdroppers listening in to mildly salacious scenes within domestic interiors. In this particular case a maid has left crockery in disarray in order to enjoy the sight and sound of her mistress giving a fearsome tongue-lashing to some hapless victim. A painted green curtain seemingly hung from a trompe-l'oeil pole is pulled back to reveal half the drama, obscuring the right-hand view of the interior and concealing the victim. The painting's theme and treatment has its origins in Dutch farces for the stage, then further back to Greek myth, to a contest of artists and their facility with illusions. Zeuxis painted grapes that were so lifelike the birds tried to eat them. Zeuxis then asked Parrhasius of Ephesus to pull

back the curtain so that he could see his rival's work, not realising that what he mistook for a curtain was Parrhasius's painting.

Museums and galleries are ritual sites, theatres in which all extraneous stimuli and invisible forces, particularly sounds, are spirited away by the thought magic of museology and its suspensions of reality. Sounds of certain kinds — noisy chatter, music, mobile phones and other distractions — are implicitly forbidden by unspoken rules of acceptable behaviour, though official lectures and gallery tours, teaching, audio guides and the noises of infrastructural machinery such as air conditioning and projectors are not. Within the frame is what matters, even though the frame, the enforcement of silence, the light, the colour of the walls and the constant presence of security are equally prevalent.

In the Kunsthistorische Museum in Vienna I sat contemplating Jan Vermeer's *The Art of Painting*, considering the trumpet held by the woman who models for Vermeer's painter, its bell half-concealed by the heavy curtain draped at the "front" of the scene. A man sat next to me, German language audio guide spilling from his headphones in ethereal whispers that could have emanated from the painting itself. He stood, ostentatiously walked back and forth to study the still atmosphere of the painting and its silent trumpet, the percussion of his shoes on the hard floor not at all silent or still. In Buckingham Palace I stood in front of *The Listening Housewife* by Nicolaes Maes, acutely conscious of maids whispering in adjacent rooms. In Dublin I walked through the National Gallery, my boots echoing from the wooden floors within their resonant halls, just as Samuel Becket's boots echoed many years before. Again in Dublin, at the Hugh Lane Gallery I peered through glass at the painstakingly reconstituted wreckage of Francis Bacon's studio, thinking it not dissimilar to the authenticism of displays in contemporary zoos in which a small patch of desert sand is littered with location-appropriate cans and bottles. Somewhere behind hides a venomous reptile.

Buried within these involuntary scenographic, sarcophagic halls of antiquity is the compulsion to build narratives from incomplete (silent) images. Yet they are not so silent. Bacon, as Deleuze noted, wanted to "paint the scream more than the horror" (as he further notes, the sensation rather than the spectacle). Mouth closed, the man is alone in darkness, like so many other men scrutinised by Bacon, half-in half-out of indeterminate masculine haunts of club, study, bar or dungeon – suit, tie, white shirt, studded chair, claustrophobic frame – its murky truncated depths a seething murmuration of creaks, air, the apparatus and effects of respiration, black resonance at the threshold of hearing. This is what we call silence, a silence both unhearable and unspeakable. Unbearable too, the kind of silence heard or not heard in recordings of memorial silences bounded by durational markers, those one-minute silences in which life is held in abeyance for the sake of those who can no longer hear them.

In these silent haunts, spaces and gaps, sound is a haunting, a spectral presence that moves ghostly and unseen through every orifice and volume. Sound is an absent presence; silence is a present absence, a sinister resonance that forces us to question our belief in the things of this world. Francis Bacon spoke about opening up the valves of sensation, as if all the separated channels and reservoirs of the body could join their various intensities, all ingress and outflow mixing together in a hallucinogenic stream (or scream).

Any student of cinema or user of YouTube knows the dramatic effect of adding sound to images, manipulating sound already associated with images (those shred videos) or subtracting sound from images. For an example, imagine swapping Krzysztof Penderecki's *Als Jakob erwachte*, used by Stanley Kubrick for climactic scenes in *The Shining*, with Mantovani's "Colors Of My Life," used by David Lynch for *Inland Empire* (thinking of the shift between Kubrick's beautifully composed interiors and the lurching fuzzy flatness of Lynch's digital video) or imagine swapping Penderecki's *Als Jakob erwachte*, used by Lynch for *Inland Empire*, with "Midnight, the Stars and You" by Ray Noble and his Orchestra, used by Stanley Kubrick for *The Shining*. But the music also changes according to the images, Mantovani's unearthly easy listening strings no longer easy, in Lynch's world overloaded with an ominous emotional tangle of tragedy and bliss.

Bring together painting and sound and two inimical theories of time confront each other. In one, the gaze claims no set duration; nothing moves under its scrutiny; its potential is to last for ever or be instantaneous, a fleeting glance. In the other, time and its complex rhythms are in a state of constant movement and unfolding. Space also shifts, the gaze of listening in motion everywhere. A piercing silence hangs in the air, lasting for all eternity or no time at all.

Gilbert & George: The Singing Sculpture

Lecture given at Lentos Kunstmuseum, Linz, Austria, 2011, on the occasion of Jack Freak Pictures, a Gilbert & George exhibition

"Art for all," say Gilbert & George. The purpose of art, according to them, is to provoke ideas but they want to reach what they describe as ordinary people. Artists fall in love with art, with the image of art, whereas Gilbert & George have fallen in love with the viewer of art. It's a game of perception played with great skill and intensity for more than forty years.

The Singing Sculpture inhabits a world which it refuses to acknowledge. This has proved to be the best possible strategy. Rather than associating with comparable art events and art movements or to its antecedents in popular entertainment, the sculpture has been presented and to some extent preserved as a mystery. It speaks directly to those who recognise its power. To everybody else it seems an anomaly, an inexplicable action that came from nowhere and whose existence is both mystifying in all respects – mystifying to the artists themselves, which they happily endorse, and mystifying as an action or object within the trajectory of post-war art.

This is what George has to say about it: "Somebody said it was rather like a force of nature, like a waterfall. People would stop in the country and park the car and look at a ravine or a waterfall and think about all the things they would not normally think about. It gives people the opportunity for thought: loving thought, difficult thoughts, and that's of course the power of culture, the power of art is to make us think and feel."

Gilbert & George met at St. Martins School of Art in London in 1967. They were, to borrow the title of the Neil Simons play, an odd couple. George was from a poor, single-parent home in Plymouth, a naval city on the

Devonshire coast of south-west England. Because of bombing during the war he and his mother moved to nearby Totnes, an arty little town closely associated with the radical education experiment of Dartington College (later attended by George) and these days a bit of a New Age enclave. Devon used to be, perhaps still is at times, a county stereotyped as a place of comical rural accents spoken by country bumpkins who are a bit too slow for the metropolis. Gilbert was similarly disadvantaged by his Dolomite village origins in the dialect speaking South Tyrol. A talented woodcarver, he had studied art and suffered bullying in Austrian and German colleges before arriving in swinging London during the first summer of love. His lack of English and difficult accent meant that nobody could understand him. Besides, these two war babies were already a little old and uncool for the mood of the time. Of course there were older aristocratic types and veteran artists of the avant-garde mixing with London's young hippies and fledgling pop stars but Gilbert & George had none of the proper credentials, either in looks, experience or attitude. Perhaps for that reason, alongside some erotic frisson, they have described their first meeting as "love at first sight."

Despite their oddity, they thrived within the atmosphere of St. Martins, working within a semi-secret department of the school. The sculpture department – headed by Frank Martin – had become famous for its dismantling of the precepts of sculpture. Anthony Caro led the interrogation of traditional sculpture while Peter Kardia and his colleagues went even further with the experimental "locked room" of 1969, in which students were shut in and forced to work only with the materials made available to them. Once they had made something, this was discarded and they were given different materials and told to start again. In his own words, Kardia "focussed on developing in students a genuine engagement with their creative process." His emphasis was on why and how, rather than what, and we can see from a partial list of graduates from the school during that period – Gilbert & George, Richard Long, Hamish Fulton, Bruce McLean, Roger Ackling and Barry Flanagan – this move away from the solidity of form to the discovery of making was hugely influential. Sculpture could be soft or uncontained, a song, a text, a pose, a process of burning using the sun's rays, a bicycle ride, a pile of rocks, a photograph or a walk.

Even within the liberal environs of the leading art colleges, such radical gestures created tensions between artists and students determined to test the limits and administrators who were equally determined to enforce them. Most notorious of all was a work by John Latham called Still and Chew/Art and Culture, 1966–67. Working as a part-time lecturer at St. Martin's College of Art in 1966, Latham borrowed Clement Greenberg's defining text of modern art, *Art and Culture*, from the library and then encouraged friends, family and students, Barry Flanagan among them, to chew the book and spit out the pages. These were processed over some months until reduced to liquid and injected into a glass vial. When he returned this transformed object – the glass vial – to the library as an overdue book Latham was dismissed from his job. Posterity has taken a different view: the work is now regarded as a key gesture of conceptual art and the dematerialisation of the art object and is now kept in the collection of the Museum of Modern Art, New York.

Gilbert & George have claimed that their first thoughts about The Singing Sculpture began in 1967. In other words, they met and their work began immediately, as if life had become a sculpture. They developed it as a performance work in their studio in 1968, then performed it under railway arch number 8 in East London's Cable Street on the afternoon of October 26th, 1969. The location is significant, Cable Street being the site of a famous street battle between police defending a march by Oswald Mosley's blackshirt British Union of Fascists and a loose alliance of 300,000 anti-fascist outsiders which included local Jewish groups, anarchists, Irish people, artists and communists. Photographs of Gilbert & George performing their new work behind a line of string show a fascinating sky-line, a snapshot of the future – a foreground of ruination that may be bomb damage from the Second World War or simply urban decay, and then in the background, clear signs of scaffolding around high-rise apartment blocks under construction. The spectators look typical of art crowds at this time – two young women wearing mini skirts, young men starting to grow their hair long. This is in sharp contrast to Gilbert & George, whose hair has grown not at all since the psychedelic summer of love. They are both wearing dark suits of a cut that is neither respectably traditional nor overtly fashionable, though I notice George's jacket is unbuttoned, which would never do now. Where everybody else is striving to be casual, they are stiff, like showroom dummies and in this respect they understood an aspect of the future that would prove to be extremely important – that it could be very powerful to defy orthodoxies, even an orthodoxy claimed to be more relevant, fashionable and free than all others.

No surprise then that hardly anybody understood what they were doing, though it has to be said that *Monty Python's Flying Circus*, first broadcast on BBC television in 1969, worked in a comparable area of English absurdity – men learning to fly on office tables, places where you could pay to have an argument, ridiculous choral songs about lumberjacks and, of course, the Ministry of Silly Walks. Another photograph shows Gilbert & George performing The Singing Sculpture at the National Jazz and Blues Festival, Sussex, also in 1969. By that time, jazz and blues actually meant rock and so the festival headliners included The Who, Pink Floyd, Soft Machine and King Crimson. Ironically, Gilbert & George have gone on to greater and more lasting fame than most of the lesser acts listed on the programme, as well as turning out to be more subversive than the prog rock headliners. Predictably, the experience was not a happy one. "It's very nice to be with the galleries," they said a few years later. "They look after you splendidly."

Their decision to locate themselves within the art world was no less bizarre than this brief flirtation with rock music. In almost every way, The Singing Sculpture contradicted the dogmas of the time. Very quickly, their piece went through a number of incarnations: originally it was performed as *Our New Sculpture* at St Martins on January 20th, 1969. One of them held a walking stick; the other a glove, though photographic documentation shows that they exchanged these items during the performance. The music was played on an ancient wind-up gramophone with the record being played twice, to give the illusion of being turned over. As befitted the anthropology of the time, concerned as it was with structural meanings of social rules, they devised The Laws of Sculptors:

> Always be smartly dressed, well groomed relaxed friendly polite
> and in complete control
> Make the world to believe in you and to pay heavily for this privilege
> Never worry assess discuss or criticize but remain quiet respectful
> and calm
> The lord chissels still, so don't leave your bench for long

Although the tone of this manifesto was archaic it was also eccentrically punctuated and, moreover, published in SHIT AND CUNT Magazine Sculpture. And what about that word "chissels," spelled with two ss's?

Perhaps God is a sculptor, busy working away on the universal project (and maybe singing as He sculpts) but the slang use of "chisel" suggests cheating, as if God is always looking for a way to undo the best efforts of conscientious workers. Right here – the beginning of their career – they have embarked on a strategy of paradox. Outwardly they are conservative and restrained yet in their art utterances they flaunt taboos against obscenity, disgust, blasphemy.

Their choice of music – a version of "Underneath the Arches," originally made famous by Flanagan and Allen – was absolutely critical to the impact of the piece. Bud Flanagan and Chesney Allen have been described as Britain's "most popular and enduring double-act of the inter-war period." As Roger Wilmut observes in his history of Variety entertainment, *Kindly Leave the Stage*, they were unusual among double acts for showing affection toward each other. Most double acts are built on the tension of funny man versus straight man. The humour of Laurel and Hardy, for example, is embodied in the disparity of their visual appearance: Ollie, a fat man, talkative and bossy, and then Stanley, a sad-faced, thin man who seems to be a simpleton but whose talent for accidents makes his partner the perpetual victim. Though there is violence and envy in the double act, there is also intimacy, shared trouble, companionship in adversity. A double act cloaks itself in anonymity to find a small space of freedom in a hostile world.

Double acts – usually master and servant dialogues – featured in ancient Greek and Roman drama, Shakespeare plays and other forms of early modern European theatre, but one of the main inspirations for the double acts of the English music hall was American so-called "Nigger Minstrel" shows. White performers wore black theatrical makeup to present caricatured vignettes of African-American life, language and music. Their disturbing popularity in Britain can be gauged by the fact that blackface variety entertainment was still being screened on primetime television up until 1978. In 19th-century minstrel shows, musicians in blackface would be interrogated by a compère without the makeup, usually to exploit the racist stereotype of the slow-witted plantation negro. This highly charged imbalance of power was adapted in the late 19th century by an Irish comedian, Joe O" Gorman, and his partner, Joe Tennyson. As Roger Wilmut described it O'Gorman "was one of the first performers to expand this primitive style [of 'Nigger Minstrel' shows] into something

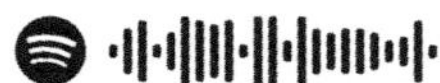

more resembling the master-and-servant confrontations of classic drama, thus setting the standard for cross-talk acts up to the present day."

Flanagan and Allen had been members of a touring show called The Crazy Gang, an ingenious response to the great depression of the 1930s. Against the backdrop of hunger marches by the unemployed, Variety entertainment in the theatre was threatened by both economic hardship and the rise of the sound film. One impresario, George Black, responded with the concept of "continuous Variety," an imitation of the cinemas of the time which allowed the audience to enter and leave whenever they liked yet still see the full programme. At the Palladium Theatre in central London, Black presented acts like Christopher Stone, the first of the BBC radio disc jockeys. Stone would walk onto the stage and say, "I'm now going to play you a very nice record, I hope you enjoy it." He would put the record on and everybody in the theatre would sit and listen until it finished.

The Crazy Gang was formed from a collection of double acts who interrupted each other's routines, showed audience members to the wrong seats and generally worked around an organised chaos of word play and physical comedy. Flanagan and Allen, specialists in free associative humorous routines and sentimental songs, were added shortly after the Gang's formation in 1931 and went on to become its most enduring and influential stars.

Co-written by Bud Flanagan in 1931, "Underneath the Arches" is one of the duo's best known songs. To the generation that had survived the Depression and both World Wars the song's lyrics came to evoke memories of hardship, poverty, people living rough on the streets, maybe busking for small change with a portable musical instrument or a song. As it true of our times, the gap between rich and poor grew disturbingly wide, but the bitterness of this reality was sweetened by the Flanagan and Allen image of affection and friendship, their gently swaying motion as they sang and the delicate lilt of their roughly matched voices.

Sentimentality and overt nostalgia were central to the song's appeal. I was born in 1949, seven years after George, so I can easily envisage him hearing the song on BBC Radio's Light Programme in the 1950s. Flanagan and Allen split up in 1945, as if the ending of the war signalled the completion of their career, but their presence was felt in various ways on the television light entertainment shows of the 1950s and early 60s. I also enrolled at a London art school in 1967, so have a clear idea of the mood of the times, what was fashionable and what was not. By that time, most people of my generation had listened to endless war stories and heard enough songs about bluebirds over the white cliffs of Dover. I was listening to soul, blues, rock, jazz, electronic music, free improvisation – anything other than the songs that had given fortitude to my parents and grandparents.

Yet there is a subconscious level at which images of the past resonate, a place from which they rise up to haunt us. Even at the level of naming, Gilbert & George connect with this mythical reiteration of cultural memory. For a British person there is a history of light popular song that begins with Gilbert and Sullivan, followed by any number of international pop references and long-vanished duos who identified themselves by a pairing of first names: Sonny and Cher, Peter and Gordon, Nina and Frederick, Jan and Dean, Paul and Paula, Santo and Johnny, Shirley and Lee, Chad and Jeremy. Introduce some complexity to the naming, as with Simon and Garfunkel or Leiber and Stoller, and the quality and longevity increases. Perhaps in search of gravitas, a little extra seriousness as a contrast to the humour, comedy double acts were more inclined to use their family names — Morecambe and Wise (both such great admirers of Flanagan and Allen that they recorded their own version of "Underneath the Arches"), Flanders and Swann, Abbott and Costello and then in more recent times, French and Saunders or Reeves and Mortimer. Peter Cook and Dudley Moore as Derek and Clive were the exception, absurdist, scabrous, filthy, not unlike Gilbert & George in fact.

Gilbert & George situated themselves in an unforeseen niche within this overripe history, a self-consciously arch "art for all" that was clearly not for all. How much forethought and planning went into the development of their unique image and cultivated sincerity will never be known. It was as if they had resolved to tap into a stream that ran deep within British culture; unable to save themselves from being outsiders – two awkward, lonely young gay men who would never wear love beads or snakeskin boots – they made the most radical statement possible, by turning to a maligned history, by embracing conservative politics when all those around them were moving further to the left, by dressing like bank managers from some forgotten backwater of provincial life and by fusing antique entertainments with avant-garde practice.

They were art of a challenging sort yet twenty years earlier could have survived as what was known as a speciality act. My father used to enjoy a speciality act called Wilson Kepple and Betty. Assisted by the seductive dancing of Betty (there were numerous Bettys), tapdancers Wilson and Kepple dressed as ancient Egyptians and to the exotica of Alexander Luigini's *Egyptian Ballet*, soft shoe shuffled on a board dusted with sand, as if arisen from the Egyptian Book of the Dead. To watch film of their weird performances is to experience a lost world of acts barely comprehensible even to their fellow entertainers – an entertainingly pointless minimalism devised only to fill a gap and yet somehow deeply engaging. Dr Goebbels saw Wilson Kepple and Betty perform at the Wintergarten Berlin in 1936, thought them indecent and declared their bare legs "bad for the morals of the Nazi Youth."

The Singing Sculpture evolved through a number of versions: first of all they walked up and down for five minutes in front of a wind-up gramophone. A 78rpm record they picked up in a junk shop crackled away, a stiff, antiquated version of "Underneath the Arches" recorded by a forgotten duo called Hardy and Hudson. Then it developed into something more formal – a cassette tape player set up on a plinth in front of the table on which Gilbert & George now stood to perform their piece. They were Living Sculptures, they declared, and so painted their faces gold, bronze, patches of colour. Like accoutrements of some strange village ritual, the glove and cane became permanent props and though the reference may have been unintentional, these accessories of the tap dancer, the song and dance man, combined with the painted faces were eerily reminiscent of the *Black and White Minstrel Show*. Margaret Thatcher's favourite pop singer, Cliff Richard, had sung about his "crying, talking, sleeping, walking living doll," back in 1959, and so they echoed this sado-masochistic desire for the doll, even implying a speciality act of two ventriloquist dummies without an operator, each singing for the other.

Then came their defining moment. In 1969, *Live in your Head: When Attitudes Become Form*, an exhibition curated by the late Harald Szeeman, was scheduled for the new premises of the Institute of Contemporary Arts in Nash House, London. The exhibition had travelled from Bern to Krefeld and then London and in every city it accommodated work by local artists. Justifiably, Gilbert & George felt they should have been included, since The Singing Sculpture ticked all the necessary boxes for installation works, minimalism, conceptual art, performance art and innovative sculptural materials. They still talk about their omission from When Attitudes Become Form with a mixture of wounded pride and glee but this is because ensuing events led to their breakthrough into Europe and beyond. Their protest was to attend the opening night party dressed in suits and metalized painted faces. Enraptured by their appearance, a German dealer, Konrad Fischer, offered them a show at the Dussledorf Kunsthalle and so between 1970 and 73 they performed The Singing Sculpture in German, Swiss, Danish, Italian, Belgian, Australian and even British galleries and museums.

This sudden celebrity was the revenge of the nerds and perhaps that is part of the fascination. The piece had expanded to become an 8 hour performance, their magnificent obsession, and in that sense it ran parallel not only with endurance marathons by artists like Joseph Beuys but with the contemporary trend in minimal music to work with simple materials extended for long durations. Terry Riley gave all-night concerts in America and Europe, La Monte Young devised pieces that could in theory last for years, and Steve Reich and Phillip Glass in their early days composed marathon explorations of just a few chords. In normal circumstances it would make sense to link this work – what we would now call sound art – with similar initiatives in experimental music. I am thinking of Alvin Lucier's *The Duke of York*, for example, a work from 1971 for solo voice and synthesisers in which the composer sang Johnnie Ray's melodramatically emotional ballad of the early 1950s, "Cry," through a barrage of synthesiser processing.

In the same year Gavin Bryars composed *Jesus Blood Never Fail Me Yet*, his now famous orchestration of a pre-recorded singing voice. The recording – an elderly man who sang alone – was taken from location recordings made by filmmaker Alan Power during the making of a documentary about homeless men in London. Bryars recalls the occasion when he made this specific segment into an endless loop. He was teaching at an art college in Leicester where he used two tape recorders to repeat the short segment of song. Going out for a break Bryars left the studio door open. Later he came back to find the art students nearby in a state of shock. All of them were silent; some were crying, and so Bryars realised the emotional potential of the song.

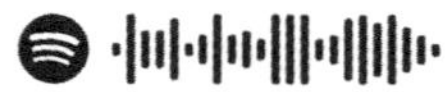

This story is not dissimilar to early accounts of The Singing Sculpture. On a number of occasions Gilbert has said that audiences fell into a trance or started to cry. This engagement with direct emotion, even sentimentality, confronted one of the great taboos of the times and so experimental music began to rework popular songs and occasionally flirt with emotional sentiments that composers like Stockhausen dismissed as the worldly superficiality of lesser beings.

The trick was to perform with sincerity but in Gilbert & George there was a paradox. Though they were sincere and apparently utterly conventional, they also gave off the air of being disengaged from human society, even human biology. There was something of Vaucanson's mechanical duck about them, automata from another era. Lost and homeless, they had made themselves the centre of their own work. They had created an art without objects but in doing so they had objectified and instrumentalised themselves. Like two wind-up speaking dolls, or an elaborate music box, they exhibited charm and at the same time, an underlying threat or horror. Beyond humour, avant-garde eccentricity and novelty, our only logical response to the work is to experience the chill of the uncanny. They might remind us of Olympia, the uncanny automaton who haunts E.T.A. Hoffmann's *The Sandman*, the living doll who plays piano with great accomplishment and can sing a bravura aria in a "piercingly clear, bell-like voice," yet whose sneezing is revealed to be the squeaking of her clockwork mechanism winding itself up.

Despite art for all, they chose galleries as the proper place for their work and yet sound in the gallery was anomalous then and still is – it makes far better sense to stumble upon the work of an artist like Susan Phillipsz under a city bridge, by accident, than it does to hear it in a gallery. The gallery was a quiet place of seeing, a library without text or a church in which God was optional, a place for the mute things of Poussin, or what Delacroix described as the silent art. In the 1960s sound was becoming more common in galleries – partly because art forms since Futurism and Dada had converged on each other but also because many music venues would no longer accommodate experimental music. Through its emphasis on duration and the creation of unique objects, La Monte Young's association with Galerie Heiner Friedrich in Munich is one example of music moving closer to becoming an installation art. Even when I was still at school in the mid-1960s, London was a good place to see and hear kinetic art, whether the slow scraping of Pol Bury's sculptures or the frantic self-immolation of Jean Tinguely's machines. If there was something excessively literal about kinetic art's desire to make sculpture that moved, then there was something uncanny about The Singing Sculpture, not only a human sculpture that moved but a meditation on the haunted nature of sound, the crackle of an obsolete gramophone, the ephemeral nature of sound as a metaphor for loss and decay, the atmosphere of Samuel Beckett whereby the most intimate and unspeakable of human desires and memories must be externalised through the medium of a speaking machine.

Sound has an uncanny quality. Is it really there? What made the sound? Am I hearing things? This is particularly true of the voice. Much is invested in the link between speech and text, as if spoken words in their transcribed state can carry the complete meaning of any verbal communication. "Read my lips," George Bush Sr. once famously said (lying through his teeth), as if the silencing of sound in the visual formation of silent text was a guarantee of truthfulness and intent, because sound, the untrustworthy medium, was suppressed and overcome.

Gilbert & George asked their audience to read their lips, their sculpture rich in the references of silent cinema, silent humour, clowning and mime – Charlie Chaplin, Marcel Marceau, Jacques Tati – reminiscent also of composer Erik Satie, with his regulated life, velvet suits, sly humour and beguilingly simple music. Identifying sources or influences in their work is a risky business, but any interpretation of The Singing Sculpture, Bend It and The Red Sculpture must take into account a long history of mechanical beings, androids and robots. They anticipate postmodern bodies in states of perpetual reinvention, transformed into machine hybrids and the posthuman utopianism of self-sculpting. Think of Madonna, Grace Jones, Klaus Nomi, David Bowie or Michael Jackson of course, or Laurie Anderson's singing cyborg of "O Superman." As if following the lead of Gilbert & George, certain strands of experimental rock music began to react against hippie free expression. Kraftwerk biographer Pascal Bussy argues that Kraftwerk may have been influenced into changing their image from rebellious improvising longhairs to besuited, short-haired science students by Gilbert & George. This seems plausible, since the first performance of The Singing Sculpture outside the UK was in Dusseldorf, the home of Kraftwerk, at exactly the time when the group was forming.

This appropriation of conservatism – the uniform, the behavioural discretion, the willingness to transform the body out of its origins, its biological determinism, into a new being that is somehow anonymous, like an industrial component, an alien, a brainwashed groupmind fully adapted to the science fiction future of virtual pop stars – became a powerful theme in 1970s music and much that followed. To varying degrees we can see it and hear it in Devo, The Residents and David Bowie, with a strange twist in Roxy Music, and in Yellow Magic Orchestra. The back cover photograph of YMO's first album, released in 1979, shows Takahashi, Hosono and Sakamoto dressed like butlers or head-waiters, each holding electric cables over a stiffly bent arm. Bleached out behind them in the workspace is a Moog synthesiser, its modules patched together like an ancient telephone exchange.

Retrospectively, YMO seem in those early days to have been as much concept art as pop band. In the same way that Kraftwerk had satirised the image of German people as identical hyper-efficient robots, slaves to industrial perfection, YMO's songs, image, name and artwork reflected back the stereotypes of Japanese people, making fun of themselves as a subterfuge for satirising the common post-war image of Japan: a land of geishas and geeks in glasses. "In the beginning we pretended to be misunderstood Japanese," Ryuichi Sakamoto once told me. For their second album – *Solid State Survivor* – they were photographed in identical red suits, playing dominoes and drinking Japanese Pepsi with a male and female dummy, which makes me wonder if members of the group or their art director had been present for Gilbert & George's Tokyo exhibition of 1975, where they performed The Red Sculpture, besuited of course, their heads, faces and hands painted solid red, moving robotically to commands spoken from a tape recorder.

The third Gilbert & George sculpture for sound and movement was Bend It. The song was originally recorded by the ultimate name band – Dave Dee Dozy Beaky Mick and Tich – in 1966. Though a big success for them in the UK, Austria and Germany it was banned by many radio stations in the USA for supposedly suggestive lyrics. One intriguing aspect of the record was its bouzouki sound, actually played on a mandolin. The song starts slowly, builds up, stops, then builds again to finally break out into an instrumental interlude in which the bend it dancers can really show

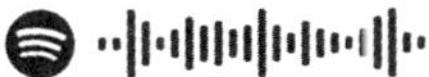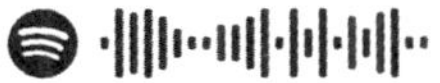

off. It's not authentically Greek but the mood almost certainly derives from "Zorba's Dance" by composer Mikis Theodorakis and the famous dance scene from *Zorba The Greek*, a film released two years before the original "Bend It." Anthony Quinn, the rough Greek peasant, is teaching Alan Bates to dance on a beach in Crete. Bates plays a repressed half-English, half-Greek writer so the scene spoke very directly to the feeling of the time, that Englishmen were so reserved, so well-mannered, so inhibited, that they had no idea how to enjoy themselves. Of course Theodorakis's music goes faster and faster and Bates finally takes off his jacket and loosens his tie.

This is the stereotype of Englishness so brilliantly worked into art by Gilbert & George. We may look like nowhere men, they say, but underneath this quiet, anonymous exterior is shit, piss and overwhelming quantities of passionate feeling. This is the meaning of sculpture.

Stick, Spit, Reed and Tubing: Writing on Seymour Wright

Blog post, David Toop, *a sinister resonance*, August 17, 2015

"Or maybe the music we are hearing tells us about the unconscious, coming from some place of archetypes or from the trauma of unspeakable secrets."
Hyperobjects: Philosophy and Ecology *after the End of*
the World, Timothy Morton.

There are many ways to think about a musical instrument. A compellingly bizarre essay published in 1976 by Alan Dundes – *A Psychoanalytic Study of the Bullroarer* – is perhaps the most extreme example of this. Through the convolutions of his argument, Dundes persuades the reader to consider the bullroarer through a miscellany of interpretations and theories: a phallus, a phallus inverted to become a womb or substitute womb, the fecundating agency of wind, fertilising breath, thunderous farts of the gods, an excremental device of shadows and secrecy, the voice of deceased ancestral spirits, an excreta hawk, shit eater, masturbation symbol and flatulent phallus. 

Perhaps this is a lot of weight for a slender strip of wood to bear, but once implausibility and risibility are set to one side, then a different kind of thinking about objects of this kind opens up, not just in relation to the instrumentality-of-the-instrument but as a loose, vast "mesh" (to borrow Timothy Morton's term) of properties, actions, conditions and futures (what I have called elsewhere "bodies without organology," which is to say an object whose extent lies far outside the constraining discourse of musicology, encompassing the deepest reaches of its composition). If what is just a simple strip of wood attached to string can inflate itself to the cosmic dimensions of flatulent gods then its supposed evolutionary position somewhere to the furthest far west of the piano becomes reversible, the piano a regression or retreat back into the cave of resonances, too timid to venture into a vibrating, respirational and unsystematic open air populated by shit eaters, excreta hawks and farting gods.

Once this was a subject of prolific anthropological debate, this complicated relationship between the playing of a bullroarer and its sounding, in which the instrument became spirit voice or mask, a collusion maintaining the structure of a society, the way in which women, men, children, non-human entities and barely imaginable beings negotiated each other's space. The object or sculpture of the playing – to whirl a strip of wood in circles – was the small spark that lit the raging fire.

At this moment I am *not* listening to Seymour Wright's *Seymour Writes Back (alto saxophone solos 2008–2014)*, partly because I have done so already and will do so again, but partly because to attend to the spark at this given moment of thinking-through-ideas is a distraction from the raging fire. It may be that he has some sympathy with this idea of bodies without organology. The physical form of the release is a folded sheet of texts and photographs on which are mounted four audio CDs, further enfolded in a wrapper reproducing a 1920s design by calligrapher Margaret Calkin James, an artist whose posters for the London transport system were both as celebrated yet as anonymizing as Phylliss Pearsall's design for the London A-Z street atlas; within this he quotes Peter Brook on *King Lear*: to paraphrase, the play (a usefully versatile word in this context) is an object, a cluster of relationships, complexes and meanings rather than a linear narrative.

If you like, it's a mythology of the saxophone, a universe inhabited by the gliding tremor of Johnny Hodges (true ancestor to Albert Ayler), Sonny Rollins mowing his lawn, the reaction of the crowd to those famous twenty-seven choruses played by Paul Gonsalves that set alight "Diminuendo and Crescendo In Blue" at Newport, 1956, Richard Wilson's *Watertable* (whereby London's agitated water table could be seen and heard through a 28-inch diameter concrete pipe sunk 4 metres into the clay beneath Matt's Gallery), the unfolding of London's spaces and places over centuries, the blurred still image of a blurred video of Willis Gator Tail Jackson screaming through a tenor saxophone without restraint on the Ed Sullivan Show in 1955. These and others.

He quotes Clarice Lispector, from *Água Viva* and *Hour of the Star* – "What am I doing writing to you? Trying to photograph perfume?" and "as for the future" – both quotes as enmeshed with spectacularly vast sets of ideas as the bullroarer; in doing so pulling aside the screen (as Daniela Cascella does also in her book, *F.M.R.L.*) that was obscuring for us the prophetic relevance of Lispector's writing to our present day endeavours in the making of an un-music, by which I mean a working in sound/not-sound that attempts to reclaim an intensity of time, feeling and objects from the emptied out rites of bourgeois music.

And if I listen I hear the vibration and resonance of a pipe burrowed through London clay into its watery substratum, a new way of listening as predicted by Clarice Lispector in *Água Viva* (1973): "I see that I've never told you how I listen to music – I gently rest my hand on the record player and my hand vibrates, sending waves through my whole body: and so I listen to the electricity of the vibrations, the last substratum of reality's realm, and the world trembles inside my hands."

And if I *am* listening then I hear the respiratory, the gustatory, the intestinal — not unlike the bullroarer whose sacredness can never be disconnected (no getting away ...) from sex, food, shit and death. And if I listen I hear the disappearance of the saxophone, lost in the woods or eaten up by circular inhalation and the voracious nature of space and its bodies. And if I listen I hear the future of a tradition. There is Evan Parker, seated at the table and photographed by Roberto Masotti for his book, *You turned the tables on me*, and there in this title and preceding titles – *Seymour Writes Back, Reed 'n' Wright*, and so on – a jazz tradition of creaking puns on names exemplified by another alto saxophone player, Lee Konitz, whose "Subconscious-Lee" and "Ice Cream Konitz" have a purpose beyond what we call word play.

Now I am listening, in-close and personal to spit, reed and tubing, to the face and mouth, to the rumble of steel through tunnels under the last substratum of reality's realm, the friction of expulsion into restless air, the softness of an instrument that gives itself up to all those vibrations to which it is subjected.

4

holes traps bat caves moths fluttering

Haroon Mirza

Alvin Lucier

trap set: Haroon Mirza

Published (first section) by Camden Arts Centre, London, File Notes, 2012, and (second section) *The Wire*, **March 2012 (Issue 337)**

A musical instrument: a crafted, singular object with monetary value, designed for purpose as an extension of the human body. Like a ventriloquist's dummy it sits quietly in a box, awaiting its other body, the voice which is both latent and lacking within itself. This is one way of thinking.

Now here is another musical instrument, spread through a place, without centre or operator. It is its own body, an autonomic body drawn from many sites: the body politic, the body of language and knowledge, the sound body, the body eviscerated. This instrument waits for no other sounding body, but draws plural voices from phenomena, rhetoric, ritual, archives, the movement of the body, divisions and articulations of space, symbolic objects, pulsations, the behaviour of sound, fragments of abject memory and all of that which may become lost.

From the first moment of experiencing a work by Haroon Mirza I felt *within* the work. Not immersed or overwhelmed, nor integrated within its field, but a clumsy moving appendage adding shifting outer edges to its already (seemingly) messy assembly. The piece was sculptural within the terms of Picasso's *Construction with Guitar Player and Violin* of 1913, the *Singing Sculpture* of Gilbert & George or John Latham's *Big Breather*. Coincidentally or not, all of these works engage with sound; they consider sound as a property that cuts into space and vibrates air, that operates first of all at the level of physiological and emotional affect, that adds complex dimensionality to the acts of looking and being in proximity to a focal point.

Clearly not just a collection of sorry materials in random juxtaposition, Haroon's work resonates with ideas. The precise nature of these gained clarity during his Lisson Gallery exhibition of February/March 2011 but my interest is not so much an exegesis of these various signs embedded within formal construction but the way they scrape against each other to make an event called music. The drum kit makes a useful parallel. According to James Blades, the drummer's "trap set" or kit was a late nineteenth-century hybrid partly inspired by one-man bands. During the early jazz era this kit was expanded to incorporate many items of exotic and domestic origin; the limits of the instrument were defined by negotiation between operator and environment. It was a contraption (hence a trap set), a somewhat pejorative term that describes a great deal of sublime music and which might also be useful for a better understanding of Haroon's *I saw square triangle sine.*

Because Haroon builds up a conglomerate of sources and objects within one work, or within the rooms of an exhibition space, there is a tendency to think about his work as an exploration of systems. That seems fine except for the fact that elements are only partially connected to each other and to some extent dysfunctional. They are part of a sound system in the more particular meaning derived from Jamaican reggae, a conglomerate of social forces, operators, musical sources and technology adapted to serve a particular passage of musical flow.

A different beginning may be more constructive. To say: this is the way music is made today. So there is turntablism, minimal techno, dub, noise, micro-improv, acousmatics, phonography, club culture, YouTube videos, loops, drones. sonification, digital audio and video sampling, pop and sound art archives and so on with all their different devices, protocols, operations and audiences. Though these specialist genres may coexist within an mp3 library as a kind of fiction-in-the-mind, it would be unusual to find within one setting (to take two late 20th-century examples) the activities of the wrk group from Japan – a near-scientific exploration of auditory phenomena pioneered by artists such as Minoru Sato, Jio Shimizu and Toshiya Tsunoda – and the minimal techno of Monolake. Despite extreme differences in their modes of "performance," they are united in an interrogation of sound, its potential forms and the environments in which sound can be experienced. In both cases, they invent new instruments. This is

one way to understand what happens in certain contemporary musics – not as a collection of separate devices which come together in front of an audience but as a "contraption" with which all listeners interact.

When I curated Sonic Boom for the Hayward Gallery in 2000 I was given the impossible task of making distinctions between many sounding artworks within a space lacking in any provision for isolating sounds. As the installation progressed I began to think of the entire building as an instrument and so the challenge was to "tune" its elements to generate a passage through sound rather than a static cacophony. Sound Spill, a curatorial research project initiated by Haroon with Thom O'Nions, was formed to pursue a similar potentiality. Their exhibition of 2009 brought together the sound work of four artists through the sensibilities not just of a "visual" curator but a composer who begins from the premise that composition is organised sound (to borrow the terminology of Edgard Varèse).

From this perspective Haroon thinks of himself as a composer, which is interesting enough in the light of what might be implicit in the notion of a 21st-century composer, but I'm also fascinated by some of his reference points: Guy Sherwin, a filmmaker associated with London's structural film movement of the 1970s, and then Fred Sandback, the American artist whose work in the same decade really pushed the limits of how a space could be articulated by the most minimal of means. Both had strong connections to sound and though Sandback's work was silent, you might say that both used sound as a way of disrupting our perception of who we are in relation to a given space. By naming the sight of the invisible, *I saw square triangle sine* goes a degree further. The instrument contraption is the room itself, all that cannot be brought to the room, all that happens in the room and all those who enter the room. The work, seen and unseen, is instrument and instrumentality.

Spike Island, Bristol: total dark, a corridor leading off into *The national apavilion of then and now*. I move forward, sightless slow and wary, hands held up defensively, right foot dragging after left (a comedy gait not dissimilar to Muhammad Ali's demonstration of why he called George Foreman The Mummy). Light flares, revealing a doorway. The corridor walls are softened and textured by grey pyramid studio foam; sound loses resonance by degrees, as if all the senses are smothering under thick blankets. Beyond the doorway lies a small anechoic room, metal grill floor laid over foam.

LED strip circled into a halo and suspended from the centre of the ceiling glows into life for 20 seconds, its wheedling hum spluttering out in decay before expiring with a pop that seems to suck all life from the universe.

Later the same day I am with Haroon Mirza in an upstairs room of this gallery/art space converted from a tea-packing factory originally built in 1960 by Brooke Bond (the company that used live chimps dressed as humans to advertise its tea). We sit in the women's toilet, now empty of all memory of its previous function but for one tiled wall whose reflections generate a very distinctive resonating frequency within which our voices swell and merge. Outside the window flows the brown water of New Cut, a remnant of early 18th-century developments to Bristol Harbour. Mirza's exhibition, flickeringly entitled /|/|/|/|/|/|/|/|/|/|/|/|/|/|, pulses with its own flow of hauntological history.

The national apavilion of then and now, shown at the Venice Biennale in 2011, is presence – the memory of a sound and the afterimage of light persisting in darkness – opposed by absence – no echoes, no national pavilions, no electromagnetic waves. The deeper reference, he tells me, was to Jacopo Tintoretto, perhaps the most grotesque analogy possible for a conservative critic like the late Brian Sewell but nonetheless intriguing. In *The Last Supper*, *The Fall of Manna* and *Storm rising while the body of Saint Mark is being transported*, Tintoretto depicts worlds oppressed in states of darkness only relieved by one burst of light, a break in the clouds, a halo.

My first exposure to Mirza's work was one piece in a group show at the Lisson Gallery in March 2009. Though genuinely perplexed by its scattered randomness, I felt a compulsion to return and figure out what was going on. In 2010 his first London solo exhibition, again the Lisson, impressed me with its rigorous economy, apparent in even the most abject, sprawling work. The elements of each piece may be worthless but each one matters, each individual work spilling into the next one. "I guess I've always followed this logic that if it's not doing something then it shouldn't really be there," he says. "If it doesn't play an acoustic role or supporting an acoustic end then it doesn't need to be there. There's an aesthetic logic. It's irreducible."

By that time I had grown allergic to the term Sound Art (and all the partial and partisan histories growing up around it) and so preferred to think of him as a composer, or at least amanuensis to a wraith composer instrumental in the operation of a score laid out through extended space and stretched time, audio technology eviscerated like a body hung drawn and quartered. References to HipHop sampling, minimal techno and the kind of microacoustic phenomenology practiced by Toshiya Tsunoda, Michael Prime, Lee Patterson and Minoru Sato were easy to spot but intriguing for their close fit, as if to say, this instrumentality, these seemingly incompatible genres through which we can work sound in the 21st-century are enfolded. This happens when they converge in a single space.

Elsewhere he has talked about being a composer; in Bristol he's reluctant. "I went to this talk about Fred Sandback recently," he says, "and Lynne Cooke was talking about when she was working with him – he was talking about space as the musical score and he was making these incisions. I don't think there's any difference between composition with sound and composition with objects in space or lines in a drawing, so the word composer encompasses all those things. I feel comfortable with that. The uncomfortability comes from standardised disciplines of practice. If I've gone to art school, studied art and show in galleries, for me to say I'm a composer to someone who has traveled around the world with the LSO, I'd feel like an idiot. It took me ages to call myself an artist. It's only in the last two or three years that I've comfortably said, yes, I'm an artist. Saying you're an artist can mean anything, so that's alright."

Fred Sandback is a recurring source in Mirza's work. Born in New York in 1943, Sandback began as a banjo and dulcimer maker, took a BA in philosophy, then studied sculpture at Yale with tutors like Donald Judd, finally combining these three strands into an extreme minimalism whereby lines of coloured string articulated the nature of a space and the presence of those moving through it. Flow and the violence with which sound cuts into a space defines the Spike Island exhibition. Each piece plays in the same key and all rhythms link. In the last of five "rooms," individual sculptures are activated by timers to play percussively in sequence, Closing my eyes I could be in the middle of a Central African funeral ceremony hearing small groups of hand drummers from different compass points, their sound truncated or blown in closer by meteorology, proximity, movement.

Some works in this room were made in collaboration with a Sheffield based sculptor, James Clarkson. The collaborative instinct is typical and to some degree heretical. Commonplace in music, of course, it suggest a readiness to relinquish control or blur the notion of ownership, at the very least to make overt the way ideas and techniques have always been appropriated, gifted, transformed and intensified in art. "The idea of this work," says Mirza, "was that it was an album. I said to James, I'm going to make an album of eight tracks and you're going to do the cover artwork."

Later, during the opening of the show, I wander round again and observe a woman standing very still in this particular room, a ferocious WTF look on her face. What does she see: a horrible teak veneer display cabinet (Parker pens, as it turns out), Mastercard and Visa stickers still attached, 3 LED lights flashing a rhythm that vibrates an upended speaker cone that in turn bounces a heavy chain suspended as if by rope-trick magic from the ceiling; a Grundig radio lying on a turntable, activated fuzzily and periodically by an energy saving light bulb suspended from one of Clarkson's black MDF sculptures; a translucent screen cannibalized from what was once the last word in TV technology, lit by coloured spots generating harsh sawtooth drones that swell, fade and pop. Added to this are more islands of utilitarian cast-offs, the kind of invisible artefact whose funeral parlour is eBay and freecycle, car boots and those shops that specialize in office furniture made threadbare by financial despair and the diurnal abrasion of indentured arse-cheeks.

One work – a flashing blue circle of light, a pale green circular thing that could be an ashtray, a line of flashing red LEDs – is possessed of surrealist mystery. Think "primitive video game from the early 1960s crossed with luminescent invertebrate." Tak-u-tak-tu-te-tun, it goes, over and over, lost creature calling for mummy, prototype autobeatboxer, discarded experiment from the workshop of Konono Nº 1. What is that blue flashing light? "That's from a bike light," he says. "It's funny when something's taken out of context. LED technology has changed so much in the last five years – you can make more efficient and complex lighting programs. They do all these visual effects but as soon as you amp up the LEDs, you get this electro-acoustic world. James made formal things, like a shelf with a drum rim, and I basically pimped them up."

Another room, a raised platform on which the empty orchestra stands: drum kit, inverted cymbal revolving on a stand, Panasonic radio revolving on a turntable, hanging light bulb, LED halo, propped up against a black MDF shelf unit a Juno 60 keyboard arpeggiating endlessly through a three-note bass line, hanging microphone running through Holy Stain Electro-Harmonix pedal, Marshall amp connected to two speakers set into a table. On the floor lies an LED display and headphones. The display is also title – *I saw square triangle sine* – running continuously (the feeling of tailending a queue in the Post Office, blankly reading "helpful information"). The headphones allow a private auditioning of the sound made by this red letter sequence, harsh and nasty as if inscription must always return to its abrasive origins. On the wall hang seven biliously psychedelic paintings of trees by Angus Fairhurst, an artist famous for his poignant gorilla sculptures, famous also for taking his own life by hanging himself from a tree in Scottish woodland at the age of forty-one.

I wonder aloud if this piece – an echo of the Fairhurst exhibition in which a drum kit was set up for anybody to play – is a tribute. Is there an emotional component buried within its post-minimal, post-structuralist economy? "I didn't know he'd hung himself on a tree," is Mirza's reaction. He learned the details of Fairhurst's suicide only after planning the piece. "Anyone can play the drums. It seemed faithful to what he was trying to do. The palette of the work came from his paintings. There's a bit of red, some yellow on the Juno, just primary colours. It's already an homage to him, to take that away would be a disservice."

All of the sounds of this piece are analogue machine sequences, repetitions that seem interconnected if only because of a neural compulsion to find pattern within cluster illusions. I could imagine the late Jaki Liebezeit stepping up onto the platform, locking deep into the groove as he did with Can, but the prevailing tendency is to self-consciously splash around on a few cymbals. A few children flail wildly and then things start to get interesting. At that point of collision between mechanical inexorability and the frailties of human "creativity," a history of automata comes more sharply into focus, a history in which the synthesis of life perpetually eludes its creator and reinforces death's inevitability. In her book, *Living Dolls*, Gaby Wood quotes a psychologist and robot specialist, John Cohen: "A robot can never commit suicide," he writes, because "true suicide implies a foreknowledge of death and some idea of its significance, and this is a privilege of man."

But these empty orchestras propose another idea, that objects and events of all kind are inherently musical. In 2007, during a residency at the Lahore National College of Arts (co-founded, incidentally by Mirza's great-grandfather), he explored the ambivalent attitudes to music within Pakistan. One of a group of pieces that emerged from this research was *Taka Tak*, based around a video of one the many cooks you can see and hear on the streets of Lahore, usually lit at night by life-threatening thickets of jerry-rigged cables and light bulbs that may suddenly go dark in a power cut. In Mirza's piece a buzzing radio and flashing LEDs mix climactically with the audio drama of meat chopped on a smoking circle of steel, a thrilling mayhem alluding to Sufism, the Qur'an, house music and the vernacular sounds of life as lived within very different cultures.

"Pakistan was an interesting time for me," he says," because I went there to find out why certain schools of Islamic dogmatic thought demonised any level of engagement with music. It was a huge contradiction because the faith itself was so musical – for example, the Qur'an or Adhan are recited in such a musical way. I guess the dogma came from the idea that music led to premarital sex or infidelity through dance so it slowly became prohibited. Spending time there I discovered that this had a huge impact because music seemed to be embedded in everyday life due to the suppression of it. Rhythm and melodic variation is part of everyday praxis.

"Qawwali and Sufism is interesting because they use music and dance as a means to get closer to God. There is a strong tradition of this in Islam that seems to have been wiped out. Things are different now but the social relationship to music is very different from how it is in the West. I guess it poses interesting questions about listening, such as whether there is a greater tendency toward reduced listening as opposed to casual or semantic listening."

The controlling of a cultural "material" as volatile as sound can be inherently provocative. One of Mirza's references is beatmixing. During his student days he earned money as a DJ, first of all consumed by house music but then opening up to Warp, Detroit, hip hop, the pop charts. Maybe this is why I can experience his "instruments" as the materialization of early house music's innards, as if "automatic" tracks like "Washing Machine" by Mr. Fingers or "Phone System" by Ricky Smith had been generated involuntarily by self-replicating systems in a Chicago foreclosure warehouse.

"Beatmixing is such an important cultural point In musical history," he says, "but it's not really addressed." He talks about using pitch control

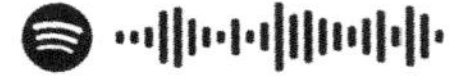

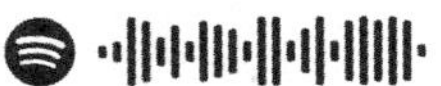

turntables, adapting the technique to his own ends, and the way pioneer DJ Francis Grasso made the claim to have discovered slip-cueing and beatmixing in the late 1960s as if humankind had suddenly discovered fire. A dedication to dance music history in all its implications underlies many of his works, but most overtly in *Paradise Loft*. As the name suggests, it evokes the story of two legendary Manhattan clubs – Paradise Garage and The Loft – sparking off thoughts about an intersection of music, technology and the search for sheltered settings in which sexual preferences, social status and race identity might become liberated, if only for one night, from those divisive, oppressive judgments that rule on the outside.

Some might consider the art gallery transplant an academic denaturing of dance music's purpose but clubs of this type already nurtured their own scholarly tendencies. In 1984 I went to The Loft on a Saturday night, then returned a few days later for an afternoon soiree in which David Mancuso and Steve D'Aquisto played a variety of records – Louis Armstrong Hot Five 78s, for example – on their incredible system. The guests, including DJ Walter Gibbons and the editor of a high end hi-fi magazine called *The Absolute Sound*, sat around, enjoyed the music and discussed Japanese hand-made cartridges.

Between materials (substances, properties, gestures, movement and time, all of which are engaged through a knowledge of artists like Sandback or Guy Sherwin, Bruce Nauman or Joseph Beuys, not to mention DJs and street chefs) and meaning there is tension. Meaning is material collected from the aether. "Artists aren't these singularities who do this genius stuff," he says "All art has to be a culmination of lots of ideas coming together. I constantly worry about having too much meaning, that it becomes about something, or not enough meaning, so it's just presenting a technological system. Working with existing objects is a way to have that balance and then not being too precious about those objects, or becoming an audiophile, worrying too much about the quality of the sound."

I wonder about this balance, looking at *Tescotrain*, three vertically stacked video monitors of different shape and type. The top one shows a malfunctioning Tesco sign, the bottom one shows strip lighting on the blink at Peckham Rye, the middle one is a explosion of light switching on and off. All of them pulse sonically in a weird J Dilla style beat. The atmosphere is of a broken world under surveillance, non-places whose malfunction has become just another phantom backdrop. They glow with desolate beauty, at the same time stirring up memories of Bristol's Tesco riots in April 2011, a portent (not that anybody in power was listening) of the August riots that swept across the UK. Though not deliberate, the reference is there to be activated. "Do you make decisions subconsciously or is it completely arbitrary?" he asks himself. "It's choices. It could easily have been a Honda sign that was flickering on and off."

Structural film of the 1960s and 70s, long condemned to the cultural dustbin, makes a comeback in Mirza's work. In *Tescotrain* he generates audio from the static of the CRT monitor. "I showed a work called *Night Train* by Guy Sherwin in the first Sound Spill show I curated," he says. "We discussed ways of presenting the film that reflect how the sound is generated. Guy suggested sticking a contact mic on the screen to pick up the static from the light. I since discovered that copper does exactly the same thing so adopted a similar process to generate some of the sound for this piece. Guy's piece amplified the sound generated by lights passing by the train he was on whereas my footage was documentation of lights malfunctioning. So it's really the process and the title – *Tescotrain* – that refer to Guy Sherwin."

Themes of interference, broken media, systems speaking for themselves run through his work with the pulsing certainty of his LED lights. Growing up in Bracknell he developed a taste for dismembering toys, examining their workings, reconfiguring them. "My sister had this doll that spoke," he says. "It was really interesting. It had a really small plastic record inside, like a rudimentary turntable with a stylus. Somehow I managed to fix it when it stopped working. I often think about my sister's doll because I felt a great sense of achievement from it. When I make work, that's how I grade it. It's only when I discover something, like if I connect this to that then that does this. It's that process."

crying: on Alvin Lucier's *The Duke of York* (1971)

Published in *The Wire*, November 2005 (Issue 261)

Johnnie Ray is a forgotten figure in twentieth-century pop music. I'd guess that most people under the age of forty or fifty don't know who he is, yet his huge success with "Cry" in 1951 was the first warning of a new era of solo pop stars, fan hysteria and incontinent emotions. "Cry" is actually a sickening song, a self-pitying message delivered with the kind of exaggerated vocal mannerisms that led to power ballads and songs like Andrew Lloyd Webber's "Memories." Ray Charles recorded a cover version but it's somewhat less surprising than the version made by Alvin Lucier. *The Duke of York* was composed for vocalist and synthesist by Lucier in 1971, and it's one of his few works for synthesiser. A lot of different ideas about memory, simulation, vocal identity, composite images and entertainment are compressed into this piece. As Stuart Marshall wrote in 1976: "Lucier intends the work to have many phantasy correlations – the tracing of ancestries, hidden family ties and ancient liaisons ... Synthesis not only takes place between successive identities but also between the performers' remembrances of the vocalist's or each others' chosen identities." What it actually sounds like is a crazy man singing "Cry" in the bath, in a bat cave, on short wave radio, in a karaoke bar, in outer space. Definitely a "What the ...?" cover version.

    note: *The Duke of York* was released on Cramps nova musicha n. 11, 1976.

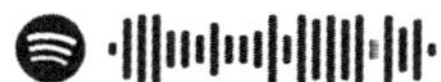

5

soft edges flickers stitches traversal

Annabel Nicolson

Alvin Lucier

Mieko Shiomi

Rolf Julius

The Woman Seen Sweeping the Sea: Annabel Nicolson escaping notice

Blog post, David Toop, *a sinister resonance*, 5 February, 2013

If a piano becomes silenced through dereliction, keys detached like so much loose kindling, is it still a piano? I asked that question, silently to myself, watching Annabel Nicolson's *Piano Film* (Camden Arts Centre Film in Space, group show selected by Guy Sherwin) and asked another more troubling question, of whether Annabel's work is still her work when she is not present? "It is what happens to things when they are not being looked at that puzzles me," she once wrote.

I had not become unconscious of her work, not turned away from it. Last summer, after lengthy deliberation and equivocation I wrote an extended essay on the subject of *Circadian Rhythm*. This concert was devised by Evan Parker as a continuous 24-hour performance for eight players – himself, myself, Paul Lytton, Paul Lovens, Max Eastley, Annabel Nicolson, Paul Burwell and Hugh Davies – for Music/Context, the festival of environmental music that I organised for the London Musicians Collective in 1978. Edited sections had been released on an Incus LP in 1980 but now Evan was proposing a release of the complete 13 hours of playing achieved on that July night 35 years ago. Paul Burwell and Hugh Davies had since died; in preparing my essay I spoke to the remaining players but Annabel's communications dwelled only on the difficulty of beginning to speak about it, then on the impossibility of the task. She was living in the Northwest Highlands of Scotland, "with gales to listen to often." Perhaps in the spring, she said. I am ashamed to say I could not wait any longer.

Escaping Notice was the title of a book she published in 1977. Prophetic, maybe? I spent time with the fragments on display at Camden – all silent amid whirring clamour – trying to find within them my own memory of Annabel, finding only tantalising wisps of her presence stuffed into that most abject means of archival display: the PVC display book. Escaping Notice also possesses that fugitive quality: its thin translucent papers through which texts and photographs are faintly visible; the events which are as nondescript as the flatness of Norfolk she describes with such cunning wit; the modesty of her anecdotes undermined by their doubtful veracity, and a detached third person self-anthropology which documents the artist, Miss Nicolson, as log rolling down a hill, or film star in the company of Mike Leggett, or sweeping the sea. As far back as 1974, she was engaged in low-key pursuit of the now earnestly fashionable practice of "walking" (or should that be "practice of walking"?), though we may surmise that these walks were not strenuous, encompassing as they did visits to jumble sales, buying postcards in the Garrigil post office or the observation of stick insects flying in strict formation, noted while Miss Nicolson lay in a cornfield above Corton Denham. In many of these works she calls upon the humble medium of local newspapers to recount the exotic life of a stranger, passing through rural communities as a woman of mystery, searching for the ineffable, the minor incident barely worthy of comment, or the more serious business of vanishing footpaths. Her observations of the woman sweeping the sea in July 1975 form a brief document worthy of the disinterested observer, perhaps a man detained for a few moments while walking his dog: "Her lack of direction was plain and she seemed to have plenty of time. After a while one realised that she was less distinct, though not actually further away. Perhaps it was deliberate this trick of making herself part of the background of being just slightly out of focus."

Now she is more than slightly out of focus, a subtle commingling of dry wit, ephemerality and modesty conspiring with her physical absence to render her almost invisible. Of course I am happy to see her represented in a London exhibition, in a context to which she belongs, yet I remember her differently, as somebody who thought deeply about convergent strategies in the 1970s and created opportunities as an organiser, publisher, writer, curator and artist to open up spaces to those strategies.

Much of her thinking seemed to embrace that which is not there or cannot be objectified, and so she was drawn to sound, to smoke, to light and dark, to silence. In a recorded conversation between Annabel, Steve Beresford and Paul Burwell (*MUSICS*, no. 8, July 1976), conducted at the old Piano Factory in Camden Town, north London, she spoke of finding a piano in the yard of the factory: "It was deteriorating and when it rained the keys started to float. It played by itself and the keys moved around quietly." Magic is always present as a possibility, quiet magic in the background, and the possibility of the artist slipping away quietly, to become anonymous as the work becomes autonomous. Phenomena are left to take care of their own work of entrancement.

For a later issue of *MUSICS* magazine (no. 20, December 1978) published after the Music/Context Festival, she contributed a page that collected together the sources of her participation in *Circadian Rhythm* but also captured the non-dimension field of its unfolding, as an event within time and darkness. So there are marks, evidence of charring, fibrous plant materials, and references to the song of women pearl divers of Taiwan, sparks thrown into water, a hidden fire, lights in trees, the room filled with smoke, and from Mark Twain perhaps, two stories: the frogs of New Orleans whose song would rise in volume when the steamboats passed, and then thick fog on the river, people in small vessels banging tins pans so the steamboats wouldn't run over them. Hidden drumming, she wrote.

Even then she was rather hidden herself, one of the only women in a cluster of male dominated scenes. Again, her anthropology came into play, particularly in the improvised music setting of the London Musicians Collective. "One of the things that puzzled me," she wrote in *Resonance* magazine (vol. 8, no. 2/vol. 9, no. 1, 2000) was just how little the musicians, all men at that time, seemed to talk to each other. Often they would meet and with barely a word prepare to play together. There appeared to be very little communication in any recognisable sense. Then somehow out of this apparent absence of communication would come the most wonderful sounds." At the same time, she was acutely conscious of her own voice. Her essay – Transcript from indistinct recording of a talk performed in the reading room of Slade School of Art 13/3/79 (*MUSICS*, no 22, June 1979) is an object lesson in what we now call reflexivity, or performativity,

the question of the voice (particularly the female voice) as social medium, performance tool, expressive and reflective marker of the self, along with the necessity of listening with a willingness to understand. "I've been thinking recently," she wrote in 1978, "that performances are almost like lectures, focussing thought as a means of sounding out what is most urgent in one's mind." Reading that again took my breath away, since it is almost exactly what I have been thinking about my own work in recent years. Sometimes we internalise a borrowed thought, unconsciously make it our own, and there it lies sleeping for years, until shaken awake by the right circumstances.

What her talk at the Slade made clear is that there was no such thing as a definable "practice" or "intact" work, as she put it; rather an evolving form of performance which might take many forms. For this there is no validation, no archive, only the ghostly trace of somebody fishing in darkest night in search of a quarry barely distinguishable from its environs. When sound art is discussed, or improvised music, or performance art, or the voice, or writing about sound, "Miss Nicolson" has somehow slipped the net, despite her centrality to the evolution of these interrelated arts. This seems to me to be a profound injustice, but also the way of things. Monuments are constructed and under cover of darkness small chisels chip away at their presumption and perfection.

MUSICS magazine (spanning the years 1975–1979) is a treasure trove of ideas and information but one of the pieces I treasure most is a conversation between Annabel and Max Eastley. They are two of my favourite artists and much-loved friends, that's one reason why, but they have a sensibility in common which is strengthened by their exchange of ideas, and the ideas are as intoxicating as they are fragile: the night-blooming sirius that opens only one night of each year; a tree shadow frozen in ice; the blazing tar barrels of Shetland Island rites; Gaelic song not as folk music but as reverence for the phenomena of its subject; a raft of straws; fireflies in cages and oily birds, threaded through with wicks, flame spouting from their mouths; the effect of Galloway dykes on frightened sheep; the projected image of a bird that was, in fact, a crack in a glass roof; the shock of twigs cracking very loudly as she walked on them. Coming and going. The presence of materials. Scattered images but potent, they exemplify the open work. In the sound of the voice they find cohesion. "Nothing else is needed, just the means you have, like your voice," she wrote. "Performance is a struggle and in a sense things are coming from far away because they are coming from something silent and making a huge leap towards being audible. Something very ancient about it."

memorising wilderness: Alvin Lucier

**Published in *The Wire*, April 2006 (Issue 266) as a review of
Alvin Lucier played by Anthony Burr and Charles Curtis,
released by Antiopic Sigma Editions**

If Alvin Lucier had chosen to be a painter or sculptor, I suspect that he would now be established as one of the greats of post-abstract expressionist American art. His work would sell for six figures a pop and be visible in every contemporary art hangar in the developed world. Sound being the reluctant, intangible commodity that it is, people are still trying to figure him out, though the figuring has lately shown signs of approaching some sort of consensus.

There are times when I regress to the figuring stage. Do we describe him as a composer, whatever that word now means, even though many of his works resist interpretation by others? Pivotal late twentieth-century works such as *I Am Sitting In A Room*, or *Bird and Person Dyning*, can be produced by others for audiences or recordings, yet they are so bound up in Lucier the person that only he can bring them fully to life. These are big idea works, memes that can be transposed, as in Bernhard Gal's *I Am Shitting In a Room*; the danger, clearly, is that homage can reduce them to one-liners. Other pieces feel elegant and unique, simultaneously dry as dust, but that response may be born of my own subjectivity as a musician and music lover: the desire for a musical sensuality where, like mathematics, there is the beauty of ideas, the revelation of natural phenomena.

Of ideas, no shortage: located in an odd place somewhere between the lightworks of Dan Flavin and James Turrell, art and technology hybrids of the 1960s like Robert Rauschenberg's *Mud-Muse*, the nineteenth-century

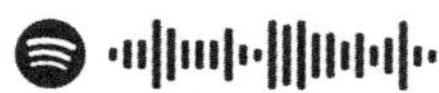

physics of Hermann Helmholtz and John Tyndall, and ancient, arcane acoustic experiments unearthed in Joseph Needham's *Science and Civilisation In China* (not that any of these analogies are quite right), they are very susceptible to analysis. The possibilities were demonstrated by Stuart Marshall, in his meticulous, intellectually expansive essay for *Studio International* in 1976, *Alvin Lucier's Music of Signs In Space*, yet still there's a mad science aura that can leave the listener little on which to dwell.

This uncertainty of placement within an existing context shifted with the compositions for pure wave oscillators and instruments. In liner notes written for a previous CD collection of such works, *Still Lives*, Lucier justified the reason for the switch in pragmatic terms. He had observed what he described as "bitterness and frustration" among his teachers and colleagues, who composed a piece and then waited for an ensemble to work up a performance. In response to requests for pieces from players of conventional instruments he began to explore the possibilities of audible beating frequencies, by pairing the steady pitches generated from pure sound waves with the less constant but more flexible tones of instruments such as cello, koto, and clarinet.

The seven examples performed by Anthony Burr on clarinets and Charles Curtis on cello are consistent in their apparent simplicity. Like a lot of post-war art, they pursue one idea with a persistence that may be obligatory for a full examination of the initial principle, but still demand a lot of stamina and patience from the listener. Their means are rigorous – pure sine waves sound throughout the performance, either fixed or sweeping up or down, and the instrumentalist micro-tunes or maintains fixed tones to draw out interference patterns.

Slow to evolve and intensely concentrated on psychoacoustic phantoms, they should be appreciated as precursors of contemporary sonic minimalism (Ryoji Ikeda and Richard Chartier come to mind). As Stuart Marshall suggested many years ago, the spatial component is central to Lucier's work. Whereas the humanly generated acoustic sounds can be located in a map of place defined by recording mix and loudspeaker placement, the hemispheric patterns of sine tones change amplitude according to the position of the listener within the playback environment, as well as appearing to spin through space and even within the listener's head.

Simple these piece may be, but they are far from easy. Both Burr and Curtis play with extraordinary control and finesse, their virtuosity channelled into the sublimation of human frailty. Given the difficulty of matching human movement and breath control to the pure harmonic motion of sine waves, the apparent plainness, scientism even, of the conception is compelled to admit bigger themes of human-machine relations. At a microscopic level another form of complexity is revealed, an auscultation of the strange phenomena that intensive listening provokes. The tiny disturbances that can be heard during *Music for Cello with One or More Amplified Vases*, for example, become eruptions as the resonant frequency of a vase encounters the pitch of the cello and is then amplified by a microphone inside the vessel.

Lucier's work could be regarded as a music stripped of stories, but I'm not so sure. There are tributes and memoriams: *In Memoriam Stuart Marshall*, for bass clarinet and pure wave oscillator, aligns the gravity of dispassionate process with the emotional gravity of loss to forge a piece that is powerfully affecting in its restraint. Poetry flickers at the edges. The title of *On the Carpet of Leaves Illuminated by the Moon*, originally written for koto player Ryuko Mizutani, comes from Italo Calvino's *If on a winter's night a traveller*, a passage in which the narrator wonders if it might be possible to distinguish the sensation of each single gingko leaf from another, as they fell onto the lawn like rain from their boughs, "as the number of leaves spinning in the air increases further, the sensations corresponding to each of them are summed up, creating a general sensation like that of silent rain …."

The eroticism of this chapter, perceptual games deriving from the keen observation of natural phenomena shading into sexual desire, and Calvino's role in Oulipo, the literary movement through which strange beauty emerged out of the observance of strict forms, might encourage us to reconsider Lucier as a clandestine Oulipian, or Pataphysicist. His fantasy, of being a nineteenth-century French Canadian fur trapper, memorising nocturnal wilderness sounds before sleep in order to match them with subconsciously heard predatory anomalies, is poetic enough to leave the question in no further doubt.

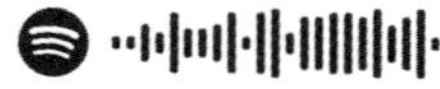

Mieko Shiomi, Spatial Poem No. 7
(At the time listed below listen to the sounds around you for a while)

from David Toop, recorded in London, N10, for 10 minutes at 01.00, 15.4.2018, "performed" for the exhibition: Orgasmic Streaming Organic Gardening Electroculture, curated by Karen Di Franco and Irene Revell, Chelsea Space, London, April–May 2018

Quiet, with a strong undercurrent of whistling and rushing air in my ears, probably because I've been rehearsing and performing a guitar piece for four days and my ears have yet to settle. Faint birdsong, high chirps A low humming from digital equipment. Aircraft noise, a roaring sound sometimes swelling to a defined pitch. Four audible footsteps from the neighbours, muffled. Passing cars, moving either left to right or right to left across my field of vision, their audibility lasting for around fifteen seconds each. Very faint sounds of children talking, then adults. When they come into view I see it is actually one woman and her son. Four cries by a crow A police siren in the distance, short duration. A wood pigeon's pulsing call, very brief. Somebody bangs their front door nearby, opens a car door starts the engine, crunching gravel, drives off quickly.

this is not a bee, it could be a kind of shadow: Rolf Julius

interview 4 February 2003, Berlin.

DT: Maybe we could start by talking about these very peripheral or normally inaudible sounds that you draw out, as if by magic, and make people much more aware of them.

RJ: Ah, so I am in Japan quite often, so most of the sounds from nature, I recorded there, which are kind of these insect sounds which I like, and then I do my own sounds, my kind of digital sounds with buzzer instruments, I think they are a kind of natural electronic – not electronic, but very simple equipment – I think it's not natural, but it has the same kind of background, because these clicking sounds from cicada, not from cicada, it's the same thing: cicadas are not singing but they're just doing this with their legs, so it's also kind of a mechanical sound, you can say that. But this is not so important, but what I found out that natural sound and my sounds, when I listen later to what I have done, when I've finished my piece, and you ask me what is the natural sound, I have problems to tell you what was exactly this sound at all. So therefore I like actually natural sounds, because when you hear sounds from a distance, anyhow, and you hear sounds which are interfering – you kind of hear atmosphere or something like that.

DT: The idea of atmosphere interests me a lot, and also the ambiguity between this idea of natural sound and – I suppose – well, it becomes very difficult to talk about, because of course you talk about natural sound and human sound, but then …

RJ: My brain maybe started to work. When I talk about natural sound, I'm interested not actually in the sounds themselves, but in the quality of having the distance, something coming from a distance I like very much. Also personally I like distance. So normally, people sit together very closely; I kind of like this distance. Also visually. And with the sounds, yeah, I like sounds coming from a corner, maybe, from there. But that means I like the air in between, and sometimes the air in between is not thick enough – so you add some other sounds, then you can kind of organise the atmosphere, I mean, the air you can organise, which is kind of sound again, or I don't know, it's kind of material. But in a more abstract way, do you understand what I'm saying?

DT: I do understand, because I think a lot of people are working in this area now, of making audible the inaudible and invisible environment and the characteristics of it, and I think actually you were one of the first people to explore this with sound.

RJ: When I do some auto-environmental pieces, I don't want to disturb the situation, but I have to do something, so I add my own sounds into the situation. But the whole situation, I call it music, because if I listen to environment sounds, I decided while I'm listening, Oh, I like that, and it is kind of pre-composed. So when I add something, I just sneak in between, and do something more, and then maybe the composition could be completed. So as an audience, if you walk through the situation, you don't hear it, because each ear is different – of course – and also the person is different, and what they think what is music is also different, so again, this person can have his own composition. Therefore sometimes I like being in an environment, because you are free which direction you put your ears. This is maybe one nice aspect. it's very difficult to come to the point. An aspect from music is silence, of course, but now, everybody talks about silence, and it's a kind of influxion. So this word silence, I more and more be careful about this word, because everybody says, "I like silence," and I say, "Oh, why?" Because silence is ... OK, silence, what is silence? I think it could be also, silence, it's a quality itself, it can be also very – not noisy, but could be loud. Silence is just not silence, but it's just a kind of form. Can be. Could be.

DT: Silence, of course, is a really relative term, because even when we have silence, our ears make emissions: we can hear our own brain working!

RJ: The older you go ...

DT: I was watching a video the other day about Takemitsu's film music. Donald Ritchie was talking about the Japanese concept of *ma*, and he was saying that *ma* is silence defined by non-silence: that's one aspect of it. As soon as you add something, then you have *ma*, which is a little bit like you were saying, something is almost finished within …

RJ: But this is the most interesting part anyhow! And then I like it, you cannot wait, you cannot work to get this. You need to know something, and then you have to forget everything, and this I like as a philosophy, or also as part of my work. So therefore I always say something, but I mean maybe this, but I don't mean this, I mean something else, so it's kind of both I like. Talking about silence may be the same.

DT: There's a Japanese poet I like, Makoto Ooka, and he wrote about *ma*, and he said that if you think of ma as being space between events, then you're wrong! A lot of people talk about silence in that way, that it's the space between events, but of course I think there's a continuum, and maybe in that sense, this kind of work explores these relative levels of intensity of sound.

RJ: [shows a photograph of a small loudspeaker on a rock] This is just photographs, and this was about silence, actually. It was in Finland, it was last summer, and actually I know this area for a long time, for more than four years. The weather conditions – it was just no wind at all, and also it was warm, and this case, it was the first time there was some clouds, therefore I like also, because the colours changed. So what I wanted to do is fake sound installations, so I put this stone into this water, and this stone already was in there, so I just put a speaker on top of it, just to make a photo, but not a photo: to do a kind of sculpture work. But I didn't connect the speakers to sounds: it was just kind of fake. But it was not fake. I just wanted to say it was absolutely nothing: it was just quiet. I was using a symbol for not being quiet, but loudspeakers say loud something, you know? I just was using it to create a piece or whatever which says, it's very, very quiet. So I'm using the opposite to create stillness or quietness. This experience took four weeks. I mean, it's the place I always go, my wife is from Finland, so I know the situations very well, therefore I like Finland because there are not so many people; they live quite alone there. Finland is actually on the Baltic Sea, so there are some islands you can see, so you see here the open sea, but because there are so many islands, it's a kind of lake. So what you see is, you have the water, which is kind of … just flat something, but it was water, it was just flat.

DT: Like a mirror.

RJ: Like a mirror, kind of, yeah. And then you see the islands, and you see the trees and you see no movement, but then you hear something from the distance, you hear one bird, or suddenly there's a fly like this and this you hear, but normally, if you hear a bird from a distance or a fly here, you don't realise: it's just, OK. But in combination with the visual part, it becomes very — not dramatic —so this fly actually was a bee that my brain said, 'Oh, this is not a bee, it could be kind of a shadow. Then again, with the bird in the distance, yeah, how can I say it, because the sign's on the microphone, actually! But this, it was very touching. So with this in my mind, I tried to make first a concept, and I tried to make a piece of music you can perform, knowing you can try that, but you cannot beat this situation. So maybe this says something about quietness in combination with audio-visual quietness. And therefore, I like Japan. Near Okoyama there is this garden which goes directly into the nature, so there is an artificial thing which goes visually into the nature mountains which are behind. I was there and I was OK, enjoying. If you are often there, then you kind of know, but still, it's very good, of course. So I was not so concentrated, realising how beautiful it is or how quiet it is, and then in the distance I heard a big noise, a very big noise, and this stopped suddenly. And after this big noise, I realised the beauty of the garden. But later I found out it was a concept which was five-hundred years ago, so they have a cage, and there were cranes, and they know the cranes every hour or once, they get crazy and they make this big noise! So I like the situation of it! So the brain was, again, ready for listening and for looking, I mean, for both.

DT: So you're saying that five-hundred years ago, if there was a crane nearby, they would expect it to make a sudden noise every so often.

RJ: Yes, there were ten there, I don't know how many, maybe twenty in one cage, so they got crazy. It was a concept! So later, I asked people, they said, Yes, you found out. But this I like also in music. You have music, and then all of a sudden, something happens, and then …

DT: I think the shakkei concept of borrowed landscape in [Japanese] gardens is very interesting in relation to sound, particularly if you're interested in distant sounds.

RJ: And also, it just comes into my mind: sometimes I look to nature and I see the situation – trees and so on – so I don't need to listen. Visually, I know it could be very different, so I don't care what's going on in reality. In the discussion about sound art, people say, OK, after so many years now, the ears are number one, but I think a combination should be very important, it's all the senses. So I like to look – of course! I do some of this kind of work, you know, OK, there is a bowl, there is some powder, pigment powder. The surface is sometimes just a little bit jumping or moving, because of the speaker underneath or inside this powder. So, also you can create with sound, because you need sound, quietness, because this movement is so kind of quiet. This I like.

DT: Quite often when I lecture, people still say, What is music? In a way this is a boring question, after Cage. Fifty years, maybe, people have been asking this question, and it still surprises me, and it also still surprises me sometimes how difficult it is to answer, despite the number of times we question this.

RJ: Actually, if you ask them the same, just look. OK, they think they see: they don't. You go to a museum, you see all these paintings or pictures there, and I myself, I like Ellsworth Kelly a lot, so I really look. I was in Basle, and I went there just to see the show. I was a little nervous, because he's getting older. Not everything is hundred per cent, which is normal! But I wanted not to say, Oh, that is nothing. I just wanted to know what are the good pieces, and of course the good pieces I knew, I could see, actually, but I was really working just to understand the language of the paintings. So later, I found out there are even better, more good ones than I saw at the beginning, as I thought at the beginning. I was just saying, I'm a professional for looking, so it's so much work. I think if you can see, also you can listen, maybe it's too short to say that, but I think something's right in that.

DT: I heard somebody say recently that in fact we hear and see too much and we should spend more time outside for smell and sensations on the skin.

RJ: It's true, then everything, then you can put it into relation, so you can skip something, and there is so much material coming, from every side. But OK, it's the same, you look at the TV or you read papers, and so on: you just have to find out your own way that you can survive, in a way. But art, in a way, and music or this kind of work, can help, to have a kind of small island to relax for a while, and maybe the whole system recovers for a while! It's like this with the birds. You always say, Oh, the birds are so endangered and if the landscape is destroyed, they will disappear, and so on. They only need I don't know how many trees and bushes, it's enough for them, so then they can recreate themselves, and it helps, the nature, I mean.

DT: I think that's also true, you were saying that so many people talk about silence, and I think, for example, many of the young generation in Japan are exploring music with a lot of silence, I think it's a retreat for a lot of them from crazy information.

RJ: Is this really true? It would be nice. I had the same feeling, but if I talked to Japanese, they said the opposite, so therefore I'm interested in what you're saying.

DT: How did you come to this work. What was the evolution?

RJ: I'm trained as a visual artist, so I did drawings and so on. And I did very minimal photographs in the beginning, and I didn't know what to do with them. I think Oh, they are good, they're not good. And usually they were not bad, I mean, they're good, but I didn't know how many, for instance, I have to use for one piece. I had hundreds, because I couldn't decide. I got some help, so they say, OK, maybe seven, but still there was a question which I couldn't answer. These photographs: as I told you, they were very minimal, and only the surface of the curve, the shape was a little bit not round but not flat. So, for me, visually, because I liked the music of Morton Feldman at that time already, so I say, Oh, it could be a score, kind of, for his music. So it was the first thing in the direction of what could be music, but I didn't know. But in that time, it was in the early seventies, I got the opportunity to listen to Cage. At that time, I lived in Bremen, there is a radio station, and the director, Hans Otte, he was very much involved in this new stuff, and he invited all the people, he made festivals so people came. So just by listening, I kind of learned. So this was La

Monte Young, for instance, from Monday eight o'clock until midnight they played, can you imagine, La Monte Young, on the radio, and that makes really an impression on me: "Oh, wow, it must be something." I mean, I liked it. I was also young: maybe now, OK, I like it too, but I don't spend so much time!

DT: It's true for me, too.

RJ: Then I realised I could do something plus my photographs, so I had a piece of iron, two pieces of iron in my hand, and I'm like this, I got one sound, click, and this I recorded with a tape recorder, a cheap tape recorder, and I thought, Oh, that's fine. More I cannot do with my own body. But then I made a copy with a reverse tape, so I had two sounds. I did some more, and so I made a tape, a reel-to-reel tape, so what you hear was, on one speaker, you hear one iron sound, and the copy of the other was on the other side. So this was, OK, this is what you could hear, and you could also see the speaker there, but in front of you, you had the six black and white photographs I talked about. I listened to that, and then something happened. Because of this click-clack, all of a sudden, the lines began to dance a little bit. It was not fantasy: we checked it, double-checked it with other people, so it began, again, to dance. So that is fine: if you have art, if you have sound, you get one more! So yes, three things. Then I made very tiny sounds, but in that time, I was using big speakers, so I had a problem. It was not in proportion. And the biggest invention I did was very similar. I just was using small speakers! Then, from that, it developed. In that time, 1981, I did my outdoor concerts. I had this Berlin concert series, each month I did one. What I did, I had eight speakers and I put them in front of a lake. I called it *Music Line*, originally because there was a line, but also the sound was kind of going from here to there, not using so much technique, it was in the composition itself, it just went like this, because Joan La Barbara, she was DAAD guest in Berlin. I showed it to her, so she liked it, and then I asked for a grant and she helped me to get this, so I went to New York and then it started! I think in New York, I did the most quiet things I ever did in my life, because the city was so noisy, so I found out, everything is the same, what you're talking about, the Japanese now. So I really just was one corner and just put that into the corner and add some sounds, that was it. But the Americans at that time, they liked it very much. When I did it in Germany, they really said I'm nuts! [laughs] So it was not so easy at the beginning, but now it's OK. And then you more and more listen, of course, because I don't need to tell you, if there is something in the corner and there is almost nothing, but there is so much things going on also, acoustically. But also there's some dust, you look closer, then you see it, OK, that's the way it is. But you kind of learn from it and then you open your mind more and more. And because of Japan, it just was – when I was young, when I went to school, I had big problems, and I had these problems since I was thirteen, because I couldn't express myself, I thought, because nobody kind of understood, and then I found out that my brain was – I forgot again, upside-down ...

DT: Oh, it was reversed?

RJ: That time, my wife found a book about Zen and she read something about Buddhism, and she has found out, OK, that all the Asian people have the same problem as I, not the problem, and then I thought, if the Asian people makes one half of the world's population, I also can be OK. I then learned about this, then I found out I don't need to learn. It's just my brain so I'm fine.

DT: One of the things I've been writing about is the development of the idea of the soundscape, from the late 1960s and R. Murray Schafer beginning to write in that time in his educational books about the soundscape. Now it's become quite a fashionable thing, you could say, to be involved in, but I think there's a realisation that a lot of Murray Schafer's work came from a sort of anti-urban perspective, he really didn't like cities or aspects of modernity and so I look at many people I know who are involved in very quiet work and concentrated listening, and they live in cities, and I don't get the impression that they're anti-city.

RJ: No, if you ask me, not at all. I mean, I say a big word, I hate, was it him or somebody else, who wants to design new sounds of cars, I don't know, or vacuum cleaners?

DT: Stockhausen said he wanted to design a kitchen, the juicer and the coffee grinder.

RJ: I think it's horrible! It's like a hundred years ago to design the horses to make another [sound], I think you can do something that is not too noisy. I like the idea that the cars are not that [noisy]. The motorbikes, the small ones, they make a terrible noise, this you don't need. But if it's a noise that's necessary, it's not a noise, it's just a sound. Actually, this is also, again, my experience with New York. You sleep on the noisiest streets or avenues, and still you can sleep, because the sound makes sense, and it's not an aggressive sound, it's just a necessary sound, not a noise, I don't know what it is, and still you can sleep, so you can even enjoy it.

DT: I've got very interested in this idea of sound memories from childhood – research on sound memories and the way sound memories can be markers of a sense of security – and so I asked people. Many people who I asked – all kind of sensitive people – all came up with incidents of silence or quiet as being frightening for different reasons.

RJ: If you would ask me, I grew up in Wilhelmshaven, I grew up in the war, and as a child, I could go to the countryside, my uncle, he was a farmer, and I remember, OK, all the animal sounds, the cows and the pigs, and also the sound, how you cut corn – not corn ... grain. Wheat, yeah, when you cut this, there's a machine and all this: this I remember very well. The old machines, they were fascinating. Very soft, I remember, and then the most thing I always remember when I'm listening to the hen, and the geese. Anyhow, this kind of stuff.

DT: But it's also a question of what feelings those sounds bring back.

RJ: I'm very happy when I think about this, even if I listen to the geese sound, I was scared, but I like the sound, it was early music, it was like Stan Kenton! [laughs]

6

ancestor voices memories distortion

steve roden

Christina Kubisch, steve roden, Stephen Vitiello

Mike Cooper

Akio Suzuki

Lav Diaz

Loré Lixenberg

Danny McCarthy

O Yama O

... I listen to the wind that obliterates my traces music in vernacular photographs 1880–1955: steve roden

Review published in *The Wire*, January 2012 (Issue 330)

A book of coffins; a book of openings to the sky. Americana but something more than that – a collection cf aged photographs, brief literary texts and 51 recordings haunted by melancholy, not just the corrosive scour of time on photographic plate and shellac surface or scenes and distant art from another world; also a record of displacement and loss from times when ancestral memories hung close and poignant, each blemish on the surface of a photographic print a scar of war and poverty, great depression, dustbowl and epidemic, Steinbeckian passages of migration and loss, and all the while a ruthless thrusting of progress to liquefy the solid ground.

I think of William Faulkner, from *As I Lay Dying*: "The women sing again. In the thick air it's like their voices come out of the air, flowing together and on in the sad, comforting tunes. When they cease it's like they hadn't gone away. It's like they had just disappeared into the air and when we moved we would loose them again out of the air around us, sad and comforting."

Rain has no sound, wind has no sound of its own – only a passing movement of water and air, their graven percussion on those surfaces vessels and volumes that erode in their own time.

A military bass drum is laid up alongside a hedge in solitary stillness omen of a future in which men have exterminated each other, those of their likeness and all that which lies around them in the pursuit of war leaving only the silenced object of death's still pulsing heart.

A banjo lies on the grass of a back garden, its neck propped on a wooden step. The woman whose face is partially hidden picks at daisies that grow around the instrument as if musical notes sprouting from the earth. Her fingers form a plucking claw, the suggestion of an occult technique whereby strings may be sounded from a distance.

Photographs so engloomed or bleached by years that the musicians centred within them recede into the void of time. A cracking and staining that overlays the texture of baking mud upon the guitarist whose instrument is dissolving into the wall against which he stands.

Spoken language was different then, those inflections and rhythms of the old country carried as if in battered suitcase and not fully yet assimilated into regional variations of a nation's speech. John Jacob Niles sings with the voice of high winds forced through a narrow granite aperture: "Now John Henry swung his hammer, around of his head, he brought that hammer down on the ground; man in Chattanooga 300 miles away, heard an awful rumberlin' sound, he heard a' awful rumberlin' sound."

Photographs in which a mystery lies, such as the tableau of six vertically displayed 78rpm discs released on the Columbia and Victor labels, then beneath them, attached to the same board in the manner of a shrine or funerary offering, four neckties of varied pattern, a cigar box, a greetings card and a copy of *Michael O'Halloran*, Gene Stratton-Porter's early twentieth-century novel of orphans surviving on the streets of a Midwestern city.

Other revenants haunt these pages, presences for whom we have attempted presently a better understanding of their obsession to collect and their vision for collecting as a means of understanding the world outside the frames of the Gutenberg parenthesis. I am thinking of the alchemist film maker Harry Smith and his Anthology of American Folk Music, of Alan Lomax and the iconologist Aby Warburg. I can also hear La Monte Young whose music grew from the sound of wind whistling through the chinks of a log cabin in Idaho and whose rendition of "Oh Bury Me Not On the Lone Prairie" (recorded with Tony Conrad at some point in the 1960s) is even more mournful than Carl T. Sprague's stately rendition included here. "It matters not I've oft been told, where the body lies when the heart grows cold."

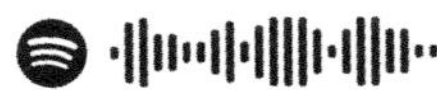

Aby Warburg's Mnemosyne, the extensive collection of images on which he worked until his death in 1929, has been described by Philippe-Alain Michaud as "the atlas in images." It is "art history without a text." One of Warburg's concerns was the representation of movement in Renaissance painting; in other words, he perceived images not in stasis but as unfoldings of time. Wind blows at the garments of a nymph and we "see" time. So it is with sound, represented in the present case by musical instruments. Twenty stringed instruments hang down from a wooden frame over unkempt grass. At once they are silent, like shaved-headed dancers of Sankai Juku suspended from ropes, yet their sound is suggestively rich and aeolian, a music that rises and falls according to the flux of nature rather than the will of humankind.

"Blue Blazes Blues" by Emery Glen, one of only a few sides cut by this obscure artist, delivers its elegant metaphor in the harsh glare of utter resignation: "I got the blue blazes' blues – they burns all night long ... blue, blue, blue ... Smoke go up the chimney, black clouds hanging low. Where in the world did my good girl go?' A string band, three men and one woman, are seated in line before a wooden wall, open window into the house revealing ornate patterns of some fabric or wallpaper. Chemical change and fungal growth have spread coral reef pink and green across cellist and fiddler, dragging them back to primordial swampland. Posthumous scratches have defaced the female guitarist who gazes dull and still into middle distance as if already lifeless. We cannot turn away from those photographs of the period in which a spiritualist version of the universe is materialised. In an otherwise pristine image, like secret thoughts which must not be uttered, a black ectoplasmic trail of song smears across the closed mouth of a frock-coated guitarist.

Pianos ruined and abandoned amid scenes of disaster, music's remnants refusing to leave a land forsaken by all others.

Photographs often feature in these photographs, a recognition of the potency of the photograph as a clustering of time and presence.

Ecstatic surges culminating in a scream, "Xango" by Roland Hayes is an incantation for voice and piano recorded in 1951, the cultured tenor of Hayes reaching back to the Yoruban *orisha* Shango, thunderbolt and lightning itself. A Georgia-born lyric tenor who studied in Chattanooga, Nashville and London, Hayes was born in 1887 to former slaves, his

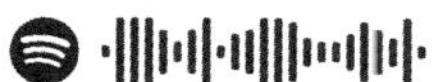

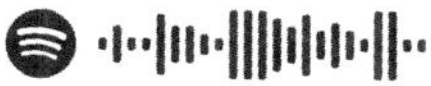

great-grandfather Abi Ougi a tribal chief captured from Cote d'Ivoire and shipped to America in 1790.

Nine cages of singing canaries are studied in complacent satisfaction by a man immaculately suited and coiffured. He sits beside a music stand and gramophone horn.

How can sound stay alive after its sounding? Steve Roden turns to writers – James Agee, Knut Hamsun, Melville, Nabakov, Par Lagerkvist (from whose words comes the title of the book) and others. Their evocations and descriptions of listening act as brief tone poems, moments of breath exhaled within the imagination, fanning air onto the thickening intensity of that which is seen. A passage found outside this book, in Flannery O'Connor's short story, *A Temple of the Holy Ghost*, sends me back to one of its photographs: "The sun was going down and the sky was turning a bruised violet colour that seemed to be connected with the sweet mournful sound of the music." Blasted by light, a man wearing a topi hat sits on a cane chair. The strings of the guitar he plays are sharply outlined against the dark circle of its sound hole. Posed by a potted palm, an orientalist study, he stares up at the sun, jaw set against the fading of his own image.

The theatre of music: two groupings, a trio whose faces are in turn resolute, imbecilic, placid, two overalled, one bow-tied and hair starched vertical by some past fright, his shiny black footwear contrasting sharply with the worn muddy boots of the banjo player who sits beside him; then a black couple, him staring at the lens, her glancing sideways at him. She is playing a guitar. They are posed as if by Picasso, c. 1913, in front of a curtain, behind a rough table on which sits a bottle and glass. He holds a pistol, as does the young man with placid face standing behind the banjo player trio. Outlaws and their ballads – Jesse James, Cole Younger, John Hardy, Kenny Wagner and Stackalee – sing silently from these stagings.

Like a Kabakov installation, an empty classroom, harmonium angled in the corner under drawings of Latin America. Sound hangs in the undisturbed air, together with the dust. A time of strange auditory technology: musical glasses and bones, stacked cylinders, kazoos, musical saws, a toy piano; phonographic horns reminiscent of blunderbuss or bat tympanum thrusting out into rooms with the intrusive force of sound itself, transmitter receivers aimed at Mars or into some invisible Swedenborgian angel realm. A one-man band equipped with bass drum, cymbal, voice horn (perhaps a kazoo), bells of the hotel reception type, played by the foot, and a guitar. The performer (I notice after some time of looking) is a one-armed man.

Out of poverty, restless invention and monomania comes the autonomic drive of the one-man band. Some of these constructions are so extraordinarily elaborate in their workings and architecture as to suggest a desire to supplant all that is communal, social and collaborative in music making. The operator stands alone, a marvellous being in control of many instruments – multiple strings, percussion and piano – all set in motion by the ingenious actions of cables, strikers, levers and foot pedals. We see Prof. Mc. Rae, Ontario's Musical Wonder, and Crawford the Musical Wonder. One operator blows trumpet and bows violin simultaneously, his piano decorated with a vase of flowers and draped with the American flag. Others sit their top hats on the piano. Foolishness and the virtue of endeavour struggle for supremacy in these scenes. Implicit within them lies a potential for the extremes of Raymond Roussel's Ponukelan instruments, a musical technology of neurotic cruelty and fanatic logic.

A woman and a microphone. She is brightness itself, cut off from her world by light. Her outstretched hands are raised above her head as if gathering in her own sound and presenting it to an unseen audience. Her long dress blows in the wind, enveloping the microphone stand.

Displacing the Air: The Marfa Sessions, Christina Kubisch, Stephen Vitiello, Steve Roden, September 2008, curated by Regine Basha, Rebecca Gates and Lucy Raven

Published in The Marfa Sessions, Texas, 2008

When an artist uses sound to activate some aspect of a room, what happens exactly? This is what interests me, even though the realisation is belated. Everything about the room looks the same, feels the same, smells the same. Aside from what constituted the room before the sound, there's nothing a bailiff could confiscate, a burglar could steal, a builder could demolish, a TV director could film, or some cheapskate member of a family could sell after a death, but when sound is introduced into a room, then nothing remains the same. If this artist who introduced the sound knows their stuff, then all these immutabilities I just mentioned – the material realities of the room – really seem to undergo some sort of transformation.

Two pieces in The Marfa Sessions forced me to consider my initial question more deeply: *Memory Room (for Marfa)* by Christina Kubisch, and *"from perfect cubes to broken trains"* by Stephen Vitiello and Steve Roden. Both were by artists I know personally, and whose work I know well. Immediately I wondered if this "enhanced knowledge" was acting as a conduit for privileged understanding, or even a kind of sympathetic resonance (better known as favouritism). Maybe so, but I think the point is that Marfa, west Texas, is a place of very specific qualities. Adding anything meaningful to them is not easy, and part of the problem is the remote location, to which one might add the profusion of hard-ridden myths already gathered within the imaginative spaces of its horizons. Sounds are not "made'; they are called, I wrote in my notebook shortly after returning to London from Marfa. That feels like a quote but I can't determine the source. Perhaps it came from Jean-Luc Nancy's book, *Listening*, which I came across in Marfa and bought as soon as I got home. Nancy's ideas about the resonance of interior spaces connected with my own thoughts of that time. "Isn't the space of the listening body, in turn, just such a hollow column over which a skin is stretched," he asked, "but also from which the opening of a mouth can resume and revive resonance?" Marfa is horizons, and a strange light, and complex histories of abandonment punctuated by various claims to an inhabitation of emptiness.

Calling sounds suggests a certain intimacy with a place, yet how is this possible for an outsider when the place is so remote, so resistant to easy diagnosis? Christina Kubisch addressed this problem in the succinct text she wrote as an adjunct to her *Memory Room*. During the customary site reconnaissance mission she realised that her preconceptions of Marfa were confounded by the place she found. She felt the sense of being "in the middle of nowhere" and used this feeling to examine her own deeply embedded sense of selfhood as a post-war European: "... we are so fixed on keeping as many traces of history as possible, so the past sometimes gets more important than the present." By "we," she means Europeans. Since I belong to that tribe and to her generation, I concur with what she says. Europe is a palace, as well as a charnel house, of memories, all on display, all priceless, irreplaceable and solid. Such memories live on in Marfa, of course, but more as traces, or islands. Artists are sensitive to such traces, so I was impressed by her integrity, which disallowed any direct engagement with the physical objects she encountered during her brief visit to the town. Instead, she filled an empty room with the sounds of digital information. Maybe this was an act of faith. I know I stood in that vacant room along with a few other people, dust motes floating in sunlight, nothing much to grasp or dissect, and felt that sound had been called, that a body was resonating, and what it transmitted by filling up the air with a kind of peculiar, delicate, frizzing energy was my first true sense of being in Marfa not through looking at arrowheads in the market, or reading pages of Larry McMurtry's novels in the bookshop, or strolling through the weirdness of Chinati (those islands of art that claims their place in the desert), but through the notion that one kind of air can be displaced by another, and in that displacement we can realise where we are.

Steve Roden and Stephen Vitiello confronted some of the Marfa myths head on, by building a shack outside of town, then using solar panels to power the audio playback. When you build a sun-driven shack you erect a monument to transience. A history of settling in unfavourable conditions is invoked. For a European of my age who grew up during the era of *Wagon Train*, *Rawhide*, and *The Lone Ranger*, I don't have much of a stockade (or firewall, to use a more contemporary expression) to protect me from the mythos such a construction flushes out of the brushwood. Again, sitting in this room, slivers of the sun's penetrating light betraying the poverty of its vernacular craftsmanship, I felt grounded by the aetherial non-specificity of the sounds introduced into the room. This reads as paradox doubled, no doubt, but I think it's a question of how place is perceived at a level below conscious thought and behaviour.

In the area close to the shack we came across a remarkable species of insect, the Red-winged Grasshopper. To the naked senses, there are three curious aspects to Arphia pseudonietana: its vivid red wings, edged in black, that become suddenly visible in flight like flags waved to signal an emergency; the erratic, seemingly doomed trajectory of these short flights; the loud clacking of its wings, more like a child's toy than a living creature. I knew, because I was told, that the distinctive clacking of these grasshoppers had been integrated into the sounds heard within the shack, yet the knowledge was superfluous to what I felt. Sitting in the shack was a vulnerable experience, suspecting some twister or freak storm might carry it off to Oz before my European sensibilities knew what was up, and yet a slow sinking into whatever this place might be about, as if I were listening to ungraspable atmospheres in the same way you could be dazzled every night by the sight of so many stars in the indigo night. One type of air is displaced by another; suddenly you know exactly where you are.

apparitions in rayon: Mike Cooper

**First section: booklet essay for Mike Cooper's *New Globe Notes*
LP, released by NO=FI Recordings, Rome, 2014**

**Second section: review of Mike Cooper's *Rayon Hula* LP, The Wire,
February 2004 (Issue 240)**

Questions hover, nebulous as the apparitions (strings whose tunings are
ancestral stones) shimmering within these distant fields of disturbance.
In a museum without walls they occupy the room labelled Machines for
Unscrolling Scenes from Hypnogogic Moments Between Somnolence
and Sleep, each one displayed for close inspection as a slippage of mere
seconds yet artfully seeming to repeat endlessly yet return and return in
diverse iterations like masks advancing, the face transforming yet always
the same face, illumination and shadow moulding the features into a
crowd of grotesques, coming forward out of the field of singing as if a
creature floating through fireflies, emerging from the broken dark of a
glowworm cave, coming forward yet never coming closer; a question
that asks, what place is this? Pulsing as if heat waves, swaying bodies,
the treefrog blink of cabled neon humming in rain. The glamour of ceremony holds us transfixed as curious eavesdroppers lured into nocturnal
alleyways or forest paths. What place is this? Across a river lights flicker
on and off, surges of power controlled by the flux of moth wings and bird
song. Mist rises from the shining water. Strings that may be bells can
be heard through the leaves, or perhaps we are seeing the abandoned
airport, a moon falling on the temple that used to be. What time is this?

Once I heard Markus Schmickler describe his feelings about loop-based music. Listening to loops, he said, can be compared to the existence of a caged mouse, running for eternity inside an exercise wheel. The analogy is apt. How many times have I stepped into some chic gift shop and heard two bar (un)variations on a looped riff lifted from the generic Miles Davis environment, locked and recycled until the end of days? Rather than uplift, inspire or liberate, this kind of music seems to dig us further into deep grooves of lifeless mediocrity. What goes around comes around.

There is a strong case to be made in favour of certain exponents of the loop, however, and this rests upon history. When Philip Jeck, Tom Recchion (or, in the past, Gavin Bryars and Steve Reich) loop, using obsolete technologies like 1/4 inch tape or vinyl rather than the formidable cloning capacities of the computer, the returns are an accumulation of sonic dust and accident, each moment in the groundhog day cycle a little different from the last. Like the proverbial broken record, time revolves in melancholy circles yet each fragile reprise of history brings a new story, a view that conflates past, present and future. As T.S. Eliot (almost) wrote, "Well now that's done … put a record on the gramophone."

This is the atmosphere of Mike Cooper's *Rayon Hula*, his homage to the Hawaiian exoticist and vibraharpist Arthur Lyman, and to Ellery Chun, the inventor of the Hawaiian shirt. Made from wood pulp, and once known as artificial silk, rayon is a fabric of human invention, though not synthetic. Cooper's fabrication of sultry Hawaiian moods is similarly deceptive. With "Mele Manu," he begins with bird song, recorded by himself in Queensland. Immediately, this references the synthetic jungle backdrop, the dream state interzone discovered by Arthur Lyman and Martin Denny in the late 1950s. A languorous overlay of loops in odd metre gradually emerges as if from early morning mist – vibes, guitar harmonics, sucking sounds from an unknown source pulling unwary travellers down towards the centre of a vortex and oblivion.

Out of the raucous bird song, slipping harmoniser loops rise – perhaps the interior workings of a tropical timepiece made from plant matter and monkey bones, operated by fire ants. Dogs bark at the black night on track two, "Ho'okani Pila," as another infernal machine crunches and grinds hallucinogenic tree bark and beetle skin into paste. Track three, "Musa Shiya," moves faster, a skipping, two chord sample of vibes and double

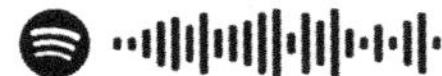 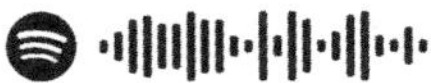

bass, cycling under drums and Cooper's lap steel solo. Throughout this album he plays with an authentic Hawaiian feel for phrasing and vibrato, yet Elmore James and Freddie Roulette have been shipwrecked on this weird tropic island and their hypnotic reptilian bottleneck slither glides through the St. Elmo's fire of "Kokoke Nalu" with sleepy menace.

With "Rayon Hula," Cooper timestretches his source material down to an ominous rumble, an archaic Balinese funeral procession moving in heavy, measured steps towards the pyre, steel blade spirits keening in the distant shadows. "Ho'ornanau nul" is sensual, relaxed, ruminative, a lap steel solo that unfolds over a gentle sunset loop of percussion and vibes. There are moments on this track when the live playing falters a little, as if uncertain about its place within the space of the created image. Personal memory seems to intervene, as if the intoxication of this exquisitely judged fakery suddenly reveals its own fragility, and the act of playing becomes self-conscious, prey to habit, technique, style, the disabling influence of thought.

The lapse is brief. "Paumalu (Sunset Beach)" moves like another bio-clock, tocking and snicking under clear green water, ejecting bubble gouts from the bowels of its tortured mechanism, tickling fish, mesmerising divers, measuring suspended time and the illusions of submarine distance. "Typhoon Inqoon" raises the tempo with a throbbing bass line, confusing polyrhythms, strange undercurrents of decomposing vocal sounds; through a haze of fast vibrato, sharp crackle and fluttering tones, "Mika Oho" trembles, radiates threat and poison, like a cache of chemical weapons abandoned by a superpower in the poisoned heart of paradise island.

Cooper writes about the Hawaiian expression – nahenahe – used to describe a musical preference. "It roughly translates as 'sweet and slow,'" he says, "and it is this quality in Lyman's music that has always made it for me authentically Hawaiian." Sweet and slow also describes Cooper's steel solo on "Caught Inside," performed over a shuddering reverse loop and a soundscape recording that mixes the "exotic" with the ordinary: rich Australian bird cries, dogs, aircraft, motorbikes, wind on the microphone, the sounds of a neighbourhood. This is the only track on the album lasting more than five minutes and his playing is very assured here, surfing effortlessly over the simple, if implicit, melody of the reverse loop.

The final, tantalisingly brief track, "The New Urban Hula Slide & the Tiki Bar Is Closed," (the title lasts longer than the music) returns us to a faux Indonesia, post-apocalyptic Java perhaps, where a woodpecker hammers vigorously in the distance and sonorous metal percussion follows its own tail at an unspecified number of revolutions per minute. This is music that will sound both disturbing and wonderful in a wide variety of circumstances: relaxing on a remote beach declared dangerous by the Foreign Office, for example, particularly if the world has just ended.

A Cat with Strange Fur: Akio Suzuki

published in *The Wire*, May 2003 (Issue 231)

In 1963, Akio Suzuki was employed in an architect's office. Working on staircases, he realised that his drawings looked like the lines of a musical stave. "How would it be," he says, "if I could make a staircase that wouldn't tire you out, that would be a pleasure to use? If I could make a perfect staircase that looked like a stave, then I could drop ping-pong balls or tin cans down the stairs and they would make a beautiful sound."

The first performance in what he now calls his Self-Study Events took place on the stairs leading to a railway station on the Chuo line in Nagoya. He tipped a large bin of rubbish down the stairs. Not only was the sound a disappointment, far from the magical sound of resonant objects tumbling down a huge musical stave, but the bystanders reacted with a frustrating excess of social responsibility by collecting all his rubbish and replacing it in the bin. Again he threw the rubbish and this time the transport police arrested him and locked him up until his parents came to the rescue. "I encountered society," he has said.

"At the time there were lots of happenings," he says, "people wandering around the streets naked and slicing canvases with knives, or putting on boxing gloves and beating up the canvas. Seeing people doing that gave me the strength to do what I did when I threw the bucket down those stairs." But the gap between his expectations and desires and the reality of disappointment has preoccupied him ever since. To fill the gap he decided to study. Then in his early twenties, he realised that music college was out of the question. Instead, he went to nature, travelling to wild places and listening to natural sounds and echoes.

The work that developed from this period was based in what he called Throwing and Following. In other words, a sound would be projected into a space and then the consequences of its resonance in time and space would be followed. In 1970 he began making his echo instrument, the Analapos. Typically of Suzuki, the artefact is simple but the principle is complex. Basically long springs connected at each end by cylindrical resonators, the instrument explores the reverberation of springs, the resonance of vessels and the transmission of signals between two physically linked terminals.

Experimenting with amplifying the Analapos in 1976, he discovered that a microphone in a cylinder will produce feedback. By using black painted cylinders of various sizes, all made from paper, wireless microphones and tuneable FM receivers designed for karaoke, he performed his Howling Objects events. In conversation with Nobuhisa Shimoda he has described these events as "kind of like performing a ceremony." Perhaps this suggests religious ritual. In reality, Suzuki's work is more concerned with materiality. Down to earth, witty, instinctual, even scatological, his performances are like audiences with a trickster. Mystifying and mesmerising phenomena emerge yet observers can see the simplicity of means for themselves: a stone flute, stones balanced on the toes of his shoes, wax paper or rubbed glass tubes.

A peculiar cycle of concentration builds between performer and audience. "This is unusual," writes guitarist Akinori Yamasaki, "because most music disrupts this cycle. If we listen to the sounds that he makes and respond, then it's already too late. The concept of "listening" in our brains actually disrupts and destroys the instinctive process of listening." Suzuki's own analysis of this concentration is a little less theoretical. "I used to think it was because I had a problem with painful haemorrhoids," he says. "I thought this communicated itself to the audience. And they started to feel the same pain. I always wanted to be more like John Cage – aware of the sound and relaxed, but aware of everything that's happening in the space and somehow creating an atmosphere in which not just me, but everyone, is aware of what's going on. This was my particular model but for some reason I can't do that. I come in, I'm relaxed and I want to be open and expansive but once I start playing I begin to lose this wide vision and begin to focus in until I'm focussing just upon the sound. That's the only thing

I'm able to put my attention on. I can't see anything beyond that when I'm performing. It's almost as if I slip into a time tunnel."

Akio Suzuki was born to Japanese parents in 1941, in Pyongyang Korea, then moved with his family to Japan in 1945. His work is unique though there are sympathetic resonances between other sound artists who are roughly of his generation: Rolf Julius, Felix Hess or Max Eastley, all of them dedicated to the minutae of sound and its place in landscapes. All of them were born during World War II. "All these people have been born in acts of destruction," he says. "I almost feel this generation was born in a gap. Sound art is perhaps born out of people who experienced the war or were born just after it. They were anxious not to be involved in mass movements. It's a solitary kind of pursuit."

Cryptically, he describes himself as "a cat with strange fur." Like a cat's paw, delicate but strong, he can activate the world without breaking its surface. One of the most difficult aspects of music and soundwork to explain is the concept of "right action." How is that music can be evaluated almost immediately, just as quickly as a fire alarm or a baby's cry? When Akio Suzuki performs, certain qualities — grace, warmth, a quiet authority of mind and action, an engagement with the vessel of nothingness through which sound can emerge — are registered through their presence in seconds.

Two Scenes: Lav Diaz

Written as part of a possibly unpublished text on Scott Walker

Scene 1: Near the end of Lav Diaz's five-and-a-half hour film, *From What Is Before*, two childhood friends talk in the water of a rice paddy. They approach old age, one near death. Cancer eats him up from inside. Filled up with this imminence he talks of the traditional Malay burial practices still extant when they were young, corpses burned on the river. When I die, he says, please send a telegram to my daughter, ask her to bury me in this way. We can revive this way of life, he says. In our dreams, says the other. The next shot shows his daughter walking along a muddy track, negotiating an armed check-point of communist guerrillas. These are emergency times, martial law just declared by the tyrant Ferdinand Marcos. The scene cuts: her father's body is alight on the darkening river, fire crackling, the bier drifting gradually out of shot. Diaz talks about regaining time, using this format of extreme slowness and long duration to reclaim film's potential from the grip of commercial studio-dominated cinema in the Philippines. Digital technology allows this dwelling of his on vegetation, weather, people whose poverty and precarity renders them victims of circumstances. They are poised between a slow vanishing world and a catastrophe of politics and in this balance soon to tip away from them the telegram comes like a flash of lightning. In our world the telegram (from a distance – a letter) is as remote a ritual as fire burial on water, never to return except in dream or cinema, yet it vibrates still with resonance of those events demanding desperate measures, concise expression, rapidity even in the tempo of a slow walk along a muddy track, radical disturbances of time.

Scene 2: In almost every Lav Diaz film there is an establishing shot of a road or path. The shot is held. Nothing appears to move though after a while it becomes apparent that vegetation waves in the breeze. After a time, figures will appear in the distance, moving towards the camera-observer. They advance slowly in real time, eventually to pass the observer who retains a fixed gaze. Life can change but only slowly; patience is a weapon. In *Elegy To the Visitor From the Revolution*, Lav Diaz is seen, face hidden, playing electric guitar through a small Marshall amp. His playing is erratic, the kind of thing you hear men playing in guitar shops, but then sliding into noise. Time, productivity and money are entangled, so in an act of disentanglement Diaz films refer back to the pre-colonial period. Malays had no clocks; time was not monetised. Watching one of his films makes every other form of moving image feel like a short cut to the emotions. In *Melancholia*, released in 2008, characters are seen only at distance, never close up. Speech is muffled, often barely audible over dogs barking, cocks crowing, passing cars, heavy rain, river flow, piercing jungle sound. There is no music, nothing extraneous to the events unfolding. These are bizarre enough, too complex to describe here, but centre on the disappearance of three Communist rebels. Five hours into the film, a scene of two bodies being found cuts suddenly to improvised noise, two electric guitarists, electronic keyboards and alto saxophone playing in a bare white room. The music feels inchoate, raw, not music at all, simply the only response to this unspeakable discovery of buried bones that has taken so long to arrive through such a convoluted route. As is usually the case, the scene is protracted, more than ten minutes. To judge it as music would be meaningless but then how to assess it in relation to music?

Loré Lixenberg: Memory Maps

Postface to Loré Lixenberg's *Memory Maps*, published by Editions AquaAvivA, 2013

A sweep of the senses; in a single glance is all of time. "No longer to imitate, but signify nature," wrote Henri Michaux in *Idéogrammes en Chine*. "By strokes, darts, dashes. Ascesis of the immediate, of the lightning bolt … out of the multiple issues the idea. Characters open onto several directions at once." For this is about time and the score, the score as defined by artist John Latham as an omnipresent, atemporal surface, lacking in temporal extendedness, and yet, the score is also all possible routes through a map viewed from the sky, paradoxically and simultaneously with the potentiality of duration, a passing into that version of time thought of as linear. And there lies the problem, because for singer Loré Lixenberg, despite all her extraordinary gifts, linear learning is so unnatural as to be almost impossible. Recalling the art of memory of the Greeks, each *aide-memoire* to a composer's work-to-be-sung condenses the convoluted paths of a song's imagery into a map of signs. *Knowledge-montage* was the term used by Philippe-Alain Michaud in his description of Aby Warburg's *Mnemosyne Atlas*, its "elective affinities" of extraordinary relationships, in itself an echo of those occult memory systems developed by Giordano Bruno or Robert Fludd, through which complex information might be drawn to the surface by being attached to strong images, "very beautiful or very ridiculous," as Frances A. Yates has described them in *The Art of Memory*, or a theatre stage of the mind, on which the elusive objects of our fragile recall might be viewed as if in a play. Populating the mnemotechnic are images that may seem abject by comparison with the pathos and beauty of the music yet

they lead the singer directly, as if in dream, into its emotional universe: a mouse that roars, conjuring the fabulous creatures of Schöenberg's *Das Buch der hängenden Gärten*; the smiling cocks that strip Debussy's sensual pan flute of its metaphoric, Edenic gentility; for the libretto of Wagner's *Wesendonck Lieder* the Walt Disney characters that rise with angels in childhood to finally tumble into the grave. This is a private language – symbolic and comic – that to some degree constructs a second, parallel score, absurd on the surface yet wholly legitimate as the distillation of a process leading by the shortest route to the sublime.

Lingering Tones:
for Danny McCarthy's exhibition –
Beyond Silence: A Bell Rings in an Empty Sky,
Crawford Art Gallery, Cork, 2017

Probably 1975 or 1976 I had been at home alone for some days, quiet, reflective, in a state of emotional pain. A long and profound relationship was breaking apart with all the effects of heartbreak, dread, isolation and impending change that such events customarily incur. Snow had fallen in the night. The whiteness of the garden seemed a projection of my mood and perhaps because of that, either to explore the mood or soothe it, I played my copy of Gorô Yamaguchi's famous shakuhachi record, *A Bell Ringing In the Empty Sky*. The recording was made in a period when Yamaguchi was teaching shakuhachi to American students at Weslyan University, Connecticut, in 1966–7. As Clive Bell has written, Yamaguchi's playing is "elegant and cool to the point of chilliness." My memory of this moment is detailed enough to recall the chilly aspect, its amplifying the severity of the scene. The cover of the record played a part also: "... the wonderful monochrome psychedelia of the LP sleeve," as Clive put it, "in which a crane appears to fly through a geisha's hair-do." There are two geisha, two cranes, surrounded by a sea (or sky) of white in which the bell implicitly rings. Whereas European bells tend to be flared at the base, Japanese bonshō bells have sloped shoulders, then either a straight line to the base or an inward curve (like the seated Buddha). Perhaps for this reason their sound is softer, more penetrating, sufficiently penetrating to be heard in the underworld. Some indefinable quality places it outside human making and sounding, as if the striking of it only sets off a feeling of nothingness inside each person within the radius of its sound waves.

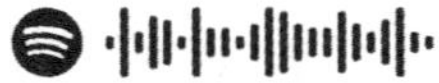

It could have been that experience – bereft through loss yet filled with the impression of snow, the thin tone of a flute, the stillness of the scene, the image of empty sky as bowl in which a single bell tone resonates – out of which emerged a new potentiality (as is so often the case when a settled situation breaks apart) based on sound detached from its physical presence.

I have another memory, of an exhibition by Danny McCarthy at Helen Frosi's SoundFjord gallery in 2013. Its title was *The Memory [box] Room*. To my recollection the works were small, like icons or the kind of offerings you might find in a remote country place. Some works lay on the floor, in their incisive scraps of memory like stained glass windows opening down into the underworld of things, people, moments, sensations that have floated into places inaccessible to human senses though still alive in that region we call the soul. We were about to play together that evening, Danny, Mick O'Shea and myself. I photographed some of these works, perhaps to contain the unsettling feeling that we would be performing on a surface that barely covered vast soundings of silent materials, the voices of ancestors and all that has gone but remains.

Hot Tea from the Spout of a Black Pot:
for O Yama O

**Sleeve note essay for O Yama O's *O Yama O* LP,
released by Mana Records, 2018**

How to think or speak about this as music, these lurching intersected rhythms, baby-songs, open-jaw buffalo roars and fox screams? As if, I want to say, new musical rites and folk festivals were building themselves up from scraps of discarded plastic, open-throated thoughts voiced in a feverish air of rummaged floors and abraded old things of wood and skin from scented workshops whose purpose if not forgotten is certainly misplaced. But then my better self prefers not to speak about music at all or sound, come to that, and think instead about people and objects working together visibly and invisibly in spaces both intimate and infinite where reverberation is not so much cosmic as a lullaby space that incites both dreaminess and the hard realities of matter and movement. Dogs bark in just the way they should, honouring the Society for Marching in Circles as it passes by, then passes by again, and again. Suddenly we are closer to music being made than we have been for many years or longer even, so alarmingly close as to feel warmth and discomfort, as if studying the sole of a foot from a few centimetres away or holding a private whisper within an enclosed hand and feeling its trembling desire to be free; but also so far away distant as to feel each vibrant, pungent ingredient within its box or jar or bowl or packet or bottle or air-tight translucent container or brown paper bag painstakingly stirred, shaken, scattered, poured into the heated cauldron of what we call recording, its imaginary rooms and its production, though my better self prefers not to speak about or analyse the notion of "the studio," this being a working up of spaces that are social, a vision of something beyond us but not quite beyond us because its existence as a listening object is real enough to make us pause and question how it was lost or never found. Was it something about a voice heard in the ear and the feel of matter activated in such a way as to make you feel the difference and sameness of beads and bones, stomach flesh and drum skin, wood pieces and finger joints? Can a voice tell us anything or does it simply persuade our own voices to tell us what they want us to hear? Hearing is a touching thing, feeling the softness, thuds, sharp edges, bitter taste, scratching rotation, fairy bells, beetle ticks and voice plain as an honest thought that speaks directly into the heart. Was it the memory of a drop of blood from a child's nose bleed, seen as a vivid image as hot tea fell from the spout of a black pot? Was it something about presence and its insistence, being so close to the breathing, smell and warmth of an entity, the force of its quiet desire to live alone and inside others, enfolded and left to be alone, to make sense and nonsense? Or was it the idea of a village somewhere remote, where the things called music, instruments and recording are unknown, a village where little is said but the paths, clearing and spaces resonate with whistles, voices, strings, drums, reeds, stamping feet, the friction and rattle of small creatures and dogs barking as they should because they are within the circle, not without.

7

kwatz graaak blap wonnnng txtspl

Nathalie Djurberg and Hans Berg

dom sylvester houédard

Jeff Keen

Bob Cobbing

mmmmmmmm, wonnnng and woh: Music and Collaboration in the Work of Nathalie Djurberg and Hans Berg

Published in *Djurberg Berg*, exhibition catalogue produced by Moderna Museet, Stockhom, 2018

Despite all the collaborative promise of Black Mountain College, Hi Red Center, Group Material and Art & Language, mixed-media dystopias by Dumb Type and group activities of contemporary collectives like teamLab and Chim-Pom, sustained working partnerships between two or more people are scarce in the art world. Artists primarily situate themselves as individuals, closer to writers than to musicians or dancers. Notable exceptions flourish – Gilbert & George or Ilya and Emilia Kabakov – though their practices are exhibited as indivisible objects. The majority of artists imagine themselves working more or less "alone" within their hidden networks, foreseeing the culmination of that process – exhibition, film, performance or whatever form it takes – as their marker of particularity. Artists have always employed assistants, technical researchers and enablers but the status of these anonymous figures is shadowy by comparison with their music world equivalents – the recording engineers, producers, hardware and software developers and session musicians who so often make things happen at a subterranean level.

For this reason Nathalie Djurberg and Hans Berg are unusual. They create a convergence of two worlds whose co-existent nature is self-evident and paradoxical. It is also mysterious. The question of why this should be so opens up to a deeper investigation of the tacit in histories of representation – the sounds a viewer may "hear" while viewing an implicitly noisy painting like Bosch's Christ Carrying the Cross – but it also connects to more obvious issues. For one thing there is a typical trend of the twenty-first century, a blurring of practices, a more fluid approach to the settings in which these ambiguous, overlapping and hybrid practices can be experienced. Gallery can become club or listening room, a development that shares strong links with some of the more radical assaults on galleries that took place in the 1960s-70s. But there is also a personal aspect to this, the way in which the two artists met, encouraged each other, initiated a collaboration and persisted in its oddity despite resistance.

Nathalie identifies art's association with the personal as a key to this resistance, as if in the orthodox view of art and its inner workings, only a single subjectivity can be articulated at any given time. Initially their idea to credit Hans as a co-creator of the works was opposed. Who by, gallery people? I ask. "No, every people," she says, "media people, collectors, curators." Gallerists were more amenable though a certain reluctance lingered, sustained no doubt by the marketing truth that focussing on a single individual simplifies and strengthens the message. Don't confuse the brand. As she points out, their reluctance was understandable, because allowing a polyphony of sources undermines romantic, semi-mystical ideas about art's genesis from within the inspired individual. So by default their insistence on being known and shown as a collaboration of two named individuals gently interrogates the politics of art production, at the same time offering refracted views of the psychology of process.

The collaboration began in Berlin in 2004. They had both moved to Berlin from Sweden and were introduced by a mutual friend who suggested they might work together. Hans watched a VHS tape of Nathalie's early films and fell in love with them. "I'd never seen anything like it," he says. "I studied art theory, so I was interested in art. And then I made some music to one film and it got really weird." This was *Désastres de la Guerre*, made by Nathalie in 2003. "It's this field after a battle," he says, "corpses on a field, dead people, crows in the trees and they're flying down and eating the eyes and it's kind of grotesque. But the music I made for it, I think we took a track that I already had made, the music is stupid, electronic – plink-plink-plink-plink – and it turned the film in a strange way."

Later in our conversation, a convoluted thought from Nathalie amplifies the feeling that this near-accidental mismatching of elements forged a key, an unlocking of their future process. Also, in its stream-of-consciousness, fractured, confessional tone, her statement circles around the process itself and why it works for them. As she says: "So what happened for me was there's something that I felt like I've failed with, that I did not achieve what I wanted to achieve with it, and as achievement, the feeling of it was just not there, but then when Hans put the music on it, it got so twisted, so all the parts that I couldn't put in were there, and suddenly I loved the work. And it was something really nice with it also because it could – I think I was a little bit greedy, in a way, with how I perceived my work, that because I'm working so much and I have so much in it, and I want it to be mine, so the more animations I gave to him, and the more I understood that I didn't know shit about music, and that the more I let go, the better it became. And it was incredibly nice to let go of the control, both of the music – like feeling that I had to control it somehow – but also for the work itself ... I'm still doing my part, is it really necessary that it's just this one name? And people discard you completely, so sometimes when we made an interview and we said, 'No, it's about both of us, and you have to have him in the picture, too,' they would take the picture, and then they would cut him out!"

So the work is one work with two oppositional aspects (holding together despite external pressures to fragment or vanish), both of which infect and warp each other to make something uncertain of all possible proliferating aspects, a description which might also apply to human psychology. In one world I am listening to music composed by Hans for *Neon* (an unfinished project), *The Secret Garden* (a large sculptural installation), *In Dreams* (an outdoor sculpture), *Worship*, *Waterfall Variation* and *Butterfly* (stop motion animations) and *The Clearing* (a large sculptural installation). All of these tracks are, in varying degrees, serene, beatific even, meticulous in their precision and detailing of surfaces, timings and dynamic relations. Descriptive of lush, sometimes vast unfolding virtual spaces, their textural density is restrained and consonant, their evolving forms and undulating tendrils born, grown, released in an almost exclusively electronic world that gleams and shimmers in otherworldly light.

In another world I am watching a stop-motion video, let's call it a sculpture. A voluptuous woman weeps in despair, her abjection awarded marks, very low marks, by three bewigged judges whose faces are as ripe,

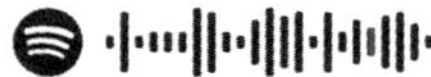

florid and distended as the lurid bird beaks and psychedelic flowers of another sculpture, *The Parade*, in which a corpulent woman stamps like a sumo wrestler, long-beaked bird clutched between her breasts. A purple-skinned woman with red nipples and pubic hair lies on her back, naked, her body shiny with sweat; a man transforms into a bird; we see pink vomit, bright within a colour scheme that seems to contain every last colour from the paint factory. Elsewhere we see a pneumatically obese woman clap like a sumo wrestler, then give birth to a rhinoceros; a woman masturbating on a giant semi-peeled banana, her scarlet lipstick matching what could be blood stains splattering the banana skin. A man caresses a glittering blue fish; a couple ride by on a golden motorcycle, their faces twisted by desire mixed with terror. The materials could be excrement mixed with cream and fruit, painted to resemble a child's dream house. Everything is somehow charming, messy and monstrous, erotically comically charged and materially fluid, as if the baby that was once us, capable of finding pleasure in playing with faeces or eating mud, is simultaneously the adult whose desire fixates on all that which is taboo and unobtainable and beautiful in our own eyes. In this world of plasticity and dream secrets, the circle of a doughnut may be sexy or cosmic, anal or faecal, tasty or sculptural, silly or profound, all at once.

I ask how these two heterogeneous worlds come together, how is their development as a unified work negotiated, tracked and shaped? "How do we talk about it? It's not like the written or spoken language," says Nathalie. "It's really not my thing, so we are talking about, "Yeah, it's mmmmmmmm ... and that makes me feel like wonnnng," and then Hans is like, "Yeah, I'm thinking of making it more like woh, like swinging," so we're not intellectualising it when we are creating." There is no reason on earth why artists should exchange high-level ideas through the language of the courtroom, the philosophy seminar, the news report, the restaurant menu, the scientific paper. Lovers and strangers communicate intuitively: a look, a breath, a scent, a private sound, the angle of an arm. Musicians can communicate through nothing other than the music itself, a type of instantaneous intra-action (to borrow a useful term from Karen Barad) that may seem to the outsider to be ESP in operation yet to participants is simply the cascade of meaning that can flow when verbal language falls silent.

Yet despite this cautionary force-field around language, a map of the process can be described: first of all a discussion in broad terms of potential themes – violence, for example – but without fixating on violence itself, considering instead the reaction people have to violence. "It's in a sensitive stage where nothing is set," says Hans, "so we just try to go deeper in these human themes." Then there is work in progress, traffic between the two areas of the studio – animation and music – and a growing consciousness of what the music might become. That might include the compilation of an inspirational playlist of music and sounds from a wide range of sources, the notebook equivalent of what film directors call a temp track, followed by the collecting of sounds, making compositional sketches on the computer, then mutual discussions on how the work is moving from within its two centres.

Whatever the final shape of the music there is a strong possibility that its presence within the art context will face problems. Where the ocular aspect of the work is likely to be absorbed through an internalised specialist knowledge of its historical and aesthetic ancestry, the aural aspect is more likely to be heard through cruder filters, threadbare histories of audio culture and the technical anomalies of visuocentric architecture. Of the latter, Hans agrees: "I would say it's a problem 95% of the times, actually, always big, empty, hard spaces." To counteract this they install as many soft, sound absorbent materials as possible. Hans has also adjusted his approach to mixing tracks. "So now I often keep the music drier with less reverb," he says, "and also listen to it with a big reverb on everything, trying to mimic how it would sound in a specific large empty room."

This question of how sound can function in spaces designed for looking is deeply problematic, as proven by the troubled history of sound art exhibitions. Sound eludes interpretation and containment, working according to a different economy based on performance and recording rather than the aura of art and its unique objects. So what exactly is the relationship between music and sculpture (film or installation) in this collaboration? Paradoxical in its resistance to the more obvious grotesquery of the sculptures, the music opens up new fields of uncertainty. As Hans says: "If it's a horrible scene and more happy kind of music, something feels more wrong than just making horrible music." This is the equivalent of a cinematic battle scene in which diegetic sounds fall silent and a clamorous soundtrack is suddenly reduced to a single voice. Through reduction and stark contrast the carnage and loss of life seen on screen is felt with greater poignancy.

In certain respects the music is acting as a soundtrack, though not a soundscape. These days the term "soundscape" – an appropriation from acoustic ecologist R. Murray Schafer – is indiscriminately used to describe almost any sound work or listening experience. In his pioneering research of the late 1960s and early 70s Schafer was proposing neologisms that identified listening environments and their properties, particularly their social and psychological effects. Soundscape described an existing conglomerate of sounds in a specific location, imagined (problematically) as the equivalent of a heard landscape. Using it to describe music not only drains the original term of its meaning; it creates confusion around the intentionality of music. Similar problems arise from the word "ambient" and its haphazard usage. From 1975 onwards, "ambient" was conceived by Brian Eno for his personal vision of music as an environmental tint or perfume, an atmosphere that shifted the conditions of a space without imposing itself on those within the space. "Ambient music must be able to accommodate many levels of listening attention without enforcing one in particular," he wrote in 1978. "It must be as ignorable as it is interesting."

The "rules" laid down by Eno in his sleevenote manifesto for *Music For Airports* are rather precise, hardly applicable to much of the music that now positions itself within the genre of ambient, but the effects of these two key concepts of the 1970s has gone far beyond the clarifications of their inventors to create multiple confusions. So the music composed by Hans for Nathalie's sculptures and animations is to some degree soundscape in the sense that it contributes to the making of a world and ambient in the sense that it perfumes the atmosphere of the space which it fills and becomes and yet, in its intention, it is neither of these things because it can never be thought of as merely subservient, accidental, background or self-contained.

To understand his music it should be read within a convergent history of electronica, ambient, techno, disco and house, particularly the shift that in some cases has transported musical strategies designed for the dance floor into art spaces. Tellingly the first experience of music that really inspired him was hearing one of the records, S-Express he thinks, that transformed something far more extreme – the late-1980s emergence of acid house from the dance clubs of Chicago – into a pop/disco phenomenon. "I was very young," he says. "I was sitting in the car with my parents and the radio was on – this super-weird music came on and I didn't understand what it was. It's like 'wow!,' and they were talking about people who had handkerchiefs they used differently to signal what kind of drugs they wanted in this strange scene. I didn't understand – I didn't even know what drugs were – it was very surreal, but the music was so strange and that's when I started noticing electronic music."

Better then to attach it to the worlds of film and television soundtrack, which considers its intentions carefully. Music is powerful in its subtleties, the ways in which emotion can be manipulated by its internal fluctuations and all those elements – tempo, harmony, texture and so on – of which listeners are barely conscious as they concentrate on what their eyes are telling them. But sound is also immersive. In its invisibility it fills a space even when barely audible and so creates a sticky or liquid materiality through which everybody must pass. This seepage of sound into all corners of a room is also durational, perpetually changing over time so that the act of engaging with a work, a sculpture, can no longer be thought of as static observation. Its existence is behind us and around our fingers, in our nasal passages and the convoluted tunnels of our ears; the events of five minutes ago have been forgotten and replaced. The sense of looking at something outside of ourselves is destabilised by this feeling of being interpenetrated, of moving through, of being in time.

For Nathalie, the impact of music and how it changed the experiencing of her work is expressed as revelation. "What I really, really like about the music," she says, "is because what I do looks physical, but when someone is watching it, it isn't appearing in them, except for like a picture, but the music you really feel. So for me, the physical part of the work we do is really Hans' music ... I had to realise it for myself, it was like revolutionary that I'm standing here and looking at something, and I don't really feel a part of it, I feel like it's over there and I'm here, but then I hear the music inside, and music touches more also the emotional part of me than looking at it, and I can feel it vibrating, there's no separation between me and the work, we are as one. So, right now, I think, for the works that we've been doing lately, it's the best thing, or the most interesting thing for me to go, 'Yeah, but it has to be like if I make an animation, can you do something with body, with bass, that vibrates,' and that becomes incredibly interesting."

This embrace of invasive physicality is a recognition that certain types of work can only be realised though the alchemy of opposites. Implicit within this recognition is the necessity of relinquishing control within a collaboration so that psychological complexities of conflict and paradox within the work are given space to develop, compete and merge on their own terms, with their own personalities and demands. This space is physical space, deeply private space and intra-active space – the personal and the public. To renounce sole control of this space is an act of acceptance and generosity that unlocks potential. Here is the paradox already signalled by the work.

dsh: apophatic wu-ness

Published in Notes from the Cosmic Typewriter: The Life and Work of Dom Sylvester Houédard, edited by Nicola Simpson, Occasional Papers, 2012

If, in the 1960s, experimental, improvised and electronic music were at the edges of British cultural life with most poets, as ever, struggling to be heard, then concrete and sound poetry were at the further edges even of those outliers. And at the edgemost edges of those rarefied scenes floated Dom Sylvester Houédard (better known as dsh), in dark glasses and the monk's habit of his Benedictine order; either Sergeant Bilko in the unfolding of a scam or a beatnik from the Middle Ages time-transported to the delirium of London's avant-garde.

Working in the early 1970s as a musician with sound poet Bob Cobbing in two performing groups and taking some part in his Writers Forum activities, meant that I was fortunate enough to encounter dsh in person a few times. He would appear, éminence grise, at various functions (an ICA event at which we were playing, for example) or periodically emerge in the shadows of theory and publishing in pursuit of what he called "apophatic wu-ness," a conjuration of theological affirmation by negation of Taoist emptiness, or what lies behind language. Immersed in words he moved through and beyond words, using a portable Olivetti Lettera 22 typewriter to build up shimmering typestracts, an architecture of letter forms. "Olivetti himself/themselves show sofar a total non interest in this fact," he wrote in the catalogue of Between Poetry and Painting (ICA, 1965). In 1970 I sat at the conference table set up as a centrepiece of APG's Inno 70 exhibition at the Hayward Gallery as he and Barbara Latham (as she then was) attempted to persuade representatives of Olivetti that dsh was the perfect candidate for placement as an artist, an incidental person within their company. They continued to show non interest.

Then in 1973 I wrote an essay for *Kroklok*, the occasional journal of concrete and sound poetry published by Cobbing's Writers Forum press and edited by dsh. In issue 2 (1971), dsh described the brief: "kroklok sets out as openendedly as possible to ingest all material that seems relevant to the label of 'soundpoem.'" My essay subject and title was *Language and Paralanguage of the Sacred*, for the first serious essay of a young auto-didact a dangerously ambitious overview of sacred and secret speech and languages, codes, glossolalia, resonance, shaman's language, voice disguisers and the nonsense languages and extreme vocalisations of artists ranging from Antonin Artaud to Slim Gaillard, Screamin' Jay Hawkins to Professor Stanley Unwin.

If there was one person in the country who shared my comparative enthusiasms it was dsh. He knew far more than me, of course. His learning was formidable, intimidating even, and when I received a 12 page edit (letter number 730817, just in case I was deceived into thinking that corre-spondence from dsh was a rare event) in the post from Prinknash Abbey, gloster (sic), I was both humbled and profoundly flattered. But his edit was also a rewrite, heavily annotated by typewriter and hand (so an archival treasure to me now), thoroughly reworked in his unique anticipation of mobile phone txtspl and lower case aesthetics, interpolated with scholar-ship and ultimately somewhat wounding to my youthful ego. Reading my essay again I can see any number of causes to rewrite it but not, perhaps, to give up my own style, poor as it was, to the distinctively personal language evolved by dsh out of e e cummings, machine poetics and boundless monastic knowledge.

Bob Cobbing, always irreverent, told me I should ignore the edit, pub-lish and be damned, and so it appeared in issue 4 with not a murmur of dissent from dsh. Damnation was not forthcoming, yet I still feel a twinge of regret to have spurned the improving interventions of such a remarkably original, erudite artist and thinker. Reading again his provocative text – *Aesthetics of the Death Wish* – published in the auto-destructive art issue of *Art & Artists* in 1966 I am struck by its prescience, its rigour. He studied the void, writing of the "stringless lute" and the white of the page, for example, and there have been times when his contribution to art seemed to have slipped into a void of forgetting. Happily there are signs that his work has new relevance to an age in which his formulation of "eyear – flickereffect – global language" has become our new reality.

blatzwurds of deepwar plasma:
the noise of Jeff Keen

Sleeve note essay for *Jeff Keen: Noise Art*, Trunk Records, 2013

Words fail me, not because the cassette recordings made by Jeff Keen (1923–2012) are exasperating or impenetrable, but because they resist analysis or explanation. They are important, not least because they were created for his expanded cinema shows, yet how exactly were they made? I can guess but to be truthful, I can't exactly tell you. All that can be said is that they have a primitive power and integrity that is consistent with the way Jeff made his films and other artworks, a labour of love in which he persevered no matter whether the world showed interest or not.

But words fail me because they are meant to. My first encounter with Jeff's work came in 1966. I visited the ICA when it was still in Dover Street, in London's Mayfair. Whatever was on show there that day is forgotten but I came away with a Jeff Keen poster in murky black and white – *Amazing Rayday*, Secret Comic Number 4 – for the cost of 9d (there were 240 pennies to the pound, so for a schoolboy like me, it was affordable). Dated June 1966, a Future City production, with Jeff's Brighton home address given in full, the poster's grubby newsprint style was populated by a mix of comic book imagery (Popeye is recognisable in silhouette), Batman-style sound-effects onomatopoeia, old scientific drawings, typewritten texts, graffiti and pen scribbles. None of these elements were formally organised: words were obscured, overlaid, run vertically, obliterated. It was as if a box of ideas had exploded on the page to leave a messy, noisy residue.

Look closer at the details and clues to Jeff's intentions, particularly his approach to sound work, begin to appear. The poster introduces Dr

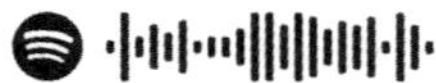

Gaz, wordkiller, clearly a descendent of fictional villains such as Dr. Fu Manchu or Fantômas. Dr. Gaz has a mission to wipe out the rationality of language. Like an alien whose limited vocabulary is compensated by the potent barbarism of its ray-gun violence, he speaks in harsh, staccato blatz wurdz and shatturd wurdfrgmtz: "gluc-c, glucuronic, graaaaaak, zap, kakakakaka, zoop zoop zoop." According to Gaz, "blatz is deepwar ryth-u-m of parallel worldsystems." The poster tells us that he was right to assassinate the poet; his techniques of sound warfare include showing films by hardman Hollywood star Alan Ladd with a variety of replacement soundtracks: the soft dance band songs of Guy Lombardo who claimed his sound to be the "sweetest music this side of heaven," Bela Lugosi's ghost whispers and "mr artode." The latter was, of course, Antonin Artaud, whose 1948 radio work, *Pour en Finir Avec le Jugement de Dieu*, was banned from transmission for its extreme vocal and musical sounds, its anti-American politics and "blasphemy." In his manifesto for a new theatre of signs, gestures, unearthly sounds and images, *The Theatre and its Double*, Artaud wrote about breaking "the intellectual subjugation of language, by providing meaning with a new and more profound intellectuality, hidden beneath gestures and signs, and raised to the dignity of particular exorcisms."

Other references to the sources of Jeff's word murdering inclinations can be found buried in *Amazing Rayday*. There is, for example, "sir ill" or surreal, which suggests that Dr. Gaz's wurdblatz was descended from Andre Breton's belief that automatism released a radically new form of poetry from the unconscious. Dr Gaz is also reminiscent of Dr Wilhelm Reich, whose controversial theories of energy and the electrical discharge of orgasm was intricately embedded within the novels of William Burroughs. A habitué of the Dover Street ICA, Burroughs also repurposed disposable pop culture such as pulp sci fi and comics, treated language as a virus and propagated the idea of mashing up incompatible cultural manifestations into delirious assaults on rational sense.

Burroughsian characters like Dr Gaz and Kamikaze Kid emerge as shadow beings out of the repetition of Jeff's sound pieces, basic delay creating entrancing loops of distorted wordsounds accumulating, decaying and exploding in the tidal surges and repeated bursts of overload that characterise films such as *Flik Flak, Marvo Movie, Rayday Film, White Lite* and

Meatdaze. Every sound is saturated by the aftermath of World War II: the madness and dread born of bomb culture; cold war folk terrors such as brainwashing; the growing media attack of the 1950s with all its accompanying fears of hidden persuaders and social disintegration. In among the cheap as chips noise of modern microelectronics swarm EVP traces of air raid sirens, aircraft engines, machine gun fire and broken transmissions from the front, or many fronts: word wars, chemical wars, propaganda wars, information wars, memory wars, comedy wars, subliminal wars, psychology wars. They remind me of the found images, halftone dots and cut-up texts of Eduardo Paolozzi, whose *Abba-Zaba* book, produced at Watford College of Art in 1970, sloshed around in the fertile mud of weird news, creature features and postwar paranoia. At the same time as echoing the pugnacious Futurism of Marinetti's Free Words, they also anticipate the lo-fi mind control aesthetic of Throbbing Gristle and all the tapes that flowed out of industrial, homemade electro-pop, loop and drone music in the 1980s. Such comparisons can be useful as context but also unhelpful, since these are deeply personal recordings, Jeff himself as the voice of Dr Gaz, using a WASP synth, an Atari PC, a ZX Spectrum, Casio keyboards, children's toys, a microphone and simple effects to persuade his listeners with the cheap promises of 1950s sales techniques: "Men, women, boys, girls, you too can create deadly power-packed blatz-poems or your brains refunded. Amaze your friends, destroy your enemies with well-directed blatzwurds."

Although Richard Hamilton made a convincing early case for British pop art, the seductive surfaces, icons and sheer scale of America always threatened to derail any artist looking across the Atlantic for inspiration. Jeff Keen's work drew from many American influences yet seemed anti-American in its rawness, its intimacy, its deep connections with the derangement of comic anarchy and anger in postwar Britain. He collaborated with Bob Cobbing and Annea Lockwood for the sound to *Marvo Movie*, a montage of texts that included Jack Kerouac's *Old Angel Midnight* (a Cobbing favourite), the daily newspaper and a scientific article; the three of them, in Cobbing's words, "read simultaneously – words tending towards abstract sound."

Then there was Brighton, and a fascination with seaside futurism and nostalgia. According to his daughter, Stella: "He would go out and record live sounds both in the cinema (he loved the hollow echoey sound of the recordings from this) and in amusement arcades. He would record the sound effects of the games being played. And in the movies he was invariably recording screams and explosions, gunfire and laser battles. What he loved was the disintegration of the sound – the fuzzier/messier it sounded the better – he wasn't going for a clean finish here!" All of his cassette recordings would be meticulously archived and labelled for future use, then copied and spliced together on a reel-to-reel tape recorder, sometimes mixed with commercially released film soundtracks or carefully overlaid with electronic effects and voices.

"He also made his own 'art brut' style instruments, like the Orpheus Lyre made from rough bits of wood and plastic," writes Stella, "and customized others, such as the electric violin with doll's body and legs included he called Orpheo Blatzo. The violin's bow is made from dolls' legs and a guitar string. Genius!" Like other UK-based artists who worked with a postwar noise aesthetic – John Latham or Gustav Metzger – the ideas were more important than the medium. For Jeff, the growing influx of American pulp combined with English parochialism to make a body of work that was unique; its range, whether film sound, graphics or constructions, was consistent – a uniquely strange collision that could jump in a heartbeat from sinister to silly, frenzied to gracefully beautiful, personal to universal. In another life he might have been Yellow Magic Orchestra or Kraftwerk (or more likely The Normal) but for a dogged individualism, a compulsion to stay faithful to the ideas that sustained him throughout his life. Thus far he has not been considered as a sound artist nor has he been acknowledged in surveys of electronic and experimental music; as sound works in their own right these cassette recordings with their densely descriptive titles — "BlatzonFragzWhitenseWasp9" or "Omozap To Plasticator" — should shift history a blatzonfragz in his direction.

raging dancing word destruction: Bob Cobbing

Bob Cobbing obituary, published in *The Wire*, November 2002 (Issue 225)

I last saw Bob Cobbing in August, one of those days when you feel that the burger merchants of Finsbury Park could switch off their hotplates and fry meat on the smoking streets. Bob appeared at his basement door as a living illustration of the phrase, under the weather. Frail and clearly in some pain, he found movement a trial, climate extremes the enemy of his arthritis. We sat at the kitchen table, surrounded by the paper architecture of a million small press publications. I had asked him to write a poem, an incantation, that would conclude the piece I was composing for the Thames Festival. He was reluctant at first. You've already got all you need, he said, when he saw the ideas I gave him. Yes, I said, but I want you to exercise verbal magic and besides, I don't have your voice. And so I pressed Record on my Minidisc and as drawn and weary as he seemed, Bob plunged into the racing Thames, calling on spirits of ooze to Rise Up.

I had hoped to see him at Lol Coxhill's 70th birthday celebration at the 100 Club but he felt too unwell to perform. This was rare enough to be a premonition. He did turn up at The Klinker for Lol's actual birthday, two nights later, but then the Fates declared enough and took charge of this rebellious octogenarian who refused to hush and be still. After a brief stay in hospital, Bob died on Saturday, September 28th.

This was a fair distance in years, if not miles, from his birth in 1920. Bob was the son of an Enfield sign writer and the temptation to draw conclusions from that is thoroughly justified. Many of us end up simply reversing the material our parents give us without actually leaving it in the past. For Bob, that was literally the case. His father wrote perfect new signs on the refurbished bodywork cf old vans; Bob took perfect letter forms and texts, then blasted them off the page, transforming them into mantric chants, roaring depths of sound, ink clouds of black night, the stutter of language returned to its source.

His father was a watercolour painter, also, and an amateur musician who played piano and fiddle. "We had one of those big old cabinet gramophones in those days," Bob told me in 2000, "and my father bought a copy of Stravinsky's *Rite of Spring*, which was advertised fairly widely. It had Mussorgsky's *Night On the Bare Mountain* on the other side. He didn't care for it much so he gave it to me and I was thrilled with it. I remember spending hours in front of it. The cabinet had doors that you put your head in and almost shut them behind you. I think from there on, an interest in music of all kinds took off."

Bob began writing poetry at the age of 11, god awful stuff he reckoned, but the first real spark came from hearing Vachel Lindsay's controversial poem, *The Congo*, at Enfield Grammar. Typically, Bob could recite it from memory, tearing into Lindsay's sonorous political incorrectness as a starving man might savage a plate of reeking stew. A conscientious objector during the war, Bob was obliged to work in a hospital. Supervising the hospital stores gave him access to a battered Roneo duplicator, the first instrument of his visual poetry. By 1956, he was cutting lines out of newspapers and rearranging them into poems. Burroughs and Gysin are heralded for that invention but Bob had the scissors out three years earlier. Rather than the bitterness you might expect, this absence from the petty chronologies of the avant-garde induces a belly laugh to mock the gods.

Breaking down the word came in the early 1960s. "It may well have been the accident of finding a lot of old Letraset in a dustbin somewhere," he told me. "It was all cracked. I thought, this is lovely, these beautiful cracked letters." Another breakthrough was his *ABC In Sound*. In 1964 he performed this new work at Better Books, the Charing Cross Road nerve centre of London's emergent counter culture. Bill Butler and Jeff Nuttall added him to their programmes at the ICA, then in Dover Street, where he performed with another poet, Anthony Thwaite. Better known as an editor of Larkin than a fan of sound poems, Thwaite expressed enthusiasm for the ABC and recommended Bob to the late George Macbeth, a

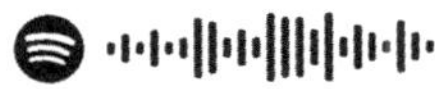

poetry producer at the BBC. This was a good time for sound poetry. Bob began to meet fellow travellers such as Henri Chopin, François Dufrêne, Bernard Heidsieck, Ernst Jandl, Sten Hanson, Lily Greenham and the monk from another planet, Dom Sylvester Houédard. He also worked with Paula Claire, who was performing poems from "texts" of old stones, or half a cabbage. Again, the belly laugh. "I thought," he said, "it's all very well to be performing these shapes from nature, but why not make the shapes oneself?"

As manager of Better Books, Bob became embroiled in the hatching of underground plots such as the Destruction In Art Symposium and two hugely successful Royal Albert Hall poetry readings. An English tour with a bunch of Beats – Ginsberg, Ferlinghetti and Corso – might have elevated his individual star but the irascible Corso objected to Bob and his ouevre so that was that. A lot more Cobbing history has filled the time between Corso's hissy fit and the Bob known and loved by contemporary audiences for his indefatigable performances with Hugh Metcalf, Lol Coxhill and Jennifer Pike. As you might expect of any 82 year old who declines to stop working, whether for Christmas or old age, Bob's story is too expansive to fit on one page.

I first met him in 1971 or thereabouts. Paul Burwell and I were playing in a duo called Rain In The Face. Paul's girlfriend, Sheila, was pregnant and Paul dutifully went to inform her father. As if in a Douglas Sirk movie about the avant garde, this turned out to be Bob. Both were so embarrassed by the situation that they discussed sound poetry, art, music, anything other than the subject in hand. Subsequently, we began playing together in a trio called abAna, touring sundry venues of the British Isles and beyond. The last gig as a trio was at The Klinker in June, a marker of our 30th anniversary together. The intensity still lingers. Somehow, I don't think the visceral effrontery of his raging, dancing word destruction will be replaced.

8

listening moving marking

Sounds Passing Through Circumstances:
Francis Alÿs

**From catalogue for *Seven Walks, London, 2004–5*,
published by Artangel, 2005**

If I am here, then where is the sound? Sound has no sight-line, no fixed
point in space, no duration beyond its own activation, no single moment
of existence, no edges, but only cumulative moments of disappearance at
the boundaries of its reach. Its place as a mark within temporal dimension
and the mapping of space can be a mixture of the precise and ambiguous: a
bell rings, the clock chimes, a cannon fires a shot. The day is divided and
the space of human relations is mapped according to the fluctuations of a
sound and its extension through air.

Call it spillage, cloud, smoke; the need for similes drawn from the tangible yet fluid world of liquids and dispersing materiality is only a lunge at
the nature of sound. So much of the world is consumed through the culture of text, in alliance with various visual forms. Urban space is divided up
according to ideas of visual drama, social connectivity, and the pragmatics
of movement, yet sound is taken for granted, forgotten, or ignored despite
its vital role as an element in urban design. Sound is not reducible to a text,
so not susceptible to "reading." Its place within the system of signs is an
anomaly, the paradox of the invisible/audible. The transience of sound, its
abstraction, its passage through time that leaves no trace, all form a resistant barrier to interpretation. Most attempts to understand sound attempt
to avoid its nature in favour of descriptions of its context, so sound remains
a barely categorised yet central element of social and cultural life.

Artists who begin with the visual are confronted with the challenge of the system within which they work, in which the visual can stand for status and containment. The space of the gallery is not designed for sound (is even designed to exclude or minimise extraneous sound), being conceived in most cases as a frame for that which has a frame (or some sort of boundary, at least), and so sound leaks, creeps or bombards in its unruly fashion. "What is still interesting to engage with and pursue is sound," artist Anri Sala has said. "Sound is one step away from language (sounds in an airport, or the sound of elevators in hotels, are already language), but once you are in the street, most sounds are untamed, they are in the act of becoming. They are part of a language that is not yet controlled. And that is why I am interested in sound: it's like an incomplete music."[1]

This follows, perhaps unwittingly, thoughts noted by John Cage in 1969. "Introduce disorder," he wrote. "Sounds passing through circumstances. Invade areas where nothing's definite (areas – micro and macro – adjacent the one we know in). It won't sound like music – serial or electronic. It'll sound like what we hear when we're not hearing music, just hearing what we happen to be."[2]

What do we happen to be? Francis Alÿs began to ask this question after moving to Mexico City in 1987. Faced with the enormity and dizzying complexity of a city whose nature is to overwhelm, he began to walk. "Certainly, at the very beginning, It was a very non-adding attitude," he says. "A let-be situation, just passing through."[3] Gordon Matta-Clark was one of the first artists who influenced him, but by literally cutting sections out of buildings, Matta-Clark was extracting from his chosen environment, whereas Alÿs was trying to build on the story of a walk and use that as a vehicle for an art work to happen.

"Being in Mexico City," he says, "it seemed vain to try to add something in that enormous and saturated situation. I was trying to affect the situation in the most minimal way. In the first walks, maybe a few passersby noticed what I was doing. I didn't want to be adding anything concrete, I just wanted to insert a story, a furtive/clandestine act."

Both Matta-Clark's anarchitecture, and this act of walking urban space, suggest a relationship to Alÿs's training as an architect and engineer, even if only in the sense that architecture is reversed out, or plotted from the external boundaries. Its occupation of space is traced in order to find transient events and human transactions within the streets.

"I don't know if there's a direct connection," he says. "but I think it's a natural state for somebody who's interested in cities or architecture in general to walk. Walking offers a very convenient space for things to happen, and it allows a certain awareness in between an ongoing chain of thoughts and a series of incidental informations around, glimpses of scenes, sounds, smells, etc." This journey between two points, encompassing the peripheral information that filters through, bouncing off thoughts to shape the piece and simultaneously question its validity, is his optimum working space.

"From early on I was intrigued by these characters I would meet in my neighbourhood," he says, 'in a sense lost souls who were inventing themselves those kind of roles to justify their presence in the urban chessboard, odd characters who do these strange acts on a repeated basis, but through that ritual they build themselves a territory and a public function which would make them a part of the neighbourhood, or a local reference. Maybe it was because I was an outsider and trying to find an entry. It was a very slow process with many mistakes."

Because of its role in urban space as a component "passing through circumstances," an element only regulated through negative action (too loud; temporally or contextually displaced; downright inappropriate) sound is assumed to leave little trace, even though its role in memory and the evolution of urban culture begins to be acknowledged. Progressively throughout the 20th century, the composer's role in regulating and constraining sounds was relaxed. John Cage was the first composer to apply this conception consistently as a methodology for organising sound. Speaking of a Mark Tobey painting, *Untitled*, 1961, Cage said: "What's so beautiful is that there's no gesture in it. The hand is not operating in any way."[4]

Cage opened a window and the sound drifted in. The remnants of a frame that always contain his works were reduced even further by pieces such as Philip Corner's *I Can Walk Through the World*, and Max Neuhaus's *Listen*. In his Lecture 1960, delivered to Anna Halprin's Dancer's Workshop in Kentfield, California, La Monte Young recalled a meeting with Dennis Johnson, when Johnson claimed to have found a piece that was "entirely

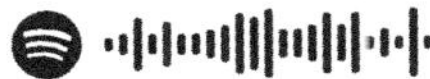

indeterminacy and left the composer out of it." According to Young, he then tore off a piece of paper and wrote on it the word LISTEN.[5] Corner's *I Can Walk Through the World*, a piece from 1965 in which the audience at New York's Town Hall was taken for a walk around Times Square, extended the implications of Cage's *4" 33."* Like Corner, Neuhaus was prepared to step out into the street. His *Listen* pieces, begun in 1966, were subtitled Field Trips Thru Found Sound Environments. Neuhaus would invite an audience to a lecture or concert, stamp their hands with the word Listen, put them on a bus and transport them to a site of distinctive sound such as a power station.

Ambiguities exist in Alÿs's work, particularly in the question of documentation that haunts performance. Just as audio recording can reduce music and sound work in all dimensions, overlaying one space with a random variety of others, obliterating the atmosphere of social and spatial sharing, so video can flatten any sense of emergence. "The image on the left is a direct documentation of the facts," he says, discussing one of his walks in Mexico City (*Re-enactments*, 2001), "and the image on the right is the re-enactment of the facts, it's pure fiction. The two images are pretty much the same, and so is their perception. As an artist who has been using performance as a medium, it's very difficult to maintain a certain integrity when it comes down to the documentation of a performance. What makes the image on the left more valid than the image on the right? What's left in both images of the experience of the live moment? And how much of the performance has been unconsciously conditioned by the prospect of its future documentation? Each time I document a performance I am trying a different take on these questions."

The mixing point of Alÿs' work comes through files of notes, which collect together drawings, diagrams, textual scribbles. The notebooks seem to function as a style of notation in which future actions can be imagined, yet they also seem to be at the heart of the work. Everything is here, and we can imagine outcomes, or sense what might be, or what might have been. This recalls the *Nature Study Notes* of the Scratch Orchestra, edited by Cornelius Cardew and published in two collections: *Nature Study Notes*, 1969, and *Scratch Music*, 1972. A loosely convened, large ensemble of composers, musicians and other artists, founded in London in 1969 by Cardew, Howard Skempton and Michael Parsons, the Scratch

Orchestra was, as Cardew proposed in the Draft Constitution, a group that "fosters communal activity, it breaks down the barriers between private and group activity, between professional and amateur – it is a means to sharing experience."

"Just as 'any activity whatsoever' could be included in the category of performance," wrote Michael Parsons, examining the group's history for *Leonardo Music Journal,* "so any kind of graphic material came to be regarded as a possible form of notation: a look at *Scratch Music* reveals a miscellany of drawings, diagrams, maps, collages, texts, photographs and found objects (even some musical notation) from the notebooks of sixteen members of the orchestra, laid out in random juxtaposition to suggest the visual equivalent of a Scratch performance. Anything that could be set down on paper, it seemed, could become part of the all-inclusive and indiscriminate category of 'graphic music.' "[6]

Text scores were central to the challenge of circumventing the emphasis on pitch and metre in conventional music notation, as opposed to alternative notations of timbres, unconventional technologies and instrument preparations, durations, actions and activities that might not be considered musical, sonic, or even connected to the making of art. They confronted hierarchies in musical professionalism, along with the systematic inequalities within society that interpenetrated the structures of music. The text pieces written by poet Jackson Mac Low were particularly inclusive and open. His instructions for *Thanks,* composed between 1960 and 1961, conclude with the following: "Anyone may submit any or all elements of this simultaneity to chance regulation by any method."

Walk – "for any number of people walking in a large open space" – composed by Michael Parsons in 1969, precisely illustrates this fracture within the conceptualisation of what music might be, as well as returning us to Alÿs. In *Walk,* randomly chosen numbers determine the speed of walking from one point to the next, and the length of time standing still at the point reached. The walking itself, the nature of the walking, the geometric patterns and intersections that can be imagined as an inscription derived from the event, and the communal spirit of its enactment as a public/private rite are what exist in the bare bones of instruction on the page. Clearly, it is also possible to understand *Walk* as a near-silent equivalent of the process of performing a symphony, in which participants begin and end at allotted moments, yet move through the work in varying rates and durations of activity and rest.

Sounds converge, through a time base; meet, merge, evoke desire, sex, the social, yet produce structures that can be deciphered and absorbed. This articulation of time, through which music and sound work emerges, has been decisive for Alÿs, and has clear parallels with the development of his scenario for a contingent of Coldstream Guards. They marched towards each other, gradually building a square under the multiple eyes of CCTV in the City of London, presenting arms, moving as blocks of colour, making sounds that were at once musical and military. In the last few years," he says, "I have been using sound as a means to destabilize time perception through the space of the rehearsal, a way of diluting time if you want ... Very much thinking of my experience of time in Latin America, a certain way of delaying the narration or postponing the conclusion. There is a progression always, but by kind of going back and forth, three steps ahead, two steps backwards, four steps ahead. It's a different take on the western concept of what efficiency could be. It took me quite some time to understand that mechanic and adapt to it."

Adaptation is also enforced by the sonic nature of a dynamic, complex city. In Mexico City, Beijing, Bangkok – cities where life rapidly mutates on the streets and legislation is informal or drastically uneven – music is heard in contiguous layers, enfolded in vernacular and functional noise. The shrill, relentless blasts of traffic control whistles, a chaos of traffic, three different styles of music heard simultaneously; such a typical soundscape of Mexico City is integral to the evolution of civic space and shape.

"Here in Latin America the function of the urbanist has been drastically challenged in the last couple decades," says Alÿs, "where I was taught that the urbanist had to plan ahead the expansion of the city, to reflect upon its future mutations etc. I saw his role inverted if not reduced to the opposite mechanism: the urbanist's role is now to react to given situations of spontaneous urban growth, to adapt to them and subsequently supply municipal services such as water or electricity to an anarchic urban phenomenon ... there is an absence of any master plan." Issues of control are touched upon here, derived from a method of making work and its specific urban context, but also illuminated by challenges to control and authority formulated in the latter half of the 20th century through chance procedures, indeterminacy, improvisation and communal music making.

There are illustrations of this in *Guards*, which developed a final form through a mixture of direction, unforeseen elements, and the decisions of the participating soldiers, their musical repertoire and the interconnection of marching, drilling, and weaponry. "A big part of the work is to provoke something beyond what you can plan," he says, "and the more the project develops, the more it becomes a game of bouncing back and forth in between all the people involved, and it sometimes can adopt a shape very far from any original intention. Now, once the action itself is launched, the development of the piece is happening within an open field of possibilities, in the sense that any outcome of the event becomes a valid answer to the premises of the piece. It's the real test on the scenario: if it isn't clear enough, or good enough, the action will deviate, and rapidly turn into something else or simply collapse ...

"I've done things where I lost total control and the crew working with me started freaking out. The one occasion I am thinking of was a year and a half ago, and it went totally wild. I was just watching, there was nothing I could do. it was totally liberating. But if the plot is simple and clear enough, the essence of the project will always survive. Coming back to the guards, we thought that there would be more different types of marches, many more notes if you want ... But they basically alternated two marches: the slow march and the quick march with some variations on the quick march. And we were expecting a game of phasing of steps, a kind of tuning if you want, but most often when they met one would stop and start marking time, and the others would join in. So again, once the parameters are set, the participants quickly imposed their own rules: These are the notes the instrument can play and that's the way we can play it."

All self-imposed challenges to the authority of score and director contain the potential to eradicate both. I ask him if this was welcomed in the enactment of *Guards*. "Yes, yes, absolutely," he answers. "They also asked if they could incorporate something else, something they thought would dramatically improve the musical potential of the piece: they wanted to carry their guns so that they could do the change of arms, ... clack, clack, clack, so they immediately translated our idea into their own skills and what they were best at. And that's precisely what the piece wanted to provoke, to see the guards display the maximum of their skills to build up this perfect human machine, and for that to happen you had to find out what the perfect cruising speed of the machine was, the one that you felt was really coinciding with their image, but also the one where that essential need for individuals to dissolve into a social group they can identify with would became physically evident."

Though not as extreme as John Cage's idea of silence, in which "the essential meaning of silence is the giving up of intention,"[7] this links us closely to the ideas of Cage and his circle. Events are initiated, but they will develop their own momentum, impose decisions, and find their own form. Ultimately, they find their own life.

Notes

1 Anri Sala, quoted www.artic.edu/aic/exhibitions/sala.html.
2 John Cage, Art and Technology, 1969, published in John Cage: Writer, ed. Richard Kostelanetz, Limelight Editions, New York, 1993, p. 111.
3 All quotes taken from conversation between Francis Alÿs and David Toop, recorded London, 6 July 2005.
4 John Cage, Musicage: Cage Muses on Words Art Music, *John Cage in Conversation with Joan Retallack*, ed. Joan Retallack, Wesleyan University Press, New England/Hanover and London, 1996, p. 127.
5 La Monte Young and Marian Zazeela, Selected Writings, Heiner Freidrich, Munchen, 1969, part 6, unnumbered pages.
6 Michael Parsons, "The Scratch Orchestra and Visual Arts," 2000, published in Not Necessarily English Music special issue, *Leonardo Music Journal*, Volume 11, 2001, p. 8.
7 John Cage, On the Interplay Between Art and Music, Sounds of the Inner Eye: *John Cage, Mark Tobey, Morris Graves*, Ed. Wulf Herzogenrath and Andreas Kreul, Museum of Glass: International Center for Contemporary Art, Tacoma, 2002.

A Ghost Travelling a Half Mile Ahead of Its Own Shape: Faulkner and Caravaggio

Extracted from lecture/performance texts and *Cold World: the deathly void of Sound*, published in *The Soundtrack*, vol. 2, no. 2, 2009

"So that at last, as though out of some trivial and unimportant region beyond even distance, the sound of it seems to come slow and terrific and without meaning, as though it were a ghost travelling a half mile ahead of its own shape," wrote William Faulkner in *Light In August*. " 'That far within my hearing before my seeing,' Lena thinks."

If we compare Caravaggio's *Martyrdom of Saint Matthew*, painted in 1599–1600, with one of its precedents, Girolamo Muziano's painting of the same subject from 1586 to 1589, we find dramatic differences. The Muziano is an example of what Poussin described as the silent art of painting: onlookers to the impending martyrdom stand mute, expressing their emotions with their hands, their mouths closed. This is also true of Saint Matthew himself, who asks for pity only with his eyes and extended hands. As for the swordsman, he says nothing, only looks down at the Saint with contempt. Everybody is still, as if a moment has been frozen in the unrolling tide of history, before violence, before consequences. In the foreground, a woman cradles her small child, holding his face into her breast; she looks away from the scene. The drama is constrained within a sensory world of seeing and touching, the world of things; silence within clamour.

What a contrast with Caravaggio, who unleashes a wide range of reactions: shock, contempt, fear. Some mouths are open – a gasp, a shout, a curse – and in the dynamism of movement, the recoiling, the turning of heads, the sweep of an arm, the awkward turn of a body, and then the

sword, poised to strike, we feel that the dark space within which drama is contained is echoing with the drama of listening. The painting is silent, of course, but at the imaginative level, hallucinatory in its intensity, sound cuts into a silent world of visual contemplation with the sharp violence of a scream.

Certain paintings guide us into a sound-world otherwise lost. In the centre of Caravaggio's *Rest on the Flight into Egypt*, painted c. 1595, a female angel is playing the violin. Joseph is holding up music for her to follow – "How fair and pleasant you are O loved one, delectable maiden" from the *Song of Songs*, set by Noël Baulduin, a Franco-Flemish composer of the late fifteenth/early sixteenth century. The notation is clearly visible, though as Catherine Puglisi points out in her book on the painter, Caravaggio decided not to show the words (Puglisi 2007). Except for the patron who commissioned the painting and those literate in musical notation, this sub-text of the scene would have been obscure or inaccessible. Madonna and Child sit to one side, both fast asleep. Although the group occupies a small space, there is a marked contrast between the ground on which Joseph sits, which is stony (he is rubbing one foot on top of the other to stay awake), and the lush vegetation in which Madonna and Child are sleeping, tall grasses, climbing plants and then behind them, fields, forest and distant hills. Just behind Joseph is their donkey, very close to the centre of the action, its nose close to the violin and the music. Concealed within the visual reality of the work is a more controversial work, a composition or sound work – a melody played (with angelic touch) on violin, the sound of grasses and leaves, the breathing of a donkey and two sleeping people, the slight friction of small stones under bare feet.

The scene is visual, but because of the drama Caravaggio brings to foreground, depth, light and movement, the painting "sounds" clearly, encapsulating the dilemma of a conflict between the symbolic and the abstract in listening. A deep symbolic language is buried within the painting, which would be very clear to those versed in its vocabulary; then the more accessible symbolic levels allow any viewer to understand the basic story, yet the words of the song are absent. We read the painting at various levels, and hear devotional musical tones without their words of devotion. This essentially abstract, or non-representative sound, is contextualized within the wider field of listening: the complex ambient sound of the scene that extends beyond the picture frame and beyond the horizon. Looking, it goes without saying, is restricted to the forward facing field of vision, whereas listening extends in all directions, connecting us to sources that have no verifiable existence other than the unstable presence of that passing moment of nothingness that we call a sound.

By reading this painting, and others, though the imperatives and filters of our visuocentric culture of looking and holding, the auditory elements (which are, after all, lacking in objective proof and resoundingly silent) are diminished in their importance. The music returns to centre stage, which is more or less where it is situated in the painting; our knowledge of the music is dependent on the visual symbols – the book of notation held up by Joseph as an aid to the angel. What is clear from Caravaggio's work is that he was less interested in talk than in emotive sounds and silences. The silence in his *Judith Beheading Holofernes*, for example, is grotesque, a black comedy: the servant looks on in grim concentration while Judith leans back, repulsed and determined, face rapt in the violent butchery of her task. Their side of the painting is uncanny with silence, yet on the other side, Holofernes screams, blood spurts, the razor grinds against bone. Aurally, if we can imagine it, the painting is even more ghastly with eyes closed. We hear (or hallucinate) these sounds. Within and extending from the frame of the painting is the auditorium: the hearing place.

Sound Thinking: Stuart Marshall

Blog post: *a sinister resonance*, July 10, 2013.

Wood striking wood, quick, hard, BOK! Impact sound sprays out, an omni-directional striking of all reflective surfaces and returning through time to the distributed centres of listening, the BOK-space of audition. This is the basis of Stuart Marshall's composition known as *Idiophonics* or *Heterophonics*, a piece performed only occasionally in the 1970s and now about to be revived.

I was present at a 1976 performance that began in 2B, Butlers Wharf, then spread out along the Thames and over Tower Bridge. Memories of the event are foggy (that may also have been true for the winter weather) but I wrote a contemporaneous account in *Readings* magazine (edited by Annabel Nicolson and Paul Burwell, 1977). In that essay I described Stuart's work with sound as being "fairly unique in this country." Of course he could be unique or not unique, not "fairly unique"; what I was trying to convey was an emergence of sound work, an engagement parallel to contemporary art practices such as structural film and video with their intellectual preoccupations (notably Lacan) that rejected the rituals of music. As an approach it was rare though not unknown.

The piece began with three people – Stuart Marshall, Jane Harrison and Nicolas Collins – striking closely pitched woodblocks, moving away from each other every time their strikes coincided. In a second phase they took up aerosol klaxons ("used in America for scaring off intruders and bears," as I wrote in the Readings essay) and moved out into the freezing night: "One of the players stayed close to the building, one moved along the river to the right and the third walked out over Tower Bridge. The sounds bounced back and forth in a most spectacular way for quite some time – after a little while most of the audience left the rather precarious platforms which jut from the doors and huddled around an electric fire. Conversations started up and the piece took on the dimensions of a social gathering punctuated by alternately mournful and strident honking from outside."

David Cunningham was present – his photograph of a murky Tower Bridge was used to illustrate my writing. Nearly 40 years later he recalls as little as me. "I remember talking to Gavin Bryars," he tells me in an email, "as the ship approached Tower Bridge which remained closed until the last possible minute. There was a possibility that the klaxons were confusing whatever signalling system the bridge uses and some speculation about another maritime disaster for Gavin to turn into a score. I have a feeling the ship sounded a klaxon too."

Within these accounts there are indications of a new way to be within the experience of a sound work. Nothing of the event could be conveyed through secondary media – you had to be there – but to behave as a conventional "audience" was clearly silly. At the time, Nic Collins recounted the unfolding of a Connecticut concert hall performance, the klaxons growing fainter in the distance until inaudible, the audience sitting patiently in silence and then applauding when the players returned.

There are things that could be said here about the development of sound art, about the influence of Alvin Lucier, about echolocation, about bats and blindness, ships and sirens, temples and time. I think always of Giovanni di Paulo's 15th-century tempera panel, *Saint John retiring to the Desert*, Saint John emerging from a city gate, a small bag of possessions slung from a stick; in the centre of the painting he can be seen again at the mouth of a mountain pass, an echo of himself dwarfing the tiny buildings depicted on the plain below. This technique is known as "continuous narrative." "The artist's intention in showing the same figure more than once was clearly to indicate the passing of time," wrote Alexander Sturgis in *Telling Time* (National Gallery Company, London, 2000, p. 22).

Each medium is limited by comparison with the body's versatility. For Aby Warburg, the inherent interest in Italian Early Renaissance painting lay in its representation of movement, an evocation of Antiquity in which the body was caught up, as Philippe-Alain Michaud described it (in *Aby*

Warburg and the Image in Motion, Zone Books, New York, 2004, p. 28), "... in the play of overwhelming forces, limbs twisting in struggle or in the grip of pain, hair flowing, and garments blown back through exertion or the wind ... He replaced the model of sculpture with that of dance, accentuating the dramatic, temporal aspects of the works."

Music has fewer problems with time but its innate desire is to rigorously control space, to bring sounds together in coherence and spatial focus, as an illusory object. *Idiophonics*, by contrast, pulls the elements apart to distribute them in space, leaving the more inert variety of audience stranded in time with only an empty object to contemplate. David Cunningham has recently noted my confusion of idio- with ideo- in the 1977 text. As he points out, the prefix idio- denotes uniqueness, privacy, a personal quality, as in idiosyncratic or idiomatic, but it also describes that which is distinct, unique or separate. Perhaps this latter meaning is what Stuart had in mind, a separating out of sounds, or like me, could he have been mixing up idio- with ideo-?

Stuart is not here to be asked – he died of an AIDS-related illness in 1993. A strange thing: we were born within two days of each other in 1949 and found ourselves assigned to the same work table, same teaching group, at Hornsey College of Art in 1967, our first year of art school. Within that year we were the only students with a developed interest in experimental music so the coincidence, from this perspective of passed time, seems marked. Our conversations about La Monte Young, AMM and Ornette Coleman helped to make the prospect of this new venture, what we now call sound art or audio culture, more tangible than it might have been had we been alone in our enthusiasms. Stuart went on to study with Alvin Lucier at Weslyan, became a pioneer of video art, then an HIV/AIDS activist and filmmaker until his death at the age of 44.

Read his online biographies or obituaries and they emphasise the latter stages of his career. This seems as it should be yet because of my personal contact with him and my predilections I consider him to be one of the unsung pioneers of sound art in the UK. In *Musics* 9 (1976) I reviewed a video screening of Stuart's work at the London Filmmakers Co-op; even from my brief descriptions it is apparent that video offered the technical means for him to explore interdependence of hearing and seeing: a bottle smashing in silence, the interior of a mouth and its ambient roar. In 1979, during the Music/Context Festival of Environmental Music at the London Musicians Collective, Stuart performed a solo version of *Idiophonics* from within a canoe paddled by Paul Burwell. I photographed the event as they glided off over the water, Stuart with an aerosol klaxon in hand. That photograph is not available to me at the moment but the memory is fresh enough, sound blasts ricocheting off the high walls lining Camden Canal. By that time, music was out of its box, sound thinking no longer "fairly unique."

Live at The Filmmakers Co-op: Stuart Marshall

From *MUSICS* No. 9, September 1976, p. 27.

In June a number of evenings were given over to non-film work by the London Filmmakers Co-operative. The first of these was taken on by Paul Burwell and myself and, it was hoped, films of our own choice. The films that Paul and I chose finally were whatever turned up out of the complete works of Harry Smith or films made by members of a Navajo Indian community [*Navajo Film Themselves*, 1966]. We had abandoned the idea of showing *Trobriand Cricket* [*Trobriand Cricket: An Ingenious Response to Colonialism*, 1973–4, Gary Kildea/Jerry Leach] since it was shown on TV just a few weeks before the Co-op evening. As it was, none of the films turned up from the USA and we had to suffer inappropriate juxtaposition with the work of Paul Sharits. The whole thing would have been much simpler if the Navajo films were kept in the Royal Anthropological Institute film library but they are excluded – adjudged of insufficient interest, in spite of – or more likely because of – Claude Lévi-Strauss's opinion that they are among the most important film documents ever made. Since the RAI carries "the book of the film" – *Through Navajo Eyes* by John Adair and Sol Wirth – then it has to be assumed that the Institute either considers the statements of anthropologists more "valid" than the statements of the observed or it considers "observation" and "commentary" more important than experience. If the latter is the case then members of the Institute – i.e. Mary Douglas – should cease the sham of paying lip service to art and the artworld. As it was, the Navajo films were shown later in the month at the National Film Theatre and caused a lot of walk-outs, or so I am told.

The second evening was devoted to performance work by Keith and Marie and Reindeer Werk – Tom Puckey and Dirk Larsen. I didn't attend so can't comment. Video work was shown in the second week. I attended the greater part of the second showing but missed Mike Leggett's *Moon Time* and most of Reindeer Werk's piece, again. They appeared to be werking. Of the three other participants – Tamara Krikorian's piece I would prefer not to comment on here although I found it quite enjoyable. Stuart Marshall and Tony Sinden are both of interest within this context since they both work within the field of music, albeit with extremely low profiles. Tony's two contributions were both self-confessedly slight – *Nothing Really / Really Nothing* being a deliberately boring comment on language. The second piece was quite amusing – the video monitor had to be tilted to show an upright picture of Tony balancing a chair. I had already seen an informal and better version – Tony balancing a small monitor on one hand – so the experience was a bit diminished.

Stuart Marshall was represented by five pieces – *Go Through the Motions, Just a Glimpse, Mouth Room, Screen* and *Arcanum*. I wouldn't propose to talk about them in the space available here. In fact, I wouldn't propose to talk about them at all without discussing them more with Stuart. I will stay descriptive for the time being. Most of the tapes posited an extreme interdependence between the visual and the audible for the video experience and proceeded to manipulate that interdependence in an extraordinary way.

A text read over and over. The lips freeze and reanimate. A bottle hits the floor in silence. You hear the smash. The inside of a mouth. Sounds like the roaring of ambient room acoustics recorded and amplified. Stuart sits, lights a cigarette seen on a monitor, within the monitor, reads a long text which is written as he reads on the studio monitor. It is a pity that his music and his writings are not available more widely. His book for Experimental Music Catalogue seems to have been delayed by EMC's financial problems. Early scores are available in the Alvin Lucier edited issue of *Source*. Writings on video can be found in the May/June issue of *Studio International*.

9

nothing silence black not-black invisible

Ryoji Ikeda

Ad Reinhardt

Picasso's Guitars

Shirazeh Houshiary

John Cage and John Latham

Lucie Stepankova

Less Is More; More Is More: Ryoji Ikeda

Published in *The Wire*, May 2006 (issue 267)

Beginning with silence. This is Susan Sontag, from her essay *The Aesthetics of Silence*, published in The Minimalism Issue of *Aspen Magazine* in 1967: "As Oscar Wilde pointed out, people didn't see fogs before certain 19th-century poets and painters taught them how to."

Twelve years ago, I was approached at the ICA in London by a Japanese man in his late twenties. He spoke little English. His project was to record short video interviews on the subject of silence. A CD entitled *Silence* changed hands, a white card box containing a booklet and somewhat surprising musical selections, released by Wacoal Art Centre at Spiral Gallery, Tokyo. Aspects of silence were encountered within, including the work of Derek Jarman, who talked of the silence that comes with an eclipse; neuroscientist and self-experimenter John Lilly, who developed a belief in superior guardian beings in the total silence of isolation tanks; the exquisite note placement, piano in silence, of Paul Bley; David Cunningham's gated feedback that returns to zero; Jan Steele's gentle swing, a hybrid of jazz composition; the drifting bamboo sho reeds of Tamami Tono and Ko Ishikawa; the circling quietude of John Cage's *In A Landscape*, interpreted for harp by Masumi Nagasawa. Coalescing like the concept of fog, the collection seemed to be auditioning past and present variations on silence then proposing a new impression of silence and its impact on the emerging post-everything sound world.

The producer of the CD, the interviewer at the ICA, was Ryoji Ikeda In the ensuing years, insistently though perhaps reluctantly, Ikeda has come to represent an absolute methodology, a rigorous aesthetic through

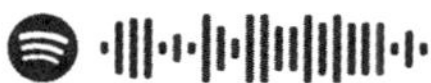

which silence is the white on white heart of lightness at the core of digital music. Though exploring significantly different presentation modes, his installations, CD releases, concerts and collaborations all contribute to a strong image of both person and work. Interpretations (of both person and work) usually settle for defining characteristics such as rigour, control, and technological formalism.

"Silence is the artist's ultimate other-worldly gesture," wrote Susan Sontag. "By silence, he frees himself from servile bondage to the world, which appears as patron, client, audience, antagonist, arbiter, and distorter of his work." This has been Ryoji's strategy, politely keeping his distance from media interviews, avoiding analysis or statements, explication, elaboration, even photographs, but now we are sitting together in London, discussing issues germane to his work, if not the work itself. Naturally, given this sudden breaking of a ten year silence, I am interested to discover if the strategy has worked.

The previous night, two of his works – *formula* and *C⁴I* – were presented at The Barbican Hall. The first, occupying the first half of the concert, was *formula [ver.2.3]*, now a familiar piece that draws from the data/media sampling overload of his 1995 album, *1000 Fragments*, the Brian Eno ambient influence audible from his contributions to *Silence*, and the more focussed clear spot of +/ -, released on Touch in 1996. Since some of the material comes from a period in which he was still a part of DJ culture – astronauts in space, stereo demonstration records, global radio and hip-hop beats – *formula* is now showing its age.

Both pieces are strikingly different, and even though the origins are familiar, I'm sufficiently influenced by more recent mediation of Ryoji's work to question the contrasts between overloaded, specific imagery and reductionist abstraction. "I'm in the generation after postmodern, after minimalism, after everything," he says, "so I just use all kinds of techniques as an artist. I'm just me." When it comes down to what he does, mostly he doesn't wish to discuss it, and this emerges when I ask him a question given to me by one of my students, Tetsumi Segawa. She wanted to know what had motivated him to make a work like *C⁴I*.

A long silence ensues, then: "It's very difficult to say because I told everything through the piece. So basically in general I don't want to explain anything about my piece because my works already are telling everything."

Oblique strategies plainly a necessity, I ask him about the period when he was a DJ, and what he played. "Abstract, from the beginning, with no beat," he says. "Ambient. I mixed your record on Obscure. I did an Obscure mix. [Brian Eno's] Obscure was a very important label. This was really at the beginning in 1990, or 1989, at the same period of house, acid house and ambient house, 808 State and so on. I was young. I loved to be at clubs with friends, just normal. I just loved parties. I was really young, 24–25, in Tokyo.

"I had just graduated from university. I had no job. I studied economics, not macroeconomics but microeconomics, like marketing. I know everything about advertising, marketing and this kind of media influence. This is very useful as an artist because I know the system and structure and concept, so it serves me a lot, but as a producer it's so useful because I know what I should do. To pinpoint, I just set the range of the listeners."

This could be construed as a lesson in how to profit from personal history and misdirected educational decisions, however tangential to future ambitions they may seem at the time. Born in 1966, into a merchant family, Ryoji grew up in Gifu, between Nagoya and Kyoto. "It's just countryside, nothing, just boring," he says. "Quite a big city but it's just industrial, no arts scene. I was there until high school and then for university I moved to Tokyo. My family is very normal. I remember my childhood very clearly, just as a snapshot. It was very normal. There was nothing creative – just my town, my school, my family, myself. I like to learn things by myself, so I like self-taught artists like Takemitsu. I don't know – maybe I have a little complex to academic as a reaction, because I never know this world. I can t write a score properly, I can't play piano, I can't play any instrument, so probably I have a complex, subconsciously."

In 1993, he took a job as audio and visual producer at Spiral, a chic Aoyama gallery, shop performance space and restaurant located close to the fashion boulevard of Omotesando. "Very soon I was fed up with everything," he says. "That period was so intense, every day, every day, every day. As a producer I organised more than 400 events in two years at Spiral, contracting with the artists, inviting, working with promoters, releasing CD, showing art films like Derek Jarman. I was totally overwhelmed and exhausted and I just needed silence. This is my silence. It's true. Then I cut all communications with all art people, music people, show business, this

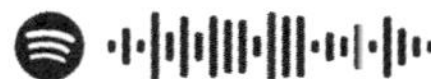

kind of superficial people. I just tried to find who is my friend and I found just three or four.

"But there was a gift. When I worked at Spiral I met Dumb Type and we quickly got married." He laughs at this point, and through most of our conversation. "I was naturally involved in Dumb Type. Then one day a fax came: 'We are waiting in the airport in Ljubljana.' There was no explanation, they just sent the fax with the reservation code of the flight. What? I just packed and went. Next day I was operating. Then I joined their tour for 12 years. It was a great experience. It was a huge chance for me to go out from Japan. And then my life was totally changed. I experienced many, many things."

Dumb Type is an artist collective, based in Kyoto and founded in 1984. Although generally considered a theatre group, their activities have included art exhibitions, audiovisual work, publications, and installations such as the spooky piece displayed as part of the permanent collection at ICC, Tokyo. In an empty room, flat screens, each the size of a person, lie in parallel on the white floor. People are visible, or not visible, in the screens; quiet sounds hum like the track of a body scan. Teiji Furuhashi described the piece as an exploration of the border of life and death, now controlled by technology but still a profound issue for the mind. Core Dumb Type members Shiro Takatani, Hiromasa Tomari and Takayuki Fujimoto all made fundamental contributions to the images and staging of Ikeda's *formula*; tellingly, Ikeda considers himself, mentally at least, to be still a member. "It's like family," he says. "It's so ambivalent. Sometimes I really hate them, like brothers."

Two years after Ryoji joined the group, Furuhashi, who was a founding member and director of Dumb Type (not that there was a director), died from causes relating to AIDS. Clearly, the political aims of the group, confronting Japanese society with unwelcome issues such as AIDS or the dystopia of technology networks, and organising themselves as an anti-hierarchical collective, seems at odds with the current public image of Ikeda's work. Here, for example, is David Ryan, writing in *Art Monthly* last year: "This listener, at least, longed for a little more risk or lack of control, a human dimension, which Jean-Francois Lyotard once described as '[t]hat analogizing power, which belongs to body and mind analogically and mutually and which body and mind share with each other in the art of invention.'"

According to Ryan, Ikeda strives for "a space of perfection." I put it to him that the pursuit of total control is deeply problematic. "If I connect the concept of control to the social aspect it's very complicated," he says, "but to me, control is just to make precision, just precise, exact, it's guaranteed. I have a big problem with improvisation. I still like many types of improvisation music – I like John Zorn, Fred Frith – but I just feel it's not my job. I never deny improvisation art and music. It's just not my job. Controlling things is just comfortable for me because now when I create a piece, music, installation or audio-visual concert my vision is so clear I need to control. This is really a short cut to reach the result. And also I'm a bit lazy. Yes, really, I'm lazy. So that's why I need to be controlled, because life is short and I want to do as many things as possible."

High modernism and minimalism are both reactions that in turn have engendered extreme reactions. Writing recently on inter-war modernism in *The Guardian*, J. G. Ballard had this to say: "Modernism's attempts to build a better world with the aid of science and technology now seems almost heroic. Bertolt Brecht, no fan of modernism, remarked that the mud, blood and carnage of the first world war trenches left its survivors longing for a future that resembled a white-tiled bathroom."

This reaction may also be provoked at a personal level, through which deep engagement with uncontrollable situations can be both inspirational and exhausting, so leading to a polar opposite. Although Ryoji has collaborated with architect Toyo Ito, photographer Hiroshi Sugimoto, choreographer William Forsyth, and in Cyclo with sound artist Carsten Nicolai, most of his defining work in music has been solitary. I ask him if he learned a lot from Dumb Type's collective conception of theatre.

"Yes, technically, conceptually, everything," he says. "It was like a kind of school for me. I was very bad in school but Dumb Type was a great place to learn any kind of thing, because there were many different kind of people in Dumb Type. Performers, they have no idea about music and computers. They just love to dance. It's more intuitive, which is great. Dumb Type has no leader or director, conceptually. The director is actually hiding. It's totally democratic, so a performer can complain about music, and I can suggest about choreography or lighting. The relationship is really healthy. That's why, if we try to move a chair from here to there we need three nights to discuss it, because it's based on ideal democracy. Fantastic, but it's so tiring. That way influenced me a lot. That's why now I'm a kind of fascist."

Then he corrects himself, realising that his use of such an emotive word can only compound his image as an extreme control freak. "Not fascist," he says. "Making a piece I have a young assistant and I'm always so clear to ask a thing. It's really concrete: can you make this image, this line, and how many pixel height? I sometimes miss Dumb Type because I'm always alone, basically, now, so I miss that atmosphere."

Nearly ten years ago, Dumb Type performed at the Barbican, which provokes some nostalgia in Ryoji. On that occasion the sound was very loud and all the sound technicians staged a boycott. "I was so afraid and at the same time so excited," he says. "More like punk attitude." Much of this confrontational approach to performance remains in his live shows. Both *formula* and *C⁴I* contain shock moments of sudden loudness, strobes, violent cuts and relentless repetition, and whatever ideas about pure white light may accrue around his recorded work, these are dismantled by the imagery, text and colour that form the visual content of *C⁴I*.

As a work that attacks everything from global inequality and environmental destruction to American imperialism, it carries certain contradictions in its powerful wake. You would have to be a right wing blimp or environmental revisionist to argue with most of the texts flashed up on the giant screen,

At one key moment, preceded by a massive audio impact, Ad Reinhardt's words – "No open book, only touch" – appear on the screen. These are taken from *Time*, a small section of Reinhardt's unpublished, undated notes, written shortly before his death in 1967. In the last ten years of his life, he painted only black paintings, and his notes explored the implications of black, both in the context of the history and function of art, and as symbol, philosophy and inherent quality. The line preceding the one that Ryoji uses is "Language serve as hiding one's thoughts," which brings us into the territory of Reinhardt's fellow students and friends, poet Robert Lax and Trappist monk and writer Thomas Merton.

All three of them wrote about silence in their different ways, and exercised versions of reductionism, a use of words to mistrust words. "The notions of silence, emptiness, reduction, sketch out new prescriptions for looking, hearing, etc," wrote Susan Sontag, "specifically, either for having a more immediate, sensuous experience of art or for confronting the art work in a more conscious, conceptual way."

This seems to sum up Ryoji's feeling: that his works should be experienced as directly as possible, rather than being instruments of theory, and he is more concerned with the individual, open subjectivity of response than with closures of meaning. Then there is the practical issue of how to present works in concert settings. In May 1997, I performed at the Stadtgarten, Cologne, in a concert organised by Frank Schulte. Also performing were Scanner, David Moss, Burnt Friedmann and Ryoji (with keyboard). This is now listed in his biography as his first concert. Since everybody at that time was struggling to find ways to translate digital recordings into live performance, I am curious to discover how many of these solo concerts there were, before he arrived at the present formulation, which is to eliminate all traces of physical human presence on stage

"I did the same kind of thing three or four times but then I had a big question," he says. "My big question was, what is a concert? What is the concept of a concert. Normally, people are going to see the concert. Not listen to music but see – watch what's going on. But for this I can't help anything. I can't dance, I can't sing, I can't entertain the people, so I was very seriously thinking what can I do? I decided to use images."

"More than anything," he says, "the piece called C^4I is how I can compose the image and sound and how I can orchestrate all the elements, so in a sense it's a kind of huge sketchbook in my head or a kind of *étude*, a study of composition for me, so that the political part is just one of the elements to be orchestrated for me. Probably, of course, because I live in New York and I feel many things in New York, these are directly reflected to the piece, but I can't analyse this point by myself. That's why some critics should analyse it. It's their work; it's not mine." Another big laugh.

Door open, door shut; theory enters. Christoph Cox, for example, published an article in *ArtForum* in 2003 entitled *Return to form: On neo-modernist sound art*. This proposed a revival of modernist abstraction in sound art, citing Carsten Nicolai, Richard Chartier and Ryoji Ikeda as leading neo-modernists. "Against the anaesthetic assault of daily life," Cox wrote, "it reclaims a basic function of art: the affirmation and extension of pure sensation." When I raise this subject of the revived interest in modernism and minimalism, Ryoji looks weary. "I really don't know about this categorisation," he says. "Of course it's quite useful but for artists it's really difficult to accept being called a minimalist. Even [Donald] Judd refused to accept being a minimalist, and Steve Reich. I really understand. People are not so simple. People have many, many aspects as a human being."

Extreme opposites are simply versions of each other. Thinking of the quotations printed in his *formula* book and DVD – Mies van der Rohe's "Less is more," and David Tudor's "More is more" – I suggest to him that the American music minimalists now sound maximalist. "Yes," he says. "Phil Glass is totally maximalist. But somebody said something very interesting. That if you listen to a Ryoji Ikeda CD you feel minimalist but if you go to see his performance you really feel he is maximalist, physically."

Perhaps some of this maximalist physicality stems from his early teenage years. I ask him about the first music that really excited him. "I have to be honest," he says. "My first really shocking musical experience was Kiss, and AC/DC. Live, so loud, it's just like wall of sound. I was 13 or 14 and I went to a big, big venue and there was a wall of Marshall amplifiers on the stage. I went with my friend and family and I was just a country boy. It was shocking, I was open-mouthed. With AC/DC the engineering on records was very good, and when I'm doing soundcheck I still use *Back In Black*. As a live person, I'm definitely rock. The experience is so amazing. If I had been a bit older I'd like to have seen Led Zeppelin and Jimi Hendrix."

At the time he disliked punk for its simplicity and even hated all electronic sounds. I ask him if he listened to early Japanese electronic music; pieces such as Joji Yuasa's *My Blue Sky (No. 1)*, Toru Takemitsu's *Water Music*, or Toshiro Mayuzumi's *Music For Sine Wave*, all of which could be heard as having a direct relationship to his own work with pure tones.

"More recently," he says, "because their music was totally abandoned. It's very complicated for them and for me. That generation experienced the war and they really hated any Japanese traditional thing, like a right wing thing, and they were so against it when they were young. Now, they returned to the basics and it's so difficult to accept for me. Joji Yuasa, why? He was so sharp and his white noise [*Projection Esemplastic for White Noise* and *<ICON>on the source of White Noise*] was a crazy idea and so fascinating. Now he's working for traditional Noh theatre and Japonesque

things and I couldn't get the point. Do you know Yuji Takahashi? He was the first student of Xenakis and now ..."

His sentence tails away in disappointment. Maybe it's to do with getting old, I suggest. We talk for a while about film soundtracks, which his father used to play every Sunday morning. The subject moves onto John Zorn and his remarkably broad listening tastes. Ryoji says he sometimes sees him in New York but is too shy to make an approach. "He was a hero," he says. I tell him the story of the time when Zorn came to my home in the mid-1980s. At the time I was working on a television series and researching British dance bands of the 1940s. I played him an early recording by Mantovani, the Italian–born bandleader whose oceanic strings hit big in 1951 with the easy listening classic, "Charmaine." Zorn knew all about Mantovani, of course, and so it transpires does Ryoji. "I have plenty of Mantovani's records," he says. "Mantovani did quadraphonic recording – it's super-sophisticated easy listening music."

Returning to silence, and Susan Sontag: "... one observes how often the aesthetic of silence appears hand in hand with a barely controlled abhorrence of the void"

"Many people can say many things about silence," Ryoji says. "There will be a thousand theories about silence – theoretically, philosophically, scientifically. It's a very difficult question. But one point I really remember. I want to begin everything from silence. Of course, I started making music as a DJ, more like a street culture, and I learned a lot about contemporary art, music, architecture and philosophy. Then I met the word, silence. I'm always still thinking about silence all the time. This is fundamental and very important for me. I can say about silence as a metaphor and actual silence, no sound. Let's say that's why I'm doing art, to find answers. It's very philosophical and of course it's obviously connected to the hardcore of Zen thought. It might be dangerous to say this but it's just a state of your mind. It's not psychology, it's very difficult."

The reference to Zen is surprising, since most younger Japanese musicians prefer to avoid the subject. "I was like this completely," he says, "so I understand. They are so afraid to be asked about that, because Zen is so deep and unspeakable. It's like music: Zen is Zen, so, it's impossible to discuss. It's a cliché, and also in the western world it's fashionable. I always suffered a lot, especially in France. Many intelligentsia ask me only about

Zen. I'm not a Zen master, and they're interviewing me as if I am Japan, and I am Zen. I have to manage the fact that people see me first as Japanese and then as Ryoji. So many cases like this, so I just shut down. It's so shallow."

Ideas of (lost) control, silence and the void are embedded in his latest CD release. One of the most beautifully refined of all his recordings, *dataplex* is also the clearest exposition of his growing fascination with mathematics. In particular, the last track, the strangely abraded tones of "data.adaplex," contains data that not all CD players can read or play. The music is there, yet not there.

"Honestly it was an accident," he says. "There's a very specific waveform so the mastering studio couldn't handle it, but the mastering engineer had no experience about it and he guaranteed me, it's OK. Then the master was sent to the factory. Nobody had any doubt about it but then there was playback error on some CD players. We tried to find the reason. Philips and Sony invented the compact disc – they researched everything but found nothing, so we decided to leave it. I thought it was a very good concept – just data. It was a very strange waveform I used. I edited the sampling rate of 44.1khz as 441000 frames. I took frames out, by accident, so there were too many manipulations and the laser couldn't read it properly. For professional CD players it's OK but for car stereos or normal stereos at home they have a buffer to read in advance. Maybe this buffer caused an error. The composition was very mathematical, and the last track was mathematically composed."

Once onto this subject, he talks rapturously about Bach, then later Merzbow. "I like the invisible phenomena in sound," he says. "Data you can see as a result on the display monitor but the concept of data is so abstract you can't touch it. This theme is very exciting to me. You just can't hear, you just feel. You don't even feel. It's like a subconscious thing, from your cells. Also, in a sense silence and white noise for me are sometimes the same. Maximum random frequency, full, paint black everything and white blank silence."

Black Is Black: Ad Reinhardt

Tate Modern lecture, 25 May 2006

This is not so much a formal essay on Ad Reinhardt's painting here, as a sequence of spoken notes on why it might matter, or what it might conceal and reveal, how it might connect to its own time, and to our time.

The painting is called *Abstract Painting No. 5*.

The dimensions are 1524 millimetres by 1524 millimetres, which Reinhardt would have called 60 by 60 square, or 5 feet by 5 feet, and the work was completed, or maybe begun and completed, in 1962. Mrs Rita Reinhardt presented it to the Tate in 1972, which is a significant date for me. Up until 1970 or so I almost lived in the contemporary painting rooms in the Tate on Millbank; then I changed course from visual arts to music and so *Abstract Painting No. 5* was not one of those works, like the Franz Kline or the Richard Hamilton, that so obsessed me as a teenage art student. I came to Ad Reinhardt a little later, and what attracted me initially was the black.

What do we see when we look at the painting? This could get silly: we see black. To talk about it in this way reminds me of a television comedy series from the 1990s — *The Fast Show* — a recurring sketch featuring an elderly, Hampstead boho couple with their easels and art materials, set up before a bucolic scene and ready to paint. Within moments the old gent has been sidetracked from this peaceful landscape by the malign power of black and has fallen once more into the slough of despond. In the same way that *The Fast Show* sabotaged all possibilities of jazz ever being given its own programme on television again, so it rejoices in the idea that for the painter, black is lunacy, obsession and depression, collecting at the edges of the light.

But the Reinhardt painting isn't black, and neither is it lunatic or depressed, though arguably it may be obsessed. According to the catalogue description, it's blue black, painted on a grid of different coloured squares divided by a green central horizontal band. Seen from the top left, and I'm quoting here, the squares are: red, blue, red, red, blue, red, with all the colours being mixed with black oil paint to give this distinctive matte surface.

A few years ago I was looking around the collection at Tate St Ives. An Ad Reinhardt was on temporary display, maybe even this one, and it was interesting, if predictable, to hear the comments made by passing viewers. A lot of them seemed to think that they could do just as well, if not better, and apart from Robert Rauschenberg's white paintings, which could potentially involve no work at all other than buying a canvas from a shop and sticking it on the wall, Reinhardt seems the ultimate lightning conductor for this form of passing critique. As Matthew Collings wrote of Reinhardt in his book, *This Is Modern Art*, "Nobody can understand his paintings and they are not popular yet."

This is neither here nor there, but painting a surface matte black with a brush, whether concealing a cross or grid of coloured squares or not, is more difficult than one might imagine. To give an example, there was the time when the blocked up fireplace in our living room was opened up. Once a gas fire was installed and a large slate laid as a new grate, I painted the cavity matt black. The time taken to achieve a flat, uniform surface was surprising, and though it works fine as a fireplace, I wouldn't presume to pass it off as anything other than background interior design. Frankly, it's boring, and maybe of more interest if the surface had been left rough and textured, though when I look at the scorched, crumpled, mud-bubbled, pitted and folded black surfaces of paintings by Rauschenberg, Burri and Fontana, or the dark excoriated greys and browns of Antoni Tàpies, they seem to be wonderfully but absolutely of their time, as fixed in an era as *The Avengers* or hula hoops. Reinhardt, on the other hand, is just here, now, just as he was there, then.

One of the problems of evaluating so-called minimalism in the twenty-first century is its colonisation by interior design. Whatever once looked radical now looks like a photo shoot from *Wallpaper* magazine. I think Reinhardt is different, still different. For one thing his paintings don't look expensive or tasteful. They don't exist to be animated by the lifestyle of well-dressed humans. They open out, or more accurately in, so are more like that portal to the other side that so often featured in 1970s horror films – not at all decorative and benign. If they are objects for contemplation, then the contemplation is serious. Reinhardt didn't live to see New Age or Zen in the art of advertising, which was fortunate for him.

In April I was in New York and visited MOMA. In the midst of that great explosion of American art from the 1950s through to the 1970s, I stood for a long time in front of the Reinhardt. *Abstract Painting* is what it's called, from 1963, and the black is not simple black but a grid of squares, a reddish tone in the corners, a cross made up of a blueish black vertical and a greenish black horizontal. Mostly I'm surprised at how easy it is to avoid looking at artworks in galleries, and perhaps this is because sound work usually provides some kind of setting in which the work can be experienced through passing time, at the very least a row of chairs, whereas galleries require you to stand and look, constantly distracted by the movements of others, before moving on. In this case, I looked for a long time, and felt myself falling at some point, into the depth of the painting. This was not a literal feeling, like falling down a well, but like passing through the surface into something more complex and infinitely rich. I came away feeling dizzy.

Ad Reinhardt was born in New York in 1913. This was the same year that Kasimir Malevich painted *Black Square*, a year before Mondrian began his *plus-minus* paintings, five years before Aleksandr Rodchenko sent his *Black On Black* canvas to Moscow's Tenth State Exhibition. I don't know if these coincidences are any more or less meaningless than astrology, or say anything about ideas, time and transmission, but Reinhardt himself seemed to find them worth noting. Judging from photographs taken in his neat studio, he looked a little bit like the kind of guy who sold brushes door to door. At the age of 54, 1967, he died, early enough to escape much of the growing commercialisation and corruption of art that he foresaw and deplored. 54 is younger than I am now, so I can't help thinking that whatever he had done to make the final act of painting a finality, he'd accomplished in less than my lifetime.

This was the way he talked. As far as he was concerned he was the last painter and this was the last art. What seems to matter now about art is that it's unique, or sort of unique, and at the same time it can be reproduced to great effect in all media. Reinhardt's black paintings are identical, give or take the important details, and they can't be reproduced. This was his intent. Neither did he want to make a business out of selling them. He taught, he drew political cartoons, he was politically active, a lifelong socialist. "This painting is unsaleable and it is not for sale except to someone who wants to buy it," he wrote. "This painting has no reason to be bought or sold or bartered. This painting is priceless, has no price tags, no markets, no buyers, no sellers, no dealers, no collectors, with few exceptions. This painting is free, and is given freely to free public museums of fine or free art, with few exceptions."

As a student of art history and philosophy at Columbia College, Reinhardt met the poet, Robert Lax, and the writer Thomas Merton. They became friends, and though very different, their work shared common interests. All of them wrote about silence. In *End*, Reinhardt wrote:

> Nonsensuous, formless, shapeless, colorless, soundless, odorless
> No sounds, sights, sensing, sensations
> No intensity

For Lax, the white page was silence, his move to Greece was silence. "Let the language fall to ashes and poetry will arise," he wrote. In 1990, 10 years before his death, Lax read his poem *bright white* into the microphone of a tape recorder. For Merton, who wrote a book of *Dialogues with Silence*, devotion was silence. He became a Trappist monk at the age of twenty-six, though Reinhardt tried to dissuade him.

In an essay entitled *The Abstract Minimalist Poetry of Robert Lax*, Karen Alexander argues that literature can highlight the differences between art media. Visual works such as Reinhardt's black paintings are concrete, she says, whereas Lax's colour poems – lists of words which she describes as "arbitrary and abstract linguistic signs" – are abstract.

As an arbitrary and abstract linguistic sign, the word black carries a lot of meaning. We don't know what Reinhardt would have thought about a fashion label like Comme des Garçons and black furniture in the Eighties, or Prince's *The Black Album*, but it's hard to imagine that he would have felt at home in the matte black dreamhome.

Black is a complex, volatile word for us. We talk, for example, about black music, and these two words strain to carry such an enormous weight of cultural information that in recent years the black part has been displaced by urban, in an attempt to neutralise it. Reinhardt, Lax and Merton were not oblivious to the racial connotations. After the Birmingham, Alabama church bombing in 1963, in which four young black girls were killed by a racist, Thomas Merton wrote that he was "tired of belonging to the humiliating white race."

Reinhardt took part in Civil Rights marches, and Lax, by then living in Greece, wrote that he was marching with him, in spirit. Black could be still, a presence in a gallery, or float, as words on a page, or in the air, and Reinhardt could say, in a seminar on *Black As Symbol and Concept* which also included the participation of pianist Cecil Taylor and filmmaker/musician Michael Snow, that the idea of black for him was not to do with "outer space or the color of skin or the color of matter," or the post-Biblical connotations of evil sin, formlessness, guilt, origins, bad guys in black hats, the void, hell, and so on, but more an involvement with non-colour, or an absence of colour, negation as a working towards perfection, yet in the imperfect world he marched in protest against the racism in America that enshrined those connotations of evil and all the rest of it. I think he may have appreciated Curtis Mayfield singing "We are the people who are darker than blue," or "Right on for the darkness," but as political statements, not as ideas that related to his own usage of black. [As a further note to this, added in 2018 after reading Fred Moten's essay, *The Case of Blackness*, I should say that Cecil Taylor became extremely frustrated with Reinhardt's universalising insistence that black is a non-colour, or an absence of colour. "Don't you understand," he said to Reinhardt, "that every culture had its own mores, its own ways of doing things, and that's why different art forms exist?"]

Composer Morton Feldman was a New York contemporary of Reinhardt. He wrote quite a bit about the New York painters but didn't seem to have anything to say about Reinhardt. I think he liked colour too much for that, though he said of his music that his primary concern was to "sustain a 'flat' surface with a minimum of contrast"; Feldman wrote about an invitation extended by Philip Guston to visit to a warehouse in which Guston's latest paintings were on view. He described his feeling, that the paintings were like "sleeping giants, hardly breathing." Both of these

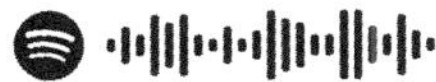

descriptions, the flat surface and the sleeping giants, make me think more of Reinhardt than they do of Guston. Feldman also talked about living stillness, which is a useful phrase for understanding where the energy of these paintings comes from.

Living stillness might be another way to describe what we call silence.

Speaking of silence makes me think about minimalism and its difficulties as a categorisation. All of these so-called minimalists are working in such different ways. Is Reinhardt a minimalist or a maximalist? Is he silence or noise, and if he's silence, is he really the final silence after which there's not a lot that can be useful achieved? Writing about two Reinhardt paintings, Donald Judd had this to say: "Their initial appearance of black nothingness is of course a precedent for a work of art really being nothing."

Critics often spoke about sound when evaluating minimalism. Reinhardt was characteristically dyspeptic on this subject. His response to an interview question was this: "Perhaps the worst thing one can say about painting is that it's poetic or dramatic or literary or musical, or like some other art." In 1967, Lucy Lippard had a article published called *The Silent Art*, and in 1964, Samuel J. Wagstaff Jr. wrote a piece in which he said that John Cage's remarks about music – "there is too much there there," and "there is not enough of nothing in it" – could represent what he called a binding philosophy of many painters and sculptors in the visual arts.

An example of this is by La Monte Young – *The Volga Delta* is its abbreviated title – performed and recorded in 1964 by La Monte with Marian Zazeela, both bowing a 4 foot diameter gong made by sculptor Robert Morris, who happened to be an ex-student of Ad Reinhardt. The cover of the record release, incidentally, was black, with Zazeela's calligraphy and La Monte's explanatory text written in grey, making this information very difficult to read.

Two histories – silence in post-war music and sonic art on the one hand; painting on the other – are intimately connected. Silence speaks volumes. In her 1967 essay, *The Aesthetics of Silence*, Susan Sontag wrote that there is "no such thing as empty space. As long as a human eye is looking there is always something to see. To look at something that's 'empty' is still to be looking, still to be seeing something – if only the ghosts of one's own expectations."

Silence aspires to transcendence. "The notions of silence, emptiness, reduction," Sontag wrote, "sketch out new prescriptions for looking, hearing, etc. – specifically, either for having a more immediate, sensuous experience of art or for confronting the art work in a more conscious, conceptual way."

This seems the point, though we have been obliged to use sound and language to get there. In a notebook entry headed Time, Reinhardt wrote this:

"Language serve as hiding one's thought
No open book, only touch."

It Is Nothing: Picasso's Guitars

**Published in *On Listening*, edited by Cathy Lane and Angus Carlyle,
Uniformbooks, 2013**

faint sound of something in its beginning

In March 2009 I was invited to give a lecture and solo performance at
the GRM Festival "Présences Electroniques," Paris. During my free time
I visited the Musée Picasso and found my ideas on the history of listening shifting as I viewed a remarkable sequence of paintings and relief
constructions depicting musical instruments, particularly guitars.

My notebook from the time records mounting puzzlement and
excitement whereby the mystery of Picasso dwelling so insistently on
guitars and violins was shadowed by close thoughts on the representation of spatiality, captured time and auditory events in an otherwise flat,
static and silent medium. At that point I had completed (or so I thought)
the writing of my book, *Sinister Resonance*, in which I explore the potentiality of a silent medium – painting for example – as auditory devices.
A compulsion to write the book came from frustration with the self-
limiting discourse of what we know as sound art, a repetitive, centripetal
parading of "heroes" whose predictability served to solidify an otherwise
fluid practice.

Perhaps the implausibility of considering Picasso as a node of anti-
history, even a primordial flash that ignited sound work in the 20th century, is an encouragement to consider the proposition seriously. I made
notes on the relief constructions Picasso made in 1926, the materials a
mix of ropes, newsprint, hessian, nails, string, a spring, tacks and canvas,
all of them named *Guitar*: "The first of these menaces, sound hole torn

out of sacking, nails protruding point-out; the second more like the soft imprint of the thought of a guitar pressed into sand, though the tension of the strap suggests a drawn bow string in both." Of *Mandolin and Clarinet* from 1913, I wrote: "In this one, the instruments are exploded, as if their inner resonance has been turned inside out."

faint sound of no outside, no inside

A music is implicit. They are listenable, these paintings and constructions. They raise a question that only becomes greater when cubism is considered as a whole: the obsessional returning to musical instrument as still life in Picasso, Georges Braque and Juan Gris. This fixation with sonic technology is unavoidable; any history of cubism will pass a mystified remark before moving on to more settled issues. With an increasing focus on the significance of the object, notably in MoMA's exhibition – *Picasso Guitars 1912–1914* – the question is examined more thoroughly. "Picasso did not play an instrument and is said to have had no patience with most types of music," writes Anne Umland in the MoMA catalogue.[1] She considers the attraction of guitar music to Picasso – "its associations with café life and with flamenco music, a contradictory genre both primitive and modern, Spanish and gypsy, fixed and improvisational"[2] before concluding that his decision to construct a guitar in 1912 was "an act that allowed him to discover what, specifically, the guitar had to offer him as a structure, or model, for a particular form of contained spatiality and for a particular vocabulary of simple, separable, iconic signs."[3]

Both of these points resonate. Lewis Kachur examines Picasso's fabled indifference to music more closely and finds a rather different story, in which his tastes (as we might expect from a painter whose vision was so forceful) ran to direct expression: the rough sound of a Catalan folk shawm called the tenora, or, at the other end of an imagined scale of refinement, the compositions of Erik Satie (a master of deceptive simplicity) and Déodat de Séverac. As an aside, Kachur draws a parallel: the embedded signs ubiquitous in cubism – the motif of the musical instrument, the scraps of newspaper and sheet music – and the strategy common to composers of the period in which folk tunes, popular songs from the music hall and cabaret and American ragtime and jazz were collaged, in the sense of being also embedded within the flow of a score.[4]

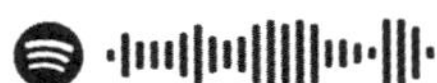

faint sound of faint sounding no sound

"Do you know how to clean sounds?" Satie wrote in 1913. "It's a filthy business."[5] His ironic wit applies itself to the notion of sonic materiality, just as Picasso's wit applied itself to the dismemberment of forms. The void of the guitar, its volume out of which issues volume, disgorges itself. The vessel of sound is opened up, emptied out and that which is nothing becomes solid. His cardboard *Guitar* of 1913 has been variously described as "a new sculptural language," and "a crucial rupture in modern art's history." As for Picasso, his reaction was a shrug: "It's nothing, it's *el guitare!*," his insouciance echoed and amplified by André Salmon: "The watertight compartments are demolished. We are delivered from painting and sculpture, which have already been liberated from the tyranny of genres. It is neither this or that. It is nothing. It's *el guitare!*"[6]

Alongside the stencilled image of a bottle of Anis del Mono, the *Guitar* of 1913 sat on a table, partial and flimsy but nonetheless "real," all suspended in space in front of two overlapped sheets of wallpaper. Also in 1913, Picasso made the more complex *Construction with Guitar Player and Violin*, a work existing only in studio photographs of the time in which a real guitar was suspended from a "wall" on which was drawn the outline of a guitar player. An arm with hand, made from newspaper, stretched down from the outlined guitarist to the floating guitar. Hanging on the wall is the paper violin from 1913 and set in front is a café table with wine bottle, pipe and cup. There is nothing in the piece that is sound in itself, no sound in the process of becoming, and yet we can listen. "Art should not be a *trompe l'oeil*, but a *trompe l'esprit*," said Picasso.[7] The eye is not deceived; nor is the ear. There is nothing, yet the mind is hearing.

Notes

1 Anne Umland, *Picasso Guitars 1912–1914*, The Museum of Modern Art, New York, 2011, p. 20.
2 Ibid, p. 21.
3 Ibid, p. 22.
4 Lewis Kachur, "Picasso, Popular Music and Collage Cubism (1911–12)," *The Burlington Magazine*, vol. 135, No. 1081, Apr. 1993, p. 256.
5 Erik Satie quoted in Nancy Perloff, *Art and the Everyday: Popular Entertainment and the Circle of Erik Satie*, Clarendon Press, Oxford, 1993, p. 83.
6 Quoted in Anne Umland, op. cit., p. 27.
7 Reference in Ruth Markus, *Picasso's Guitar, 1912: The Transition from Analytical to Synthetic Cubism*, Assaph, Studies in Art History 2, Tel Aviv University, 1996, p. 238.

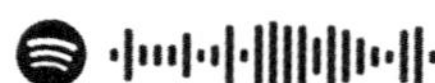

no boundary condition: Shirazeh Houshiary

Published in *No Boundary Condition*, Shirazeh Houshiary exhibition catalogue, Lisson Gallery, 2011.

Presence

Breath through its emanation is an unfolding and upholding, the smoke of life whose boundary within the space it sounds lies beyond seeing and ordinary description, beyond the conception of boundary, beyond human perception itself.

We look into the face of Piero della Francesco's *Madonna del Parto* (circa 1450–70). Her eyelids, curvilinear, asymmetrical, appear to drift. The left eye, in particular, seems to be moving sideways and downwards simultaneously, as if she is, despite her dramatic unveiling, distracted by some object beyond and below the picture plane. The eyelid itself is an opening, a cut; the soft entrance to the tent in which she stands, drawn aside by angels, is an opening; the vertical slit at the front of her dress is suggestive of a Caesarean opening into her swollen belly. Time is compressed here, as if by some miracle the imminent opening of her body to give birth to the child inside her is anticipated by openings of all phenomena.

We look closely into her face in an atmosphere of silence, because this is a painting that generates its own aura of silence. Talk is silenced (caesura) by a look, or by the eyelid itself which threatens to drop down over seeing, the pensive, grave torpor of her face perhaps revealed only for a moment before the angels bring down the curtains again in a refusal of our gaze. We may consider another history, that of cinema and the camera shutter which opens and closes, and Tarkovsky, whose *Nostalghia* (1983) begins within a chapel in which the focal point is this same Madonna of

Piero (in actuality a reproduction of the painting). Out of the reverberation of footsteps a dialogue ensues between a central character – Eugenia – and the sacristan of the chapel, a question of faith, a woman's role, and creativity, and in some other time (deceptively, because Tarkovsky makes us believe that both events happen within the same time frame even though physical space makes this impossible) a procession of women carrying lit candles and a statue of the virgin assembles in devotion before the painting in a sounding of bells, a murmur of heterophonic chant. The dress of the statue is pulled open; many birds fly out, their songs pouring forth as multitudinous life that collects beneath every surface which invites its own cutting. The camera dwells on the candles, on Eugenia's face as fleetingly she seems to hear some faint residue of the birdsong, then the face of the Madonna in close up. From this tighter focus and concentrated mood the song of the birds now takes flight from its avian realm to become transformed into a sound at once electronic and alien. In their isolation under the gaze of the Madonna, they are calls that resonate in extra-human space.

We had been looking at Shirazeh Houshiary's painting, *Sigh*, in which the two words repeated over and over in pencil, one of affirmation, the other a denial of affirmation, overlap to create a space within a space. She had spoken in terms of pulsation and breath: "From that overlapping we get this movement. It's like something that is coming into existence and this is like absence of breath, there's a hole in the web of breath." Writhing within the pulsating blue there are curvilinear cuts, as if incisions into mist. They echo the *Madonna del Parto* eyelid, the device with which Piero expressed a continuous narrative that neither moves nor remains static. Another space is proposed, a dimension of multi-dimensions and movement beyond this flat stillness. "It's almost like music in a way," she says as we look into its depths of words that cannot be read, "knowing if you stop that the silence is more important than what happened." For her, this comes from Piero's expression of silence through movement, a barely perceptible sense of motion that gives his paintings such radiant stillness. "It comes from the eye of this Madonna," she says. "I tried to imitate it to see if it was possible and realised that actually what is very powerful about this is that it gives a sense of stillness to this Madonna. It's so heavy, she's so still."

She had been talking about an epiphany, an encounter in 1977 when she was a first year art student only recently arrived in London from Iran: "I went to a show of Fred Sandback at the Lisson and I saw one piece of string from the floor to the ceiling. That was my presence. It was the first time I understood what it means to have presence in a space because that one piece of string described my boundary as a human being. I was very shocked to observe such a thing. That one string made me aware of my own movement."

Fred Sandback began as a musical instrument maker, building banjos and dulcimers, moving almost imperceptibly into his mature work by impermanently drawing the shape of a large board on the floor with string and wire. It was as if the craft of stretching soon-to-be vibrating strings over silent vessels, surfaces, openings and cavities had rewarded him with an understanding of how volume, temporal differentiation (what John Latham would have described as not-nothing upon nothing) and states of being and not-being may be articulated through such simple means. The string itself disturbs a silence, a disturbance so fugitive as to be almost nothing, an opening as still and profound in its imperceptible vibration as the Madonna's eyelid. "It is not just that silence allows memories of the past to be invoked, or the bells of Easters yet to come to be heard," wrote Vladimir Jankélévitch, in *Music and the Ineffable*. "Silence also develops the infinitesimal sounds of a universal multipresence."

For this subtlety of exploration into presence we may look back – to the work of Fluxus artist Mieko Shiomi, for example, whose *Boundary Music* text-score of 1963 gave the following instruction to performers:

Make the faintest possible sound to a boundary condition whether the sound is given birth to as a sound or not. At the performance, instruments, human bodies, electronic apparatus or anything else may be used.

… or we may look back further to Jun'ichiro Tanizaki's 1933 essay on Japanese aesthetics and modernity, *In Praise of Shadows*, in which he wrote of the effect of indirect light falling through paper screens onto the neutral hues of a traditional Japanese house:

The hue may differ from room to room, but the degree of difference will be ever so slight; not so much a difference in colour as in shade, a difference that will seem to exist only in the mood of the viewer.

Tanizaki also wrote of those scrolls housed in dark alcoves within the temples of Kyoto and Nara, the age of the painting and the gloom of the alcove uniting in a challenge to perception itself, so that "the painting here is nothing more than another delicate surface upon which the faint, frail light can play."

Cage talked about the Rauschenberg works as repositories of light, shadows and dust, a dwelling on that which is not of the work that becomes the work, as in Man Ray's celebrated photograph of Marcel Duchamp's *Large Glass*, laid flat in his Manhattan apartment as a collector or breeder of dust, the *Elevage de Poussière* whereby dust (that mysteriously ubiquitous phenomenon that doggedly collects itself from the detritus of largely organic life, skin, hair, seed fragments and other bodiless airborne particles shed from ourselves and our close and far environments) was fixed with varnish by Duchamp as a record of its passing through an unstable universe.

This is the universe to which Shirazeh Houshiary works such as *Dust, Sigh, Black Light, Between* and *Deep* address themselves. "No Boundary Condition comes out of physics," she says, "directly out of quantum theory. No Boundary Condition is the condition of true human beings. We know in quantum fields, when we shoot an electron through a concrete wall it can go through the wall and land anywhere in this universe. That is really the reality of perception. If our perception belongs to the Middle Ages, that's not my fault. The fact is the universe has many layers so why do I have to have a boundary? I really believe a true artist lives in no boundary condition."

In the latest of a series of works begun in 2005 – *Breath, Veil, Shroud* and now *Dust* – that are difficult to categorise or even describe, being accretions and manipulations of intangibility, visual materials that move and sometimes sound from within the digital domain, this question of where the work exists is addressed. Through the accumulation of soot from a flame, an elusive cycle of images moves in transitions and transformations from soot to smoke to soot. Emanations, they interrogate what it might mean to be without form and aspire, perhaps, to the human voice. "The voice is formless," she says. "It has a different form. As a visual artist I know what form is but I'm fascinated by the voice as a different form. My whole work is based on this. It makes you aware of your own presence more than any form that we know. It's a living form. It carries emotion, it carries depth of feeling."

This idea of a fragile remnant, the trace of an aerosol that is itself a by-product of fire, is reminiscent of breath. In particular it recalls the story that she has told previously, about wanting to know who she was, moving close in to a mirror to see her own face. "I couldn't see any face," she says. "I went in closer to see if I even exist and my breath left a residue. That is exactly what life is like: you're there and you're not there." *Dust* was imagined as a projection of multiple images, like the Buddha's cave in which there are hundreds of images of Buddha. Nothing is precious.

I am reminded of Morton Feldman's music, which can be a form of writing of short breaths – a piano phrase or violin – extended over long durations so that the composer, the performers, the listeners can get closer to what he called "Time in its unstructured existence," or it can hover in air with the presence of suspended combinatory colour, a pulsing exhalation-inhalation that transforms unpredictably in stillness. Feldman structured his music through pattern. "I'm being distracted by a small Turkish village rug of white tile patterns in a diagonal repeat of large stars in lighter tones of red, green and beige," he wrote. "My music has been influenced mainly by the methods in which colour is used on essentially simple devices. It has made me question the nature of musical material. What could best be used to accommodate, by equally simple means, musical colour? Patterns."

"Is there repetition or is there insistence?" Gertrude Stein once asked. Within Shirazeh Houshiary's work, words are buried, an insistent patterning that cannot be read or even seen with the naked eye. "All my work is made out of text," she says. "It's like you listen to a language you don't understand. Most of the time when I listen to vocalists I don't understand anything. It's beyond. That's more powerful to me. I have tried to recreate this experience in the paintings too. The foundations of these paintings is text, not text, it's word, because it's just two words. The two words are "I am," "I'm not," and they're crushed upon one another so it's not about being able to divide them. It's about being able to experience this tension. This tension exists in the way we are as human beings in the world. We are not so definable."

All around our feet as we talk are books on art from Italy and Spain: Piero della Francesca and Francisco de Zurbarán. Of the latter's work, *St. Francis in Meditation, St. Serapio,* and the *Ram with Tethered Feet,* all show figures

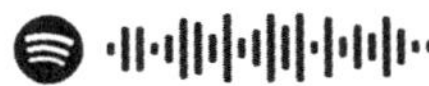

under duress, alone in deep blackness vibrating softly amid infinite depth as the low C from a bass flute might sound in the recesses of a lightless cave. The presence of Zurbarán's paintings is overwhelming. Through their intimate dialogue with the void they breathe audibly, a rising and falling stillness.

Artist Steve Roden has described Shirazeh Houshiary's works as quiet but human, "not manipulative or too technique-y, just whispering ..." The paintings are large, exhibiting extraordinary skill, yet they do whisper, and like the digital video works they challenge their own materiality. Working both slowly and quickly, from many different viewpoints, in the ambiguity of their tensile presence in space they are suggestive of, even fascinated by skin as a supple surface that moves, almost coming to life, then finally cracks, dissolving into an empty frame: a matter of life and death. I think of boundedness and unboundedness, a striving toward invisibility or form-lessness, and the question of physical presence and affect. Discussing Zurbarán's miraculous (in all senses) *Veil of Saint Veronica* (1631–6) in *The Book of Skin*, Steven Connor has this to say about such works that represent "the miraculously marked skin or fabric, untouched by human hand":

Since colour deposited on top of the skin is so associated with corruption and deceit, being both too speciously superficial and too pore-cloggingly piled, the painter of flesh must strive, through all the delicacy of his means and technique, to give the impression of a flesh which is in fact immaculate, untouched by human hand, and therefore illuminated by its own light and by radiant, rather than pigmented, colour.

The presence of presence is a constant in Houshiary's work, raising questions of physical presence and affect (these were discussed exten-sively by Mel Gooding in his catalogue essay of 2008: *A Suite for Shirazeh Houshiary*). "A lot of it comes from Zurbarán," she says, "*The Veil of Saint Veronica*. My fascination with that painting is the cloth itself. The cloth is the face itself. That painting is one of the most extraordinary paintings I've ever confronted. It is a sense of identity that has no identity. It's the skin. It's the cloth. The face [of Christ] is irrelevant in a way. I know people think the face is more important but I think the cloth is important and the unification of the face with the cloth, so maybe there was never a division as we think there is. I try to challenge people's perception of this duality that exists in the core of the way that we perceive the world. I try to do it in abstraction, not in a literal sense, which is not easy, because I found abstraction for me is very powerful, because it doesn't allow you to make recognisable connections."

Quotes

Vladimir Janekélévitch, *Music and the Ineffable*, 2003; Jun'ichiro Tanizak_, *In Praise of Shadows*, 1991; Morton Feldman, *Give My Regards to Eight'ı Street: Collected Writings of Morton Feldman*, 2000; Gertrude Stein_, *Lectures in America*, 1935; Steve Roden, personal communication to the author, 2011; Steven Connor, *The Book of Skin*, 2004; Leonardo da Vinc_, *Notebooks*, 2008.

All quotes from Shirazeh Houshiary from recorded conversations with the author: 4 April and 6 May, 2011.

Nothing Hear: John Cage & John Latham

Published in *Sounds Like Silence: John Cage 4" 33" Silence Today*, exhibition catalogue, ed. Dieter Daniels/Inke Arns, Spector Books, Germany, 2013

A recent discovery in the archive of artist John Latham caught my attention in January 2012 for two reasons: first of all because of the ongoing significance of Latham's work within my own research and music practice; also because I had recently developed a music theatre project based on the notebooks of Leonardo da Vinci and had been struck by Leonardo's refrain on the unfinished: tell me if ever anything was finished? "Nothingness has no centre," he wrote, "and its boundaries are nothingness."[1]

Latham's text was prepared for the catalogue of a London exhibition by Graham Stevens entitled *Nothing*, held at the Seven Dials Gallery, London, in 1984 (though there seems some doubt as to whether it was published). It begins with a quote from Leonardo:

Among the great things which are found among us the existence of Nothing is the greatest. This dwells in time, and stretches its limbs into the past and future, and with these takes to itself all works that are past and those that are to come, both of nature and of the animals, and possesses nothing of the indivisible present. It does not, however, extend to the essence of anything.[2]

In less than two pages, Latham expounded a theory of least event as the converging position of both art and science from the beginning of the twentieth century – the discovery of sub-atomic particles, Quantum theory and the monochrome paintings (black or white) of Aleksandr Rodchenko and Kazimir Malevich made in Russia between 1915 and 1918.

"On the basis of actual 20th century trajectories," he wrote, "both science and art each arrived at the conclusion

'everything = nothing.'

"Lacking a reasoning as to how to interpret this finding the next orthodox step has been backwards in both science and art, a retreat. But on line and forwards if unofficially, it has led to a defining of 'event' structure. The dimensional framework of 'event' embodies this equation."

Latham went on to discuss his idea of Event Structure as a paradigm distinct from material/mental dualism, an escape route from the "common sense" reality through which it becomes so difficult (and so undesirable within bureaucracies) to reconcile measurable quantities and observable phenomena with intangibilities, immeasurabilities and qualities (the purpose of art, for example) that resist definitive explanation or quantifiable usefulness. He also proposed the point of nothingness as a starting point for this rethink: " 'Nothing' is referring in [Leonardo da Vinci's] mind's eye to a *nonextended state of everything,* a dynamic component in the cosmic *EVENT.*"[3]

Nothing to Say

In 1974, artist Marie Yates and I added our names to a letter written by John Latham and sent to *The Guardian* newspaper. Printed on the letters page under the heading – *But who deserves patronage?* – it addressed a contemporary debate on the efficacy and relevance of the Arts Council of Great Britain and its recently published report, Patronage of the Creative Artist. "Sir," the letter began, "A short glance at some journal of current art would show immediately how unrealistic it is to think 'artist' means simply 'poet,' 'painter' or 'composer.' However much it might suit administrators to restore those neat distinctions, it is decades since John Cage and Ad Reinhardt (among many) began to reorientate and rephrase the serious capital A activity as quite a different kind of consideration."[4]

Latham had very little time for Cage. They met at one of Cage's events in London (date unknown) and what seems to have occurred is either a monumental misunderstanding or one of those titanic clashes between strong artists in which a kind of negative energy is generated. Latham was impressed by the elaborate setting up of a performance which led to what he described to me as "just a one-note song." Latham went up to speak to Cage afterwards and was apparently rebuffed: "I asked him, it was obvious that there was an ordering, that he understood an ordering principle and didn't it have an event kind of structure to it, and he just stared blankly at me and said, 'I don't know what you're talking about.' He could have said, 'No, I don't, I don't think about it,' but if he didn't think about it, he couldn't have been that precise about what he did."[5]

This unpropitious encounter did little to enhance Latham's view of Cage's ideas. "Yes, well, I've got a book of his," he told me during the same conversation, "but it's nonsense. It's worse than nonsense. It's not good nonsense, it's just plainly boring nonsense. I don't know what it's called because I don't read it but I was very disappointed to try to find, well where's Cage in all this? And he just didn't show up at all."

Very few people have read all of James Joyce's *Finnegans Wake*, appear to have understood it and have not only incorporated its implications into their own work but actively promoted it as a key work of twentieth-century literature and art. Cage and Latham are two of the most prominent (and two of the only) exceptions so their mutual incomprehension is tinged with a degree of pathos.

Despite the famous cross-talk between Rauschenberg's white paintings and Cage's *4' 33",* perhaps artists who stood on the brink of this mid-century void could only properly exist in exclusionary spheres. *Finnegans Wake* is one of the works described by Latham as a "non-spatial continuum." There is narrative but it appears to tunnel down into deep time or mythical time in which all events can happen simultaneously at the beginning of time, in all possibilities and means of communication and at the most present, microscopic fleeting instant of the now. After this momentous step other writers could only ask themselves, what now? The monochrome paintings of Malevich, Rodchenko, Reinhardt, Rothko and Rauschenberg, Latham's *One Second Drawing*, Nam June Paik's *Zen For Film* (a loop of blank leader film, first projected by Paik in 1964) and John Cage's *4' 33"* stand alongside Joyce's final work as finalities in themselves, end point in the eschatology of art, into great silence.

From the point of view of the composer, what could possibly be committed to manuscript paper once nothingness – no action, no

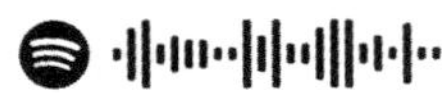

sound – had become a spectacle? This was, of course, an illusion based on a number of myths, exaggerations and misunderstandings, plus a fatal flaw in Cage's own work, though these are not the reasons why composers decided to forge ahead with their work despite its apparent futility and irrelevance (the pursuit of career offers a more plausible explanation for that). The exaggerations and misunderstandings are too well rehearsed to need reiteration here. As for Cage and his flaws, he was no different to other revolutionaries, one foot stuck in the past. The whole business of *4' 33,"* its rituals and formality, its hierarchy, its irritatingly fussy timings which mean nothing whatever to the audience, its nostalgic deference to the conventions of the concert hall, ensure that Cage kept tight control on sounds.

Look again at his *Lecture On Nothing*, from 1949,[6] in which he begins, notoriously, by saying: "I am here and there is nothing to say," and then continues, "What we require is silence but what silence requires is that I go on talking." Did silence demand any such thing or did Cage just love to talk? Later in the same lecture Cage expresses allegiance to music structured through the twelve tone row (though not because it is twelve tone), repudiates the phonograph as a musical instrument and advocates the destruction of gramophone records (my archaic terminology is deliberate here). He imagines listening to Japanese shakuhachi music or the Navajo Yeibitchai (or Yeibichai, songs performed during the ninth night of the Nightways ceremony) for any length of time, or sitting near Chinese bronzes (perhaps as a way of listening to a form of silence) but then admits that such proximity inculcates the desire to possess. There is much here with which to sympathise and much here to dislike, particularly a kind of folksy nostalgia disguised as revolutionary rigour. Is breaking gramophone records any more acceptable, for example, than burning books? John Latham burned books because he believed that books have dominated and warped our way of thinking about the world; Cage felt that a record was an abomination when you could have a person singing in real time and a particular place. He has a point, though it is deeply conservative. I am reminded of a passage from author Norman Lewis's *A Dragon Apparent*, his account of travel through Laos and other South-East Asian countries in 1950 (a coincidence, but striking for being contemporaneous with Cage's *Lecture On Nothing* and no doubt indicative of a certain remorse among men of a certain age, that those same devices that allowed

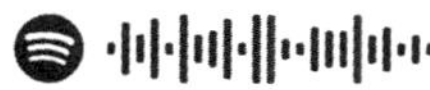

them to ply their trade – magnetic tape, microphones, passenger aircraft and motor cars, were destroying a world that they idealised):

As I arrived the organizers were having trouble with the microphone – an indispensable adjunct to any social occasion in the new Far East. A young man chanted a soft, nasal melody which could only be heard in the boun enclosure itself [a boun is a Laotian festival]. But suddenly the electricians were successful with their tinkerings and all Vientiane was flooded with a great, ogrish baying. The electricians hugged each other, and, enchanted by the din, the audience began to drift away from the theatre and make for the dancing floor … the microphone is an infallible sign of what is to come. Nothing of the kind will survive the era of materialism, under whatever form it arrives.[7]

Indifference and Estrangement

My own trajectory with Cage since I first read *Silence* in 1968 has become a slow downward spiral – the writings that I so admired as young musician now seem riddled with difficulties and contradictions, whereas those few pieces of his music I still enjoy, *Prelude For Meditation*, for example, do the work of the theory far more succinctly and completely without the notoriety of *4' 33."* This is not entirely Cage's fault. Posthumously he is undergoing transformation into a saintly culture hero (even though the status of hero seems no more agreeable than coveting ancient Chinese bronzes – in the present moment auction rooms are growing very rich on that particular trade) and is gradually sinking into a state of orthodoxy, partly because he represents a golden age of the avant garde which has now passed, partly because conservative institutions are now beginning to recognise him as part of the canon.

In his introduction to the overly luxurious 50th Anniversary Edition of *Silence*, Kyle Gann tackles some of these difficulties. "Personally, I have tried, at Cage's urging," he writes, "to enjoy a baby crying at a concert, not letting it ruin a piece of modern music; so far I've failed. But that's why I keep coming back to Cage, because I keep thinking that if I *could* evolve or relax a little more, I could enjoy babies crying and fire alarms ringing, and feel as comfortable with the universe as he always seemed to be. He thought his way out of the twentieth-century's artistic neuroses and discovered a more vibrant, less uptight world that we didn't realize was there."[8]

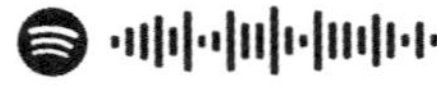

Was it really true that Cage was less "uptight"? His persistent critiques of improvisation (hence the entire history of African-American music) suggest not. Of course it may also be true that Cage was indifferent. That would lead us to the arguments presented in Garret Keizer's book – *The Unwanted Sound of Everything We Want: A Book About Noise* – in which he explores the intractable ethical problems of noise in a world increasingly filled with noisy devices and increasingly characterized by claims to personal freedom that are sublimely indifferent to the discomfort and sufferings of others. Keizer's book was published at the same time as other manifestos of silence, all of them arguing for a quieter world and in the case of Sara Maitland's *A Book of Silence*, documenting a retreat from human habitation and company.

But as Keizer pointed out, personal silence can be illusory, a kind of luxury sustained within a bubble of detachment. "Silence, even the innocent silence of an hour's silent reading, can lie," he wrote. "It can tell us that we're quieter than we really are. It can tell us that our seemingly 'quiet lifestyle' disturbs nobody. Noise, on the other hand, has an uncanny way of telling the truth. Much of the truth it tells is political."[9]

This could also be addressed to Cage's *4' 33."* A contemporary audience arrives en masse to hear a mixed programme of Cage's works, including the infamous "silence." The piece is performed with a knowing self-congratulatory irony in which token, painless participation in the crawling historicity of avant gardism deflates its original claims to be an end point of art. The listeners may congratulate themselves also on their silence, during which they heard their own ardent listening within the arid context of the concert hall. Then they return to their noisy cars to drive home; the many lights, the heating system, air conditioning and all other utilities that allow the concert hall to function are switched off, the wasteful nature of the exercise is forgotten.

Perhaps this is too cynical? Yet as an affable pioneer of the forbidding, Cage is well suited to our contemporary taste for heroes and celebrities and so we can afford to be critical of the way his ideas enable a form of high-minded hypocrisy, both his and our own. Could it be that one of the difficulties of thinking about Cage in the adulatory atmosphere of an anniversary year is that he is now the man who "invented silence," or, better still, the man who "invented listening"?

Again, this is not entirely his responsibility. Certain stories which even Cage admitted to telling many times over have passed from anecdote into myth. They have come to form an unshakeable, largely unquestioned foundation for so-called audio culture. A perfect example is the famous story of Cage's visit inside one of the anechoic chambers in use at Harvard in 1951, in which he heard two sounds, despite the total absence of reverberation in the room. One was a low pulse, the other a high-pitched singing tone. Being disturbed by these, he was told by the engineer that they were the sounds of his circulation and nervous system respectively. This is so close to the experience of the mole creature in Franz Kafka's short story, *The Burrow* (in which the creature builds a secure burrow underground only to become disturbed by sounds indicating an unseen intruder), as to be uncanny, as if Kafka had struggled with the perpetual disturbance of these same externalisations of interior body processes.

One of the key texts of twentieth-century music, sound art, and American minimalism, the anechoic chamber story may also be incorrect in its details. Cage may have been hearing symptoms of tinnitus, or spontaneous otoacoustic emissions from his own ears, rather than the sound of his brain at work (or as Susan Sontag put it, confusing the issue still further, the blood in his head). These faint sounds of otoacoustics, produced by the expansion and contractions of hair cells within the outer cochlea, could not be measured until the development of sufficiently sensitive low noise microphones in the late 1970s, so the Harvard engineer (and Cage) would have been unaware of their existence.

The origins of the sounds heard by Cage do not affect the sense or impact of the story, but these uncertainties emphasise an estrangement from the emissions of the body. We are left with the conclusion that Cage was a less diligent listener to his own body than Kafka, or indeed those writers whose explorations of listening preceded or were contemporary with his own: Joseph Conrad, Edgar Allan Poe, Virginia Woolf, James Joyce, Samuel Beckett, Herman Melville, William Faulkner, William Wordsworth and others. The conception of silence as an external phenomenon that can be heard (as opposed to metaphorical, mystical, philosophical or political silences) presupposes an absence of the body, a neutralisation of space as an active presence. My own experiences of anechoic chambers have emphasized the artificiality of this manifestation of silence, a theoretical construct that can only be achieved through extreme measures. In all other environments in which sound waves can meet resistance and be reflected, silence is only a potentiality, aerial yet substantial: the sound of the listener; the sound of space and the air with which it is filled.

The Deep Time of Listening

With his references to figures such as Erik Satie, James Joyce, Gertrude Stein and Henry David Thoreau, Cage began to sketch in a continuum of listening practice of which he was just one part. He was a composer of ideas, not a scholar, not obliged to be in any way comprehensive, and so the compatibility of certain Cageian ideas to the conditions of contemporary life – his arguments against government (borrowed from Thoreau, which might now find followers on both left and right extremes of the political spectrum, his formalizing of the principle of silence, his desire for simplicity – have ensured his centrality within a music scene which might be more true to itself with no heroes at all.

The other factor that consolidates this position is the dearth of composers who continue the work of Cage and his generation with the same impact, or even a similar sense of purpose. As John Latham suggested, after an end point, ' the next orthodox step is backwards." These sentiments are echoed in Gabriel Josipovici's *What Ever Happened To Modernism?* Josipovici interrogates the conservative literary culture of the present day (particularly that of Britain) and wonders why the innovations of Modernism – its fragmentation, its assault on the subject and its bold experiments with language and time – have been displaced by unchallenging literary fiction.[10]

He argues that Modernism is not an unprecedented phenomenon of the twentieth century, discussing William Wordsworth, of all writers, as one antecedent of the moment when self was swallowed by an abyss. Intriguingly, many examples given from the poems arise out of listening. In *The Prelude* Wordsworth compared the human mind to music, in both a "dark invisible workmanship that reconciles discordant elements, and makes them move in one society."[11] Cheek pressed to a mossy stone he listened to subterranean waters as if their murmuring echoed the resonation of his own unconscious. "He used rock to orchestrate the sounds of water," say the authors of *Wordsworth's Gardens*. "He was accustomed to using the flow of water, among other sources of natural music, to balance with the ear what he referred to in the *Prelude* as the domination of the eye."[12] An example of this urge to orchestrate can be seen in a photograph of Wordsworth's garden, taken by Herbert Bell in 1958, *The Well – Dove Cottage garden*, which could be mistaken for a pond in a Kyoto garden. The water continues to pour and we might imagine its sound as a murmur unchanged from the early nineteenth century but as Heraclitus of Ephesus famously wrote: "We both step and do not step in the same rivers. We are and are not."

Dying Away Upon the Ear

Nothingness, to return to Leonardo da Vinci, is not empty as an ending, rather, a potentiality through which other beginnings can become born. No wonder music seems to have gone backwards. "Nothingness, one might say, has no properties" wrote Vladimir Jankélévitch, in *Music and the Ineffable*. "One nothing cannot be distinguished from another nothing. How could they be distinguished without having qualities or a manner of being; that is, without, at least, being something? Two nothings are only a single, same nothing, a single, same zero. But silence has differential properties: and as a result, this particular nothingness is not nothing at all – in other words, it is not (like Parmenides' nothingness) the negation of all beings: it is not a nonbeing that totally annihiliates or contradicts total being."[13]

Silence is not a "thing," a fixity or common state to be defined by timings or setting but a condition of constant flux subject to the subjectivity of the listener. "The air smells like sulphur," William Faulkner wrote in *As I Lay Dying*. "Upon the impalpable plane of it their shadows form upon a wall, as though like sound they had not gone very far away in falling but had merely congealed for a moment, immediate and musing."[14]

This was an intensity experienced by Wordsworth; *On the Power of Sound*, composed as a poetic essay in 1828–late 1829, begins with the ineffability of sound passing into the body to register as emotions, sensations, signals of great import that vanish into air at the moment of their becoming:

> Thy functions are ethereal
> As if within thee dwelt a glancing Mind,

> Organ of Vision! And a Spirit aerial
> Informs the cell of hearing, dark and blind.[15]

Through listening and its decomposition, sound's presence oscillates alongside absence with the potential of a return (through reflection and echo), Gabriel Josipovici reiterates a question raised by Wordworth – "Is there a way of interacting with nature which is not destructive?" (as Josipovici frames it) – supplying the answer with that section from *The Prelude* in which Wordsworth describes a boy mimicking the hooting of owls through cupped hands. He calls out to the silence of the owls and they respond to his call, the exchange falling away into a deep silence into which rushes another form of echoing, "to a complete incorporation of the landscape into the boy and the boy into the landscape."[16]

The drive toward nothingness, to silent music, is perhaps more "natural" than we think, or natural in a world divided according to the mental/material dualism that John Latham hoped to resolve with event-structure. Humans are adapted to sight, to looking, to touching and holding, so the listening world is disconcertingly abstract, ambiguous, always to a greater or lesser degree disconnected from objects and sources. These same attributes can become qualities associated with freedom – like a dream of flying – particularly when experienced through music and even more particularly when connecting that which seems natural with that which seems musical, as in birdsong. This is what Cage sought, perhaps, in his desire to listen just to the Japanese shakuhachi. As a man of his time he interpreted such music as an experience of being not being, an art closer to nature than culture.

Easy to see why it should be so – my own response to this particular instrument and its repertoire is similar – yet the Japanese shakuhachi is embedded in theory, history, craftsmanship, a hierarchical lineage of performance and schools. In that sense it is little different to any other music: sanctioning free movement in a sightless, weightless domain, at the same time offering compensatory structures as a form of invisible making. Music is apparently object and order yet always intangible event and so through that tension, pleasure is always underscored with fear in that music reminds us that order may be illusory and that solidity is undone by loss and decay. To privilege a form of listening not wedded to music is to enter a perpetually dynamic space both unmeasurable and precise.

Listening represents instability, a blurring of boundaries, the feeling of moving out through multiple spaces into sounding events and at the same time drawing sound inward to the place in the self that has no place except as the listening place, the act of listening.

Writing listens to the self in silent discourse with the self, detached from performative sounding. Speech is stilled in writing, even though the formation of writing is a form of speech or song; even though writing emerges from a confusion and flow of inner speech. Think of all the words invented by James Joyce in *Finnegans Wake* to describe intensities of listening: quiet darkenings, flitmansfluh, hushkah, soft belling, amossive silence, lispn. I would say that the ears turn inward, being disengaged from the act of monitoring speech and its effects in outer air, but I'm suspicious of all this ear-talk in the discourse of listening. Maybe R. Murray Schafer started it in 1967 with a book called *Ear Cleaning*, which as a metaphor entraps us within the wrong part of the body. Better to draw upon James Joyce and his shell, in which the imaginary and external is joined with the pulsating vessel of the whole body.

Gestures of the hand and other soundings enact this inner listening to the unknown formations of a stranger-self which emerge into resonant space and light to become listening in waiting, a silent text calling for reply, a music without music. Nothing hear: a nothingness falling to which there is nothing other than (as James Joyce would say) to lispn.

Notes

1 Leonardo da Vinci, *Notebooks*, Oxford, 2008, p. 260.

2 Leonardo da Vinci, *Codex Atlanticus* 398v, c. 1500.

3 Latham, John, unpublished text, 1984, source: John Latham archive.

4 Letters to the editor, *The Guardian*, Thursday July 18, London, 1974.

5 John Latham interviewed by David Toop, Flat Time House, London, Friday October 8, 2004, published in the catalogue to *The Body Event*, David Toop, Flat Time House, September 2009.

6 "Lecture On Nothing," printed in John Cage, *Silence: 50th Anniversary Edition*, Wesleyan University Press, Middletown, Connecticut, 2011 (1961), pp. 109, 125–26.

7 Lewis, Norman, *A Dragon Apparent*, London, 1987 (1951), pp. 251–52, 254.

8 Gann, Kyle, "Introduction" to *Silence*, op. cit., 2011, pp. xxv–xxvi.

9 Keizer, Garret, *The Unwanted Sound of Everything We Want: A Book About Noise*, New York, 2010, p. 46.

10 Josipovici, Gabriel, *What Ever Happened To Modernism?*, New Haven and London, 2010.

11 Wordsworth, William, *The Major Works*, Oxford, 1984.

12 Buchanan, Carol & Buchanan, Richard, *Wordsworth's Gardens*, Lubbock, 2001.

13 Jankélévitch, Vladimir, *Music and the Ineffable*, Princeton and Oxford, 2003 (1961), p. 137.

14 Faulkner, William, *As I Lay Dying*, London, 2004 (1935), pp. 68–69.

15 Wordsworth, 1984, p. 358.

16 Josipovici, 2010, p. 55.

Folding: Lucie Stepankova

Blog post, *a sinister resonance,* March 2018

Nine people sitting on the basement floor folding paper into origami birds, four microphones hanging from the ceiling, a loudspeaker pair at each end of the room. A sound going on, unmistakeably but ambiguously emanating from this activity, suggestive of the palpitations of a locust swarm, the feeding of insect eaters biting their way through a bounty of desiccated wings and bleached bones. The white cranes accumulate, piling up in earthbound flocks next to their makers. I am conscious of furniture in the room, the chair on which I sit, the movement of hands, a thin garment hanging loosely on the wall, a vivid red teapot.

Gradually, patterns emerge in the sound, lulls falling mysteriously, overtaken by industrious surges. A Max patch is at work. Now the sound piece thins, leaving a sparse acoustic crackle that exactly matches the quick, concentrated effort of the folders. Their number has grown to fourteen. This is a durational piece – four hours at this point – so some of them have returned from a break. The atmosphere of dedication is the focal point that holds it all within its shape and volition, no obvious breakage points other than the sight of doing and making, the growth of birds.

Upstairs we speak in low voices, respectful of the crackling quiet below. For a moment I think of Chim–Pom's installation pieces – *Non-Burnable, Real Thousand Cranes* and *The History of Human* – all of which refer to the vast quantities of paper cranes sent from all over the world to the city of Hiroshima each year and to the practice of Senbazuru, folding one thousand paper cranes connected together by strings. According to Japanese legend, a person who folds one thousand paper cranes – one for every year of the mystical crane's life – will be granted whatever they wish for.

But then I think of the sounds of labour: the physical impact of an axe cutting into a tree, the making of objects by hand, a typing pool (as only seen in old films) or the agricultural workers in Suffolk who would ease the monotony of threshing by mimicking the patterns of bell-ringing, their flails beating the same rhythm on the elm floor as the bells in a church steeple.

There are those records in my collection devoted only to songs and sounds of working: a Folkways 10-inch LP, *The World of Man: His Work*, which, notwithstanding the title, includes examples of women working: a Norwegian woman calling cattle to the barn to be milked, a Japanese woman spinning thread, women waulking, pounding and pulling tweed in the Hebrides, singing to make the work go with joy and pace. Then more grim than that, Alan Lomax's recordings of prison songs made at Parchman State Penitentiary, Mississippi, in 1947, and Bruce Jackson's *Wake Up Dead Man: Black Convict Work Songs from Texas Prisons*, made in 1965–6, the percussive thud of axes and hammers resounding in hot air as they rise and fall in unison, beating the rhythm of songs like "Rosie," "Grizzly Bear" and "Early In the Morning."

Lucie Stepankova's idea for *Fold* was to bring together a spatial composition with this physicality, the working of paper and legend, "[exploring] the sonority of the ancient tradition of paper folding (origami), its ritual aspects and meditative potential. It values collectivity, simplicity and the transcendental quality of repetition over a long duration."

At the beginning of Yasunari Kawabata's post-war novel, *Thousand Cranes*, a young woman serves tea to the male protagonist. She becomes known as the girl of the thousand cranes, simply because she "carried a bundle wrapped in a kerchief the thousand-crane pattern in white on a pink crape background." The image of a thousand cranes haunts the text. Starting up in flight or flying across the evening sun, their flashes of brilliance momentarily cut across guilt and suffering. "The sound of her

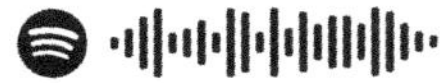

broom became the sound of a broom sweeping the contents from his skull, and her cloth polishing the veranda a cloth rubbing at his skull." Happiness is a wish.

Fold, a listening environment, was performed at Hundred Years Gallery, London, during the afternoon of Saturday March 24th, 2018.

10

clocks hearts nervous tic stutters

Christian Marclay

Carsten Nicolai

Robert Ashley, Cevdet Erek, Mary Butts, Dorothy Richardson

Painting in Slang: Christian Marclay

Published in *The Wire*, October 2011 (Issue 332)

A scene from *Gulliver's Travels*, perhaps. We are situated within a circular piano keyboard of grotesque proportions — piano gargantua — and in the centre, raised on a dais and encircled by an audience, sits a tiny piano played by an equally diminutive pianist. The pianist is Steve Beresford, a human of average size (as is his piano), but in this context he is dwarfed by the immensity of his own hands. I think of Alex Steinweiss's cover design for *Boogie Woogie*, a record released in 1942. Two huge hands, one black, one white, loom over a small grand piano on which rests a cigarette, its smoke rising in one elegant series of waveforms. 

This is Christian Marclay's *Pianorama*, created for designer Ron Arad's *Curtain Call* at The Roundhouse, North London. Arad has created a circular curtain of hanging plastic filaments for this famously circular building, then invited a selection of artists to make works which exploit the format. Marclay's response was to film Beresford's hands playing piano, then editing the images into a seamless, circular keyboard that surrounds the audience and allows hands to appear at any point in the circle. Hands that dart and retreat like nervous birds, a single note here, a short run, an alluring hint of Cuban montuno; ghosts of a piano music that is as abstracted as the black and white keys of the piano itself.

Two days earlier I was standing in the same circle, this time with Marclay. He frets about sound, the way the budget always narrows when it gets to the sonic point of the wedge, the way the sound should emerge at exactly the point where each finger strikes the keys, the compromise solution that has small speakers outside the cylinder of the curtain and

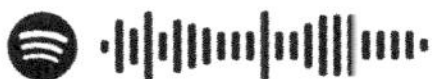

then a big speaker pod hanging overhead to destroy any sense of precise placement in space. He also frets about the matching of this muffled sound with the clarity of the live piano. "Visually, [*Curtain Call* is] extremely sophisticated and complex," he says. "But on the opening night the sound system was so damned loud that my ear is still bothering me two weeks later, ringing. I find it criminal. People just don't understand sound, they're so careless about it."

Of all the artists who are decidedly not careless about sound, Marclay has risen with gradual, inexorable grace to pre-eminence. In June I found him at the opening of the *Gone With The Wind* exhibition at Raven Row gallery, still in shock from winning the Golden Lion as best artist at the Venice Biennale. I'm surprised that he's surprised. *The Clock*, the 24 hour video piece which won him the award, is both artful and crafted, a simple but smart concept that has proven to be nonchalantly engaging for a casual audience but as weighty as you like for cultural commentators whose deliberations on time unwind until their clock gets stopped. But later for *The Clock*: now Steve Beresford sits at the piano within his two-and-a-half-times cylindrical piano and deals, as musicians must do, with the immediate problems of improvising a response to his own playing, to the muffled sound from the speakers, and to the prospect of drawing a line that connects the repeating fragments and pauses of the video work.

A line is what seems to unfold, an exercise in concentration and poise. At one moment it feels like reverse karaoke, in which the video sound is reacting to the live playing, trying to knock its implacable invention off course. The hint of salsa nudges and pokes, follow me, develop me, but Beresford is too seasoned an improvisor to capitulate into cliché. He plays with clarity and care, waits, occasionally erupts, mirrors a descending scale but then next time walks it in the other direction. Certainly it's more demanding than the shorter experience of sitting within the video piece. The listener needs to remember the unfolding, the variations, yet stay with its taut outstretching. As Marclay had said, two days before: "Live, you're in a balancing act and you can fall any minute."

While we were talking on that occasion, Ron Arad appeared at our table wearing that funny cloche hat of his. What can you say about Christian's piece, I ask. "There's things you can't get your head around," Arad says. "When I thought to ask Christian to do something, stupidly

I thought there's 24 columns, so we can watch *The Clock* in one hour I regret all the hours I missed when I watched it, because I didn't stay for the whole 24 hours." Instead, he got *Pianorama*, which has more than one relationship to *The Clock* but connects more closely to a whole other world of improvisation and ideographic scores. The imperfect nature of performance remains as important to Marclay as the fanatically precise editing exemplified in these works (and earlier examples such as *Telephones Video Quartet* and *Cross Fire*).

From talking about the specific technical challenges of making *Pianorama*, Marclay begins to expand on similarities and differences: cursor cuts and digital frames versus slamming a swiftly chosen vinyl LP onto a turntable in half-darkness and hoping for the best, or the malleability of sound against the specificity of image. "I do enjoy editing," he says. "That's the fun part. With video, found footage, it is and it isn't like DJing, mixing things, but it's definitely connected to that approach of just dealing with what you have, what you've found, what you can do. There's more freedom with just sound because things blend. With images you can blend them but it'll always look like a soft dissolve, have a cheesy quality."

There was the feeling of a vast living archive, music, art and film. One summer in 1980, Marclay's punk/performance duo — The Bachelors, Even — played a San Francisco dive called Club Foot. Bruce Conner would hang out there, once inviting the group to accompany a night of films using multiple projectors, home movies, industrial films and found footage. "He was notoriously grumpy and paranoid toward the art world but somehow he liked us," Marclay recalls. "I stayed in touch. He was someone that I really liked right away for the found footage. It was always set to another soundtrack so he was a bit of a precursor to MTV. I remember seeing his early films - *A FILM* - the heat of a radiator makes this feather go up in the air and it becomes an atomic bomb. These beautiful transitions. Another film maker I liked was Maya Deren. There's one scene in *Meshes Of The Afternoon* when she's walking and every footstep is a cut to a different landscape surface on which she's walking, and so there's the momentum of the body yet it's always cutting to another thing. These are the moments that you remember because they make you aware of what you can do with editing. It's a whole poem, going from one texture to the next."

I suggest to him that his recent work can't be fully understood unless you know the music backstory. "The DJing satisfies something more raw," he says, "improvised and direct, hands on, that incorporates mistakes and accidents. It's a live process. The viewer/listener becomes involved and sees the process but with video you end up with this product that doesn't necessarily show how it's made. It's a different experience for the viewer/listener but it's different for me also. In the editing there's more a cerebral process. It's a struggle and then you have these little moments of pleasure when you get a smooth transition, something exciting happens, you find a narrative connection, a symbolic connection, a musical connection. It keeps you going, though there isn't that joy of the moment that you get when you're on stage, you lose balance, you catch yourself. I miss that because I don't do it as much anymore."

What I also feel, and this is more contentious, is that works like *The Clock* are given greater culture weight than the unpredictability, sensuality and quick wit of improvising and turntablism. Maybe it's Pan versus Apollo all over again. At different times in history one will triumph over the other and for the moment Apollo has Pan's face in the dirt, but I'm thinking of an album like *Moving Parts*, recorded by Marclay and Otomo Yoshihide in the late 1990s. What is implicit and distilled in Steve Beresford's *Pianorama* improvisation — a fluid, inclusive music that can move in many directions simultaneously, working with the possibilities of silence and chaos — is given its head in tracks like "Blood Eddy" and "Fanfare." Untidy and splintered, they have a delirious appetite for life as it is. Or was.

"I don't know," says Marclay, "it's maybe a different time. My reason for pulling out of DJing, or 'DJing' quote, is that I feel that records have lost their meaning in a strange way. They're not the objects that we all used to interact with and that brought us all the music we enjoyed. Now it's retro. Culturally it has a different signification. Being a DJ means something else. It doesn't have that same relevance it had in the 80s or even the 90s."

Something similar could be said of clocks – they are being superseded by mobile devices. Rueful, he nods and starts to tell me about a forthcoming exhibition in San Francisco in which he will use cassettes combined with cyanotypes, an obsolete 19th-century photographic printing process (used famously in the early 1840s by Anna Atkins, who laid various species of British algae directly onto the coated paper to produce mysterious near-abstract blueprints). There are two or three ideas here that pinpoint the themes of Marclay's work. One is circularity: the records, the clocks and now a circular piano keyboard. You might call it an instinct for design, which incidentally helps him make compelling sound work for the visual art world, or something deeper, whereby circles and cycles tell the universally understood story of loss and return. On the one hand you sense a ruthless streak — it's over, move on — but then there is a recurrent attraction to obsolescence or that which is caught only in the corner of the eye, to the salvage of peripheral, abject and transient cultural objects, scrap discarded in the inexorable churning of evolution.

"The gramophone record becomes a form the moment it unintentionally approaches the requisite state of a compositional form." This was Theodor Adorno, from *Opera And The Long-Playing Record*, published in 1969. "Looking back, it now seems as if the short-playing records of yesteryear — acoustic daguerreotypes that are already now hard to play in a way that produces a satisfying sound due to the lack of proper apparatuses — unconsciously also corresponded to their epoch: the desire for highbrow diversion, the salon pieces, favourite arias, and the Neapolitan semihits ... This sphere of music is finished: there is now only music of the highest standards and obvious kitsch, with nothing in between. The LP expresses this historical change rather precisely." Views on that gap between kitsch and high art may have changed but the point remains still valid: technologies shape cultural expression and mirror the politics of their epoch in ways that can only be fully grasped when the technology becomes obsolete.

In 2003, Marclay exhibited photographs, drawings and notations at White Cube gallery. Let's say they were all notations because they indicated alternative methods for instructing musicians. Writing a catalogue essay at the time I mentioned Max Ernst, a reference that Marclay accepted. "He found the fantastic in the ordinary," he said. I remember a photograph of tin cans tied to the back of a marital get-away car and thought about indeterminacy, improvising percussionists, the charivari noise rites for domestic percussion enacted during threats to cosmic or social order, not to mention Marclay's own *Guitar Drag*. Marclay likened his photographs to the kind of sketches made by travellers, hasty recordings of memorable scenes as fragile as dreams in their tendency to slip from memory. "It's like

these little moments of recognition that we encounter every day," he said. "They're small and insignificant. They need to have enough excitement to pull out the camera and press the shutter but most people pass them by."

In another life, Marclay might have been a professor of linguistics. He is fascinated by the language of signs, constantly returning to ideographic languages, semiotics, codes, translations and transliterations. The severely abstracted, extended and enlarged piano keyboard of *Pianorama*, for example, plugs us straight into a history of what Alfred Appel Jr called jazz modernism — that Steinweiss design I mentioned earlier which showed up in the record collection of Piet Mondrian — or a wonderful Walker Evans photograph from 1935, *Sidewalk And Shopfront*, a New Orleans barber shop where the helical stripe of the traditional barber's pole has been applied to the entire shop front, as if blood and bandage transformed into the ebony/ivory binary of Jelly Roll Morton's piano keyboard. "Rauschenberg's collaged 78 rpm record label shard of Monk's most famous composition, 'Round About Midnight' (1947) is a jagged fragment," writes Appel Jr in his book, *Jazz Modernism*, "a phrase that telescopes the essence of Monk as minimalist piano player and composer, especially his dissonances, like the nerve-wracking way the lines of the trumpet and alto saxophone refuse to mesh in the ensemble passages of 'Round About Midnight,' enough to crack or shatter a 78 disc."

Thelonious Monk and "Tea For Two" were nested within Steve Beresford's performance alongside the eruptive piano styles of Charlie and Eddie Palmieri (the cover design of Charlie Palmieri's *A Giant Step* album shows a giant piano keyboard played by a shoe, while Eddie Palmieri's *The Sun Of Latin Music* has a disembodied red hand clawing at a cluster of vertical piano keys). This language owes as much to the iconography of film and cartoons as it does to cubism, Futurism, surrealism, constructivism and the De Stijl movement of Mondrian and Theo van Doesburg. Like Fernand Léger, who wanted to "paint in slang," Marclay makes sound works which draw on blurred memories of vernacular culture, maybe a *Tom And Jerry* cartoon in which Tom's efforts to perform Liszt's *Hungarian Rhapsody* are thwarted by Jerry hammering at the same note over and over, slamming down the lid to flatten Tom's white-gloved fingers, laying a mousetrap on the keyboard, dancing on the hammers to play jazz or running up and down under the keys to make a Mexican wave.

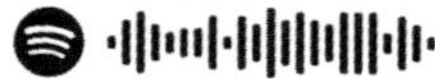

A car passes in the street outside, A-ZOOM painted on the side fo: some unknown reason. Marclay gets excited momentarily because these are exactly the kind of signs he has collected for an improvising score — *Zoom Zoom* — writtǝn for New York singer Shelley Hirsch and first performed by them both in Théâtre de Grütli, Geneva, in September 2008. "Using video to prompt a performance is something I've been interested in," he says. "Shelley has a certain skill, she's a great improvisor, but she has this stream of consciousness way that she can get lost in the stories and react to sound and image. I send these photographs that I've been taking of onomatopoeias that I find in everyday life, advertisements, the name of a store or a restaurant — Ping Pong, which is a London reference, or Crush. These onomatopoeias are sourːd first but they become little hints at story-telling. She embroiders them as a springboard for her own crazy stories. I project the images oːe at a time. I'm on stage with my laptop — I have this grid of images that I can click on so I'm reacting to her as much as she's reacting to the images. The images are projected big and she stands facing the audience bːt she has a monitor that shows her the image, so she can still project towards thǝ audience but she's reacting to the images that the audience sees. What has always been for the private world of the musician — the score — becomes now public. The audience is creating relationships in their own minds which are maybe different from whaː Shelley is thinking. TⱵat connection between the image there and whaː Shelley's doing becomǝs part of the perception of the piece. These are new experiments for me."

What interests me in this account is what it reveals about the insistently collaborative nature of Marclay's work. No longer in thrall to movements the art world now feeds on individuals who are strong and distinctive enough to survive as internatioːal commodities without the collective anɔ critical mass of an "ism," a manifesto or national identity. While this is true of Marclay, a peripatetːc Swiss-American living in London and not aligneɔ to any particular group or style, he also has a background in music, more particularly in improvised music, a form which survives and develops because it is based oɳ interlinked and evolving communities of players spread across the globe.

Those photographːs and notations mentioned earlier, the onǝs exhibited at White Cube gallery in 2003, may have seemed like amuse-bouches at the time; retrospectively we can see them as interim stages, gradually developing into notations for a type of playing which has its roots in 1970s improvisation. In New York, John Zorn's game pieces — *Lacrosse, Pool, Archery, Cobra* and so on — were innovations not just in ensemble improvisation but as an ingenious solution to problems of notation, organisation and stylistic diversity. Marclay took part in some of these pieces and played turntables with Zorn on tracks like "Forbidden Fruit" (from *Spillane*, 1987), "Battle Of Algiers" (from *The Big Gundown*, 1986), "John Zorn Présente: Godard" (*Godard Ca Vous Chante?*, 1985), the 1986 studio version of *Cobra*, the "Trip Coaster" section of *Filmworks VII* (1988) and the first side of *Locus Solus* (1983). The turntables are beautifully integrated with other instruments on *Filmworks* tracks like "Hysteric Logo" and "Coaster 2" but with track timings of 0'24" and 0'47," the music places extreme demands on immediacy, invention and physical dexterity. Out of curiosity, I asked John Zorn about how players like Marclay shaped the game pieces and how the game pieces shaped them. "The game pieces were a very fast operation," he says, "with cues and signs and split-second timing. Christian's ear was always very sharp and all the players loved working with him. Christian has an uncanny sense of what follows what — you can see that in his film work as well."

"It was very important and very influential," Marclay says of his "apprenticeship" with Zorn. "To me it was my discovery of improvisation. These early game pieces were liberating for me, realising I didn't have to try to repeat on stage what I had rehearsed. I also realised that the records didn't allow me to do that because they were so fragile and so temperamental, wanting to do what they wanted to do. John is a really important influence on my development. He also introduced me to every musician I ended up working with in the 80s. He introduced me to a social network I didn't know. It's through encouragements from people like John, who somehow saw something in me that I couldn't see because I didn't have the musical knowledge or the experience to see that what I was doing had any value. That's one of the great things about John, how he brought all these people from such different backgrounds together in his projects, from classical musicians who were eager to get away from the rigidity of classical music, jazz musicians, rockers and even amateurs and weirdoes like me."

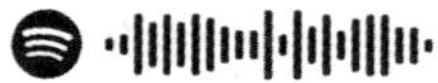

One of the issues Zorn was thinking about was time. In his notes for the 1981 release of *Archery* he wrote about eliminating set amounts of materials whose completion was a requirement of a score; his compositions were pulling "free from the idea of Time as a linear progression of amounts (size of information, length) to a more vertical conception: as an energy that appears immediately everywhere, and can be collected, balanced and regenerated in 'pockets' of information/material." Though they are all different, Marclay compositions such as *Screen Play Ephemera, Shuffle, Zoom Zoom* and now *Manga Scroll* have grown from this dynamic of omnipresent energy. They attack the normal hierarchy of perception whereby what is seen is taken to be the core reality of the event. The fluctuations of dialogue I experienced during Steve Beresford's performance of *Pianorama* — fixed and unfixed elements seeming to exchange roles, stabilise or destabilise each other, ebb and flow as focal points — is true of all these pieces. Sound pulls them away from the frontal flatness of the image, its fixity in space; improvising adds that element of unpredictability, the loss of balance, the catch.

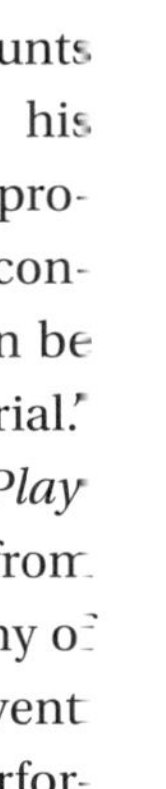

As well as distributing the formal elements of a composition, they also distribute the composer to some extent. Last summer's Christian Marclay Festival at the Whitney Museum in New York acknowledged this blurring of identities. More of a Festschrift or carnival than a typical one-person show the festival threw light on his musical collaborators as well as Marclay himself. Yet scores for improvisors (or improvising, to be more precise) can be a treacherous area. Who gets the ultimate credit? What about those notorious, sometimes celebrated "composers" who ask experienced players to "play something sustained in a high register," then walk away with the accolades and perhaps the money. Marclay's scores are different — they are highly sophisticated visually and conceptually. Audiences find them entertaining but they also open up potentialities for the musicians. For this reason I asked three of his collaborators — John Butcher, Joan La Barbara and Steve Beresford — to reflect on the question of why these pieces are stimulating to perform.

Steve Beresford has acted more or less as musical director for some of Marclay's more recent performances. For him, the scores engage parts of the brain that lie dormant in other situations: "Playing *Ephemera* [a piano solo in which the soloist works from twenty-eight folios that collectogether musical notations printed on confectionery wrappers, clothing, record covers and other found materials] is like improvising with a great musician — you get a sense of overwhelming urgency and you don't want to miss a trick. Sometimes the difference between improvising freely and playing (or can we re-use the 60s term 'realising'?) Christian's pieces is obvious: readable bits of simple tunes in a piece like *Ephemera* may get played very literally; over-repetitive phrases that looked good to the graphic designer might get played past the point we'd normally do them."

"I've played a few now," says John Butcher, "the film ones, *Screen Play* and *The Bell And The Glass* and *Shuffle* and *Ephemera*, which are both based around photos of items or places that contain images of musical notation, ranging from a snippet of Mozart on a chocolate box, to stuff that makes little sense — like adverts with three-line staves. With the films, I think some people approach it like Foley artists, and at the other end, some play quite independently. As they're very carefully edited — he has a great sense of flow and rhythm in this — I think it works best to follow the action quite closely but keep the material pretty abstract. You can invent your own schemes for responding to the films, but it pays to rehearse and think through this in advance. I think it's wrong to look at it as the musicians 'interpreting' the work. The improvisation is as much the work as the visual material that's stimulating it. It's unusual to have an artist prepared to operate this way. I guess it comes from the fact that he himself values the intrinsic possibilities and meanings of improvised music, and isn't interested in just using it for colour or effect. Of course, he's a great improvisor himself, which must be why players like collaborating with him."

In 2009 Marclay invited Joan La Barbara to perform a work for solo vocalist, *Manga Scroll*. He speaks of her with a note of awe — "She's like the queen of vocal experimentation." The piece has been performed by La Barbara, Shelley Hirsch, David Moss and Phil Minton and will be performed in Aldeburgh later this year by Elaine Mitchener. *Manga Scroll* is another project derived from signs, transliteration and onomatopoeia, a 60 foot long scroll of graphic sound effects lifted from North American translations of Japanese manga cartoons – basically a long strip of pfff, skrrrr, krak and blech.

I asked La Barbara how she approached such a daunting task. "When we discussed it, I said that I hadn't decided where to start or which direction to go," she says, "and Christian said that he wanted the scroll performed from beginning to end. Other than that, he gave no specific directions as he wanted to experience how each of the performers who took on the task would deal with the material. There was a great deal of motion, for me, in the visual material, and so I progressed through the score with somewhat of a sense of 'urgency,' moving along with the energy required to perform what I felt the score implied. I worked as close to real-time reading/performing as possible, spending the amount of time necessary to perform the fragments immediately in my gaze, while looking somewhat ahead to the next set of events in preparation. So, each performance was both improvised in the moment but informed by the decisions I had made about certain sounds and sets of material in my rehearsal sessions. When the visual material circles back on itself, I tried to create real-time loops without electronic support. When it splits apart and goes in several directions simultaneously, I tried to create the feeling of multi-directionality by choosing one stream, progressing along it and then leaping back and choosing another. I created complex patterns by fragmenting material and utilized some of the extended vocal techniques that are my personal vocabulary, which often sound as if there are multiple voices performing simultaneously."

In *The Sound And The Fury*, one of William Faulkner's characters passes on his grandfather's watch through the paternal line to his son, Quentin, describing it as "the mausoleum of all hope and desire ... the reducto absurdum of all hope and desire." As Quentin prepares for his suicide, he lies in bed listening to the watch: "It was propped against the collar box and I lay listening to it. Hearing it, that is. I don't suppose anybody ever deliberately listens to a watch or a clock. You don't have to. You can be oblivious to the sound for a long while, then in a second of ticking it can create in the mind unbroken the long diminishing parade of time you didn't hear."

Time is unmentionable, always ticking down to cliché unless you are Faulkner, Joyce, Woolf, of that calibre. But it can be shown, or better still, find its more complex, unfettered and subtle expression through music, through sound. Marclay discusses editing and writing, yoga, the small platoon of researchers who found and delivered material for *The Clock*, this beautifully constructed timepiece of collective memory which nobody will ever fully experience, if only because its parallel loop of time, each minute of a 24 hour period shown in real time on found film, is a constant reminder that life goes on, things to do, places to go. As Joe Hinton once sang, "Ain't it funny how time just slips right on away?" Although developed very slowly, the concept of *The Clock* emerged from a video score. Working on *Screen Play* in 2005, Marclay realised that the musicians would need some indication of time in order to prepare for changes in the found film footage they were seeing. During a residency at Eyebeam Art + Technology Center in New York, an intern assistant researched film footage showing clocks. "He started bringing in all these clocks," he says. "I thought, wow, I wonder if it's possible to find every minute of every day in the history of cinema?"

Everybody is in crisis, I say. Who knows how to release work or make a living from it except by flinging some lo-res giveaway out into the digital ocean and hoping for a better tomorrow? *The Clock* is something else, a media work enhanced by its limited availability, too big to download or buy in a shop so you must go to the work, sit with it and, yes, give it time. He tells me the economics, that *The Clock* is being sold as an edition to institutions or sold jointly to museums in order to avoid the squirrelling habits of private collectors.

"I welcome the crisis," he says. It's not as if the old structures were particularly easy or helpful anyway. Marclay's work is not just about his own salvation — his collaborations uncover new audiences and strategies for a large group of musicians and the success of *The Clock* amplifies that effect. "Sometimes it opens a door for audiences to learn about the music," he says. "When we did the Whitney they organised a dinner for all these funders — a crowd of uptown people who would never want to go down to Cafe Oto or the Knitting Factory. They're fine with Vito Acconci jerking off on the floor but going to listen to some crazy music somewhere is not acceptable. I asked Shelley Hirsch if she'd perform and people were completely blown away. They gave her a standing ovation and she was the hit of the party. It's hard to cross these borders so I'm hoping *The Clock* might help a little bit with that but I'm not worried about it so much. I'm happy making music for people who enjoy it. If you strive toward a certain perfection or something you believe in then people will see that it's for real, it's not a joke. You can't control that. You can't even worry about it too much."

pink glisten for Carsten Nicolai

**Published in Carsten Nicolai: *Parallel Lines Cross at Infinity*,
published in conjunction with the exhibition, Carsten
Nicolai: *unitape*, Chemnitz, 2015**

for Carsten Nicolai: hot flat haze in a pulse of quickened weather / stop / flick on/off dark blitz of switching then open out stab the pulse-tone colour into the cathode for disturbance mode not -sit back relax- / stop / flat black to ice white point field of sharps / stop / filament vibrations through which space / stop / cubic air / stop / is shivered to slivers point needle pierced membrane of air punctum viewed through a tiny hole thin rays of streaming lights all visible spectra colour the area in spider web shuddering / stop / spectrum the area / stop / pink light white noise pink noise white light black noise black light the area a flat haze of pulsation relaxing taut and flaccid tension / stop / lightly breathing secretly withholding handfuls of this viscous medium through which passes the trembling-flickering-pulsing that we call colour-sound-light / stop / for the second time of asking how can we comprehend time? 00000010000000? / stop / beyond our edges a temporal question of slow quick low high / the lunging gait is a respiratory movement open + close signal + gate breathing its last and first / stop / cuts slicing into cobalt darkness leave the pink glisten of a luminescent wound whose edges close with supernatural grace only to flutter open close open in the gill breath of a shark undulate abyss / stop / then left only with the buzzing tremor of consciousness rips and tears of movement the firing trailing of sharded snips whose thin splinters illume a distant silence black as cloud before the static rain is poised to fall to start and / stop / from David Toop

Automatic Writing: Robert Ashley, Cevdet Erek, Mary Butts, Dorothy Richardson

Blog post, *a sinister resonance*, March 2014

Robert Ashley's death gave me the odd feeling that I should have been listening to more of his music. Absurd really, to self-impose a kind of obligation to consume. The truth is I loved his work but never felt much of a compulsion to listen to the recordings. They seemed beautiful shadows of something genuinely new, something he spoke about, wrote about and enacted: a different way of being within music.

He was one of the few composers of his generation (and subsequent generations) who fully understood that after Cage, electronic music, free jazz, pop music, happenings, proliferating media and all the rest of it, you couldn't just carry on as if nothing untoward had happened, making your "experimental" music with all the formal constraints and solemn ritual still in place. There's an interesting essay he wrote for the CD release of Alvin Lucier's *Vespers* and other early works. You can't hear *Vespers* on a recording, he says, because the experience of hearing the music comes from the space in which it's performed. He also talks about attempts to subvert the concert hall or redesign concert halls specifically for experimental music, as if it's going to stand still for another hundred years to please the architects. He talks about fire marshalls. Basically, if you think you're being subversive but at the same time pleasing the fire marshalls then you're not subverting much at all.

A lot of composers and musicians shared that sentiment for a while but then turned away into less problematic territory – back to the 19th century or to a vision of jazz bars as they were prior to the 1940s, wherever

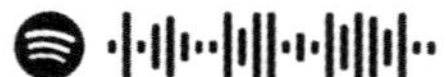 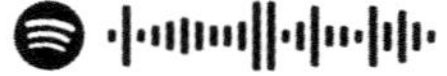

their spiritual home might be. Robert Ashley worked out ways to make a piece that could be heard on television, or heard in a public space as if you were in a hotel room alone, hearing some other guest's strange yet mellifluous speech rhythms coming through the adjacent wall and by putting a glass up to the wall and listening hard you could eavesdrop on th_s man's muffled monologue about the particles of life, recited to the accom-paniment of a radio broadcas by Liberace or one of those early New Age composers from Los Angeles, but Liberace after his audience has departed the building and his secret deepest vision of a cosmic music is finally given a silence in which to float like a voluminous sun bed on a Hollywood pool.

There was a stillness or stasis about Ashley's pieces that demanded some new venue for listening that just doesn't exist. In a way, television (arguably the most important medium of the twentieth century but a wasteland for most composers) was the best way to encounter what he did, a box in which dreams spewed forth as if mind itself; now television is pretty much dead so there's a moment that won't come back.

In what setting should music be experienced? The question resurfaced, as it does in every exhibition that demands listening, while I walked around Cevdet Erek's *Alt Üst* exhibition at Spike Island in Bristol (15 February to 13 April 2014): the video of his fingers attempting to tap out a sonic transla-tion of a timeline of "life-related events"; the measures, markers and cycles alluding to sound's temporality; the rooms called Üst and Alt – day and night, up and down, high and low, heaven and underworld – in which blue LED lights and bass beats measure time in the murky claustrophobia of Alt. In the dazzling sky light of Üst, a cardiac pulse of a low beat emanates from under the feet as if seeping upwards from the underground below. All of the ideas within the exhibition came together in that moment of being within the shock of daylight, the emptiness of a room, sound coming from elsewhere.

Later that afternoon (04.03.14), I gave a talk at Spike Island about music and time, playing tracks like Joe Hinton's "Funny How Time Slipped Away," Sly Stone's "In Time," Felix Hess's frog recordings from Australia, an exceedingly slow, sacred Javanese gamelan from Jogjakarta ("Sekatan Kyahi Guntur Madu"), Arpebu's "Munsta from Kavain Space" and Ryoko Akama's "Jiwa Jiwa," created on the Max Brand synthesiser during a composer-in-residence programme at IMA in Austria. Ryoko's recen

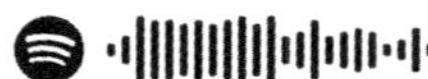

CD – *Code of Silence* – gives no information about the latter piece other than its title so I asked her for some thoughts. Her return email spoke about sine waves and beat frequencies, sustained tones, other worlds with the emptiness of sonic "surfaces" and an aesthetic that arises not from duration but the complexity of listening and its context and conditions.

She translates "Jiwa Jiwa" as "slowly but certainly happening," giving the example of finding a water leak coming through the ceiling, the stain gradually growing in circumference: "You might say – it is getting jiwajiwa there, water is permeating jiwajiwa." So the sound is a type of sculpture (maybe like the slower kinetic sculptures of the 1960s by Pol Bury, Gerhard von Graevenitz and David Medalla); change is taking place but at a rate that is hard to discern, closer to stasis than movement.

Last week (07.03.14) I gave a talk in the Royal College of Art's *Vocal Dischords* symposium, using the technique of automatic writing I'd tried once before at Bristol Arnolfini's *Tertulia – Writing Sound* event, a set-up in which I write without a script and whatever I write can be seen on screen by the audience in real time. The question of whether automatic writing is possible in these circumstances becomes part of the performance, not least in the sense that fluidity of so-called inner thought is hard to realise except in private. The pace is slower than speech, more stilted or inhibited by observers, subject to error and revision. At one point I played an extract from Robert Ashley's *Automatic Writing* and in that heightened, stressful atmosphere heard it as unvoiced thoughts bubbling out of the eyes like soap, seeping slowly from the pores of a face.

For all I know this live performance of writing may be painful to watch but it comes, in part, from an active questioning of spontaneity in improvised performance, along with a questioning of the voice-as-sound, the droning seducer that transmutes ideas into theatre (or so it hopes). What are voices in the head? In pursuit of the origins of spontaneity in 20th-century music I have been reading largely forgotten writers who experimented with automatic writing and what William James described as stream-of-consciousness. Mary Butts is one of them, an author whose short and turbulent life included the thankless task of assisting Aleister Crowley with the editing of *Magick in Theory and Practice*. Her novel *Armed With Madness* (1928) opens with a sentence that makes you want to love it – "In the house, in which they could not afford to live, it was unpleasantly

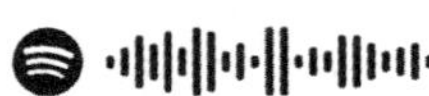

quiet." A description of listening and silence as an uncanny, occult experi-ence follows, not dissimilar to passages written by Virginia Woolf at almost exactly the same moment in history.

Dorothy Miller Richardson is another. Her *Pilgrimage* series of thir-teen novels, the first published in 1915, was a meticulous, if highly selec-tive recording of a life, each instalment given an enticing title: *The Tunnel*, *Pointed Roofs, Honeycomb, Deadlock, Revolving Lights, The Trap, Dawn's Left Hand*. The protagonist – Miriam – lives a modest, unspectacular life. In *The Tunnel* (1919) she is ecstatic to be renting a dingy room that gives her some measure of independence. Time barely seems to move, yet the cycles of life, day and night, work and time off, drudgery and tea, the tasks to be performed at a given time within the patterns of her job, her walks through a London that feels both hostile and magical, the surging and ebbing of feelings, convictions, confidence and often silenced opinion open out, fold upon fold, light and dark as she learns how to live and finally to write. The reader is caught in the stream of this interior monologue (as Richardson liked to call it), absorbed, like Robert Ashley, in the particles of life: "As she began on her solid slice of bread and butter St. Pancras bells stopped again. In the stillness she could hear the sound of her own munching. She stared at the surface of the table that held her plate and cup. It was like sitting up to the nursery table. 'How frightfully happy I am,' she thought with bent head. Happiness streamed along her arms and from her head. St. Pancras bells began playing a hymn tune in single firm beats with intervals between that left each note standing for a moment gently in the air."

An ordinary life; a dull life even, yet the polyphony of emotions and sensations is hallucinatory in its precision and accumulation: "The lec-turing voice was far away, irrelevant and unintelligible. Peace flooded her." Why do we have to spatialise time, sound and thought, reducing all three to a manageable linearity and locus that has nothing to do with the way we think or hear? Because they are elusive, everywhere and nowhere. The pouring of thoughts may take place in a dark room as if a kind of ectoplasm gushing out of some hidden spring and dispersing into nothingness, into the blood or becoming a sound recognisable as audible words, the marks of writing or some other signs on or from the body.

11

agitations surfaces decomposition

Instruments of Non-Existence

Hugh Davies

Tomás Saraceno

Tania Chen, John Cage and David Tudor

Rie Nakajima

Instruments of Non-Existence (through which Heaven and Earth Seek Reconciliation)

Published in exhibition catalogue *Art Or Sound*, edited Germano Celant, Fondazione Prada, Milan, 2014

In the *Sound (Acoustics)* section of Joseph Needham's *Science & Civilisation In China* (written by Kenneth Robinson), a strange instrument is described.[1] A single-roomed building is built, in fact three rooms built concentrically, like a Russian doll. The walls are plastered to cover any cracks; doors prevent the ingress of draughts from outside. Pitch-pipes are mounted on wooden stands, further sealed from external air within a tent of orange-coloured silk, each pipe arranged around the compass points in their proper positions. The upper ends of the pitch pipes are packed with the ashes of reeds. Once prepared they are watched according to the calendar.[2] As the Neo-Confucian philosopher Tshai Yung, an expert in acoustics and music, wrote during the later Han Dynasty: "When the emanation (*chhi*) for a (given) month arrives, the ashes (of the appropriate pitch-pipe) fly out and the tube is cleared."

(noting that in the 21st century, cosmological machinery such as this might crassly be described as "sound art") can this be described as a musical instrument? Of course not, no, but then again its existence as an instrument of observation, of "waiting for the *chii*," was predicated upon the use of bamboo tubes to determine tuning in relation to the proportionate, cyclical, physical world and to *chii*, the mysterious, rarefied emanation which in its rising from earth to the heavenly ancestors and from ancestral heaven back down to earth could give life to all phenomena. So bamboo tubes were divinatory instruments, "humming-tubes" as they are called in Needham, used by shamanic music-masters to predict outcomes and influence events. "Every man has within his body his own *chii*," writes Robinson. "The diviner uses his to set up a disturbance in the outside

world when he blows through his humming-tube. One *chii* will then "by a kind of mysterious resonance" react on another *chii*, just as one musical instrument will touch off another which is in tune with it."

So *chii* (*chi*, *ch'i* or *qi*, as it better known in our time) and sound can be understood as closely related emanations within ancient Chinese belief, just as such divinatory actions can be contextualised within a wider category of spirit or ancestral voices "speaking" through voice disguisers, flutes and other esoteric instruments, a form of extra-human communication once widely practiced in regions of Papua New Guinea, Central and West Africa. Even in 1930, when Henry Balfour was collecting material in Nigeria for his paper, *Ritual and Secular Uses of Vibrating Membranes as Voice-Disguisers*, such "mysteries," as Balfour described them, "had already been shorn of much of their former significance and terror,"[3] and yet the principle remains consistent throughout the global history of sound producing devices.

The demonstrably physical yet alluringly intangible properties of sound can be shaped, modified, projected, transformed and given life through constructions that belong to the prosaic earthliness of matter on earth. No matter how beautiful to all the senses, they remain abject. Though fetishised, monetised and anthropomorphised, musical instruments are the waste products left behind after sound. By representing sound through matter as a static design for the eyes they represent the betrayal of music, particularly from the radical Taoist point of view, that "the greatest music has the most tenuous notes." In James Joyce's *Ulysses*, Leopold Bloom ruminates on sound and the outer limits of the musical instrument: "a blade of grass, shell of her hands, then blow."[4] Such an instrument can be found illustrated in Jaap Kunst's volume, *Hindu-Javanese Musical Instruments*, described as a rice-stalk aerophone, the instrument that has no archive, becomes without sound, air, lips, the cupped hands, simply what it was before its sounding – described by Kunst as "a remarkably powerful shawm-like sound" – abject wisp of discarded organic matter.[5] In Bloom's museum of streaming consciousness there is comb and paper, hunting horn, shepherd's pipe, policeman's whistle, the bestiary of the instrumentarium – "Brasses braying asses through uptrunks ... Semigrand open crocodile music hath jaws" – the holes of the female body to be played like a flute, the chamber music of Molly Bloom pissing in her chamber pot: "It is a kind of music I often thought when she. Acoustics that is. Tinkling. Empty vessels make most noise. Because the acoustics, the resonance changes according as the weight of the water is equal to the law of falling water … Drops. Rain. Diddle iddle addle addle oodle oodle. Hissss."

(while noting that even Joyce's attempts to transcribe sound into language are simplistic) Rembrandt's etching of 1631, *Man and Woman Pissing*, gazes without judgement on this history of abject music, one of the origins of music perhaps (though never mentioned in those speculative texts that imagine early humans listening to birds and imitating their cries).

Organ(ology) Without a Body

"Composing means: building an instrument," Helmut Lachenmann wrote.[6] In response: Building an instrument means composing (if composing is taken in its broadest sense, a gathering of materials).

(noting the previous line of thought) an instrument may question the fixity of boundaries between all that it is to be human and all that it is to be not. Many musical instruments are based on animal forms. They are channels, reservoirs, surfaces, chambers out of which sound is projected into the world through varieties of violence – friction, expulsions of breath, shaking, striking – as if empty animals whose cries can be transmuted into human desire. In dreams, hallucinations, the generative machine of Raymond Roussel's fiction, alchemical symbology and in paintings we find ensembles which broach the boundary between human, animal and heaven, proposing a reversion to the garden of Eden: a hybrid creature depicted in the 14th-century alchemical work, *Aurora Consurgens*, its instrument a lobster bowed by what may be a snake, a harmony created from the elemental chaos of Materia Prima.

This silent music is closest to the ideal of an instrument that has become dematerialised to fully allow music its heavenly status. Music without sound was a Taoist ideal. One note would only spoil the last one or the next; the most exalted of instruments, the ch'in (translated by Robert Hans van Gulik as "lute")[7] could reduce earthly instruments to silence, its heavenly sound only audible to supernatural beings: "Now when the lute is played, the sounds of the instruments made of metal and stone die out, and the breath blown into those made of gourd or bamboo ceases. Wang Pao breaks off his chant, Ti Ya loses his gift of taste." In this resounding silence a more subtle silence rouses immortals, a phoenix pair and heavenly fairies

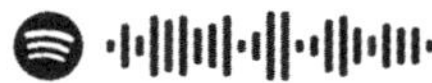

In search of regeneration (or heaven), music apostatises its instruments, relinquishing agency to the forces of nature or automata, hypothesising (to reverse the Antonin Artaud formulation of a free body without organs) a (free) organology without body. Mississippi-born R&B singer Bo Diddley, creator of guitars whose shapes and textures – rectangular, rocket-shaped, covered in pink fur – celebrated the un-guitar nature of the solid body electric guitar, identified the catalyst as an instrument without body: "There was this dude called Sandman that used to carry a board, an' a bag of sand an' a swish broom, you know, one of them 'sweep-ups.' He'd put the board down on the ground, an' dump this sand on it an' spread it around, and then he would sing an' sand – it was kinda like a shuffle, but it was called 'sandin.' It carried a rhythm with it, like a tap-dancer taps, an' not too many people around here could do it. When he finished his act, he'd sweep up the sand an' put it back in his little sack."[8]

This modulation between the (literally) earthly and immaterial is reminiscent both of the sand mandalas created by Tibetan llamas over days, then ceremonially destroyed and discarded in water, and the Japanese dōtaku bronze bells displayed in the British Museum. These bells were made by the Yayoi people between 500BC and AD250, often buried at the edge of fertile agricultural land. Decorated with images of animals valued as predators on the insects that destroy crops, their ritual absence from sight and activation suggests a profound relationship between the invisible, hence mysterious energies of sound, the observable workings of nature and the unfathomable proclivities of the gods. A disappearance is enacted; the presence or physical sounding of the object is unnecessary for the magic to take place.

Such sounding devices and actions-at-a-distance cling precariously to the outer edges of our contemporary idea of the instrument. They are closer to an extra-human zone beyond culture, to the involuntary and uncontrollable, to meteorology, bioacoustics, cavernous reverberations and inexplicable submarine and subterranean noises, to supernatural conceptions of auditory phenomena. They arise as if from the deep, often treated as living beings, symbolic of afterlife and speaking their own language. The huge garamut slit gongs recorded by Ragnar Johnson in Papua New Guinea in 1976 were decorated with long-nosed ancestor figures carved as a projection at each end of the garamut slit. Made from single logs, these wooden gongs were used both to send messages over long distances and to accompany the sacred flutes whose voices were the cries of spirits. "In the past,"

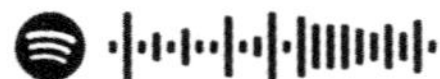

wrote Johnson, "when a new garamut had been carved and brought into a village, the villagers would dress up and hold a feast, decorating the garamut with feathers and leaves, and give it offerings of sago and coconut milk."[9]

Wooden slit gongs played a pivotal role in Bamum society in Cameroon, their function to send out signals in times of crisis from within the royal court.[10] By the time German colonists reached the Bamum kingdom for the first time in 1910, they found eight of these spectacular instruments, each hollowed out from a central slit, decaying on the palace ground like felled giants from a mythical past. Two French colonial administrators stationed in the Bamilike region noted that these gongs were carved when a king was enthroned; when he died they were abandoned in the market square and left to rot as if undergoing the same process of decomposition as the king himself. One Bamilike gong photographed in 1911 was carved in the shape of a buffalo, bigger than life, its body decorated with relief carvings of a chameleon, a lizard, a toad and the iron double gong used by secret societies. Consecrated by sacrifice, a gong like this might be used by all-male secret societies to summon members for a funeral. Torn from earth, the sound of these tree gongs gave voice to the dead and the otherworldly of the deep and so it is with these unknown devices that they speak in chthonic tongues from James Joyce's "rambling undergroands" of the lower depths.

noting that:

Notes

1 Joseph Needham (with Kenneth Robinson), *Science & Civilisation in China, Volume IV: 1*, UK, 1962, pp. 187–88.

2 Ibid. p. 132.

3 Henry Balfour, "Ritual and Secular Uses of Vibrating Membranes as Voice Disguisers,' *Royal Anthropological Institute Journal*, UK, 1948, p. 47.

4 James Joyce, *Ulysses*, UK, 1997, references taken from the Sirens section.

5 Jaap Kunst, *Hindu-Javanese Musical Instruments*, The Netherlands, 1968, p. 29 (and illustration, after Curt Sachs, Fig. 103).

6 Helmut Lachenmann, *Über das Komponieren*, 1986, in *Musik als existenzielle Erfahrung*, Germany, 1996, p. 73.

7 R. H. Van Gulik, *Hsi K'ang and his Poetical Essay on the Lute*, Japan, 1968, p. 118.

8 George R. White, *Bo Diddley: Living Legend*, UK, 1995, p. 27.

9 Ragnar Johnson, *Sacred Flute Music from New Guinea: Madang, Quartz 001 LP*, UK, 1977 (reissued on Ideologic Organ/Editions Mego, 2016).

10 Christraud M. Geary, "Slit Gongs in the Cameroon Grassfields: Sights and Sounds of Beauty and Power," pp. 63–71, published in Marie-Thérèse Brincard, *Sounding Forms: African Musical Instruments*, USA, 1989.

Everything Is Vibrating: Hugh Davies

Liner notes for Hugh Davies: *Tapestries*, Ants CD, 2005

Ubiquitous in the performance of contemporary experimental music and electronica is the table, an essential piece of equipment that may support anything from laptop computers and record decks, to curious collections of contact microphones, invented instruments, and amplified domestic appliances. Hugh Davies, live electronics pioneer, improvisor, instrument inventor, composer, educator, and researcher, was one of the pioneers of this approach in the 1960s. He even invented an instrument that embraced the table as a connective element linking performer with sonic sources – the Solo Performance Table (1969–72).

Hugh Seymour Davies, musician, researcher, and instrument inventor, was born on the 23rd April, 1943. Even while studying music history, harmony and counterpoint at Worcester College, Oxford, with Frank Harrison and Edmund Rubbra from 1961–64, Hugh was exploring an unconventional career in sound. In 1962 he was invited to work for a weekend with Daphne Oram in her Oramics studio, the first private electronic music studio in England. As a teenager, living in Exmouth, Devon, he purchased his first recording of electronic music: Karlheinz Stockhausen's *Gesang der Jünglinge*. The record featured in a lecture he gave to the Oxford Contemporary Music Society; then shortly after completing his degree, he moved to Cologne to take over Cornelius Cardew's position as Stockhausen's personal assistant.

This was a time of breakthrough for Stockhausen, in which some of his most important pieces were composed. Fluent in German, open to experiment, yet possessed of a rigorously analytical mind, Hugh was given

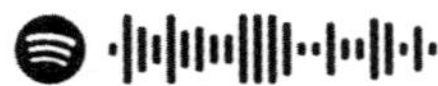

the responsibility of preparing a listening score for *Gesang der Jünglinge*, writing new performance material for *Momente*, and, in 1964, performing in the ensemble that recorded Stockhausen's exploration of live electronic music, *Mikrophonie I*. With the expertise of Jaap Spek, the technician at Cologne's WDR radio, Stockhausen had devised a method for using contact microphones to amplify sounds made by activating the surface of a large Paiste tam-tam. "The tape recording of the first microphony experiment constitutes for me a discovery of utmost importance," Stockhausen has written. "We had not pre-arranged anything; I used several of the laid-out implements at my own discretion while probing the tam-tam surface with the microphone, as a doctor auscultates a body with his stethoscope … actually, this moment was the genesis of a live electronic music with unconventional music instruments."

There are interesting parallels with space probes, oceanic exploration, anthropology, microscopy, or spying. Others were fascinated by this potential of the microphone: "Sound is vibration," John Cage told a reporter from *Newsweek*. "Everything is vibrating. So there is no earthly reason why we can't hear everything. If we push beyond the limits of perception, there is a chance that perception itself will be extended. Mushrooms are making sounds and we should be listening to them. When I went into the anechoic chamber, I could hear myself. Well now, instead of listening to myself, I want to listen to this ashtray … I'm going to listen to its inner life thanks to a suitable technology. Imagine people bringing objects to a central place. You would be able to listen to their offerings."

"Stockhausen also used contact microphones to amplify metal percussion instruments in *Mixtur*, which I also performed in," Hugh wrote to me when I was researching my book, *Haunted Weather*. I had asked him about his early influences. "Jaap Spek was the expert on all these things and gave me advice. Contact mics were also used, as far as I remember, by Johannes Fritsch (in Stockhausen's group) on his viola, and in pieces by a couple of students on Stockhausen's composition course. In autumn 1965 Max Neuhaus played a solo concert in Cologne, with a programme similar to his solo LP, including contact mics on percussion and use of acoustic feedback."

After two years in Cologne, Davies returned home to England in 1966, armed with a small array of electronic equipment and plans to develop

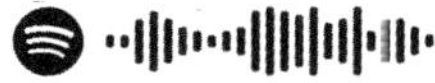

his own electronic projects. "Living at my parents' home for a few months in the summer of 1967, before I moved to London," he told me, "I finally started. There was a radio set in their bedroom with an extension loud-speaker, so I wired up a connection from my room so that I could plug my mixer into the radio and work in my room using the extension loudspeaker with the door closed. In that period I was still thinking primarily of making tape music, and produced a few small pieces using a considerable amount of splicing, including a short piece based on modified musical extracts taped off the radio that would now be described as plunderphonic."

Although he still perceived himself as a composer of tape music at this time, Davies was drawn into a more spontaneous world in which the compositional control and expensive equipment of electronic music became increasingly irrelevant. Typically, he converted his lack of resources into a virtue. "Wanting to extend my sound sources, I ordered a sine/square wave generator from Heathkit in kit form, which I assembled (excellent practice for my soldering, and it worked first time!), and I started putting contact mics on found objects – including a quartet of combs mounted in holders, and an upturned tea-tin with several small springs stretched across a wooden 'bridge.'"

This recycling of everyday objects, and the detritus of a wasteful society, developed into a deeply felt environmental awareness. In 1974, for example, his article *Sounds Heard at La Sainte-Baume* (published in *MUSICS* no. 5), listed various activities that both encouraged deep listening to the sensorium of the natural world, its sound, echoes, and resonances, and suggested non-invasive methods of engagement with this world: "In a forest, listen for a woodpecker. Quietly approach the tree that is being pecked, and listen to the resonances produced inside it. You may find that these are most audible with 'bone-contact' – place a finger in one ear and hold the knuckle against the tree." Then in *MUSICS* no. 20 (1978) he published a number of environmental music projects, some devised as early as 1969, that analysed the sounds produced by varying road surfaces on motorways, and advocated the building of acoustic parks in cities.

These were pieces that could be related historically to Fluxus works, to Scratch Orchestra pieces, to conceptual art and land art, and to the sound ecology concurrently being developed in Vancouver by R. Murray Schafer, yet they wore their influences lightly. Perhaps this was because Hugh belonged to no faction. A pragmatic, down to earth person unaffected by the political and mystical storms that uprooted some of the charismatic figures in experimental music, along with their disciples, he worked where and with whom he felt comfortable, with a strong sense of individuality; many of his ideas were original inventions, rather than a reflection of modish techniques and procedures. At a time when people chose camps, he enjoyed a certain amount of control, certainly never rejecting composition, yet also relished accident and the unforeseen.

Even in the late 1960s, few musicians were able to move confidently between the divided factions of experimental jazz, classical composition, and rock. Crossing boundaries could be interpreted as a lack of commitment to a cause. People took sides: composition versus improvisation; art versus politics; conventional notation versus graphic scores; humour versus seriousness; electronic versus acoustic; live electronics versus studio. Any of these disputes could turn into long-term warfare, yet Hugh managed to negotiate the calls to arms with great tact. In fact, he pursued all of these activities and possibilities without ever becoming the subject of an ideological purge. Equally problematic was his determination to balance the life of a gigging musician with serious academic research. In 1968, his comprehensive *International Electronic Music Catalog*, compiled during two years as researcher at the Groupe de Recherches Musicales of the French Radio, was published by MIT Press. In December 1967, at the invitation of composer Stanley Glasser, then Head of Music, he set up and directed the Electronic Music Studio at Goldsmiths College, London. This was the first university studio to be founded in Britain. From 1968, Hugh led workshops in which he described his own knowledge as progressing just a small distance ahead of his pupils.

This period has been described as a golden age for experimental music. Hugh's activities included the organisation of concerts at the Arts Lab in Drury Lane, and in November 1968, having worked in a duo earlier that year with Richard Orton, he joined Orton, Richard Bernas, Patrick Harrex, Graham Hearn, Stuart Jones, and Michael Robinson, in one of the first groups dedicated to performances of live electronic compositions, Gentle Fire. Hugh's account of the group's history in *Leonardo Music Journal* 11, "Not Necessarily English Music," is typically comprehensive, detailing performances at the first Glastonbury Festival, on a roundabout in Shiraz, Iran, and at the Royal Court Theatre with Marianne Faithful. Their only LP, released in 1974 and featuring performances of pieces by Earle Brown, John Cage, and Christian Wolff, is virtually impossible to buy now, but still sounds remarkably contemporary. The group can also be heard on the 1976 release of Stockhausen's *Sternklang*, and on the double-CD I compiled for *Leonardo Music Journal*.

Another major step came in 1968 with an invitation from Evan Parker and Derek Bailey to join their trio with drummer Jamie Muir. This became Music Improvisation Company, later supplemented by vocalist Christine Jeffrey. Given his attention to precise detail, Hugh's playing was surprisingly rough and visceral, even combative, and the group's two LPs on ECM and Incus are important documents of this formative period of improvised music history. In notes included in the Incus CD release, Richard Leigh recalled many evenings at London's Little Theatre Club, listening to the group. "I remember it as a democratic music with no stars," he wrote, "no impassioned soloist turning the others into a backing group – a music based on respect and sensitivity."

By this time Hugh was playing invented instruments such as amplified springs and shozygs, defined by him in the little book I edited in 1974, *New/Rediscovered Musical Instruments*, as "any instrument (usually amplified) built inside an everyday container, such as book-covers, breadbins, accordion files, radio and TV sets, card tables." His shozyg instruments of that time included a 3-D photograph, sounded by amplifying fingernails running across the grooves at different speeds, amplified springs of various lengths, some of them pitch adjusted by means of key rings, a plucked or blown egg slicer, and a cardboard accordion file interleaved with doll squeakers of varying pitch. A solo album, *Shozyg: Music for Invented Instruments*, released by FMP in 1982, is an excellent way to discover the sounds of these instruments in detail, and to hear the way Hugh developed improvisations from their possibilities and limitations.

He even argued successfully for the institution of shozyg as an instrument category (one exponent) in the Musicians' Union Handbook, and through this entry was invited by the 1980s pop band, Talk Talk, to play on their *Spirit of Eden* LP in 1987. In collaboration with John Furnival, a lecturer at Bath Academy of Art, Hugh also made Feelie Boxes, which were version of shozygs with an extra tactile element. In his article, *Hugh Davies: Instrument*

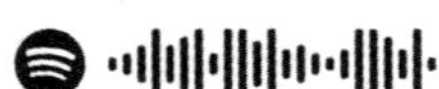
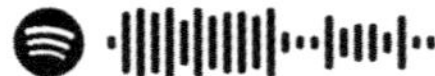

Maker (*Contact* No. 17, Summer 1977), David Roberts described Feelie Boxes as follows: "These are designed for exhibitions, though ideally they should be installed at bus stops, railway stations, dentists' waiting rooms, hospitals, etc.: anywhere that people have to wait with nothing else to do. Their general principle is to have a number of objects built into a box and amplified; these are explored with hands and fingers through holes in the sides. The Jack and Jill Box is for two people, four hands: "fur is thoughtfully provided in case the two people wish to hold hands in the middle."

I can't remember exactly how I met Hugh, but it must have been at the beginning of the 1970s. We sat together in a restaurant, maybe in Birmingham, and I was tremendously impressed by the sophisticated ease with which he ordered a vegetarian meal (meat-free dishes were rare on menus in those days). At the end of the meal, I was even more impressed when he noted down all his expenses in a small notebook. Recently, Jim Sauter of Borbetomagus (another of the many groups Hugh could list as collaborative projects) fondly related a similar story from one of Hugh's visits to the US. "I took him on a 'spending spree' to a local orchard to buy a small container of maple syrup to bring back to England," Sauter wrote to me via email after learning of Hugh's death. "He recorded the purchase."

Such meticulous documentation of minutia, along with the broader themes of life and art, was typical. A stickler for detail and factual veracity, Hugh became an authority on many hitherto neglected subjects relating to twentieth-century music: Futurism and the art of noises; musical instrument invention and building; the Theremin; the work of electronics pioneer Daphne Oram; sound sampling history; electronic music of all kinds, notably its early, murky history; sound art; hardware hacking; and environmental sound. Many who listened to him talk on these subjects would experience a sense of awe at the depth of scholarship and accuracy of information, and his entries to *New Grove Dictionary of Musical Instruments* (1984) numbered over 300. There are difficult aspects to being filled with such erudition, and Hugh's mind sometimes seemed to be a massive filing system running out of control. On occasion, his lectures had a tendency to collapse into disconnected and indigestible facts as the need to link everything to everything else overwhelmed the clarity of good communication.

But Hugh's work was a riposte to poor scholarship and lazy history. Many composers regarded him with great respect. After the publication of

New/Rediscovered Musical Instruments he sent a copy to Toru Takemitsu. I was immensely touched when a message came back from Japan, addressed with great warmth of feeling to Hugh. His dedication to the work of Daphne Oram was particularly impressive. From 1958 Oram had been a member of the BBC Radiophonic Workshop, but left to pursue her own compositions and technological inventions. Given the scarcity of women in the largely masculine, technocratic domain of electronic music, along with the scarcity of her recorded output, she seemed condemned to obscurity. After two strokes in the mid-1990s, she became unable to work, and so Hugh became the manager of her archive. Following her death in 2003, Hugh wrote obituaries, and planned a retrospective CD of her music. One piece – *Four Aspects* (1960) – was included on the *Leonardo Music Journal* CD, and later included on a Sub Rosa compilation, but sadly, Hugh's desire to establish her as a British equivalent of Raymond Scott was sabotaged by his own premature death.

Hugh and I played together in a variety of situations in the 1970s, either in duos or larger groups. I particularly remember playing in duo at the Unity Theatre in London, on a Musicians Co-op evening that also featured the Three Pullovers trio of Terry Day, Steve Beresford, and Nigel Coombes, and at Riverside Studios in 1978, during an Artist Placement Group week of events. Bill Furlong recorded the latter concert and released it on his Audio Arts cassette label. For the notes accompanying that release, I wrote: "The fact that this duo has not worked together often combined with the differences in approach and types of sound-producing technology highlights a quality of so-called free improvisation. The structure of the music derives from an immediate listening interaction rather than from externals such as composition, choice of instruments, cosmic schemes, preferred mode of dress and so on."

Yet with our shared interest in ethnographic recordings of global musics, lo-fi electronics and invented, self-made instruments, similarities were greater than differences, though I felt that I was always learning from Hugh's greater experience. Because of his association with Parker and Bailey he was considered to be a first generation UK improvisor, but he responded with great openness and generosity to second generation players such as myself, Paul Burwell, John Russell, Roger Turner and Max Eastley. His solo release from 1997, *Interplay*, demonstrated his long-term commitment to these musical relationships by featuring duos with Russell, Turner, and Eastley.

In 1975 I invited him to play amplified grill harp on one of my tracks on *New and Rediscovered Musical Instruments*, recorded for Brian Eno's Obscure label. Hugh earthed himself to prevent hum by connecting a wire to his ankle. It was a peculiar sight, even for those times, but he antic-pated by decades the common practice of computer technicians who earth themselves with Velcro wristbands. This is just one of the many anecdotes about Hugh that can be told with affection by musicians. He could be funny, erudite, gentle, diplomatic, terrifyingly bad-tempered, open-minded, pedantic, alarmingly obsessive and unfailingly courteous. Every one of our encounters and conversations, from first to almost last, ended with his sign-off: "Thanks very much." Inevitably, there's too much to say about a person whose character, career, and collaborative projects all added up to something complicated, resolutely unconventional, indif-ferent to the negative career impact of mixing your genres, strange, yet very human.

Whether as a member of Artist Placement Group, working with Ian Breakwell and Bill Furlong on a reminiscence aid for the elderly, or giving workshops to children, often with his close friend and colleague Max Eastley and once with Don Cherry, Hugh was dedicated to commun-cating beyond the avant-garde audience. This was a utopian mission and he never lost sight of it. In a symposium that I organised in 1978, during the Music/Context Festival at the London Musicians Collective, Hugh talked about his experience of exhibiting environmental sound objects. "People come and have a lot of fun," he said, "and they don't ask, is this music or not ... Perhaps we have to do more things like that. More educational, in a sense, a subtle way of educating people to listen in a different way."

Sadly, despite an exemplary lifestyle of healthy eating and modera-tion, Hugh was diagnosed with cancer in the latter half of 2004. His decline was distressingly rapid, and he died in a north London hospice on January 1st, 2005. One of his priorities at the onset of his illness was the comple-tion of a solo CD, *Tapestries*. At an Artist Placement group meeting in the summer of 2004, we discussed the writing of sleeve notes and agreed to meet for a conversation. This never took place. Hugh became too ill to fulfil his many commitments, even before he realised what was happening; the

next time we met was when Max Eastley and I visited him in hospital. Even during a bedside chat in the hospital ward, Hugh engaged in animated conversation, telling us about the lack of consideration given to sound in this environment. A rubbish bin, for example, squeaked when opened. Suffering from painfully heightened senses due to his illness, Hugh was woken frequently by this noise in the night. His good humour was still in place at that time, and the irony of finding a squeaking bin lid too painful to endure was not lost on him.

Although strongly influenced in the 1960s by techniques pioneered by Stockhausen, Cage, David Tudor and Gordon Mumma, Davies stood out as an idiosyncratic inventor of singular originality. His unique vision of an accessible, humorous approach to live electronics – a way of making music too often hidden behind the technocratic alienation of mysterious processes, expensive standardised equipment and an atmosphere of remote science – has threatened to marginalise him from the electro-acoustic mainstream. The music heard on *Tapestries* has an informal feel, drawing on sampling techniques and concrete sounds derived from the same humble sources he used in instrument making and live performance: toys, stones, a plastic bread bin (one of those squeaky lids), or actions integral to the transformation of a tree from living organism to art work. The mixture of both hi-tech and lo-tech always leads him to an evocation of the natural world. Hugh's approach to electronic music was non-doctrinaire within a milieu that thrived on factionalism. Again, this did his reputation few favours.

Yet over the 40 years in which he has been active, his influence on younger generations has grown noticeably. In London, the Bohman Brothers, for example, have continued his explorations into found objects, using home-made string instruments, spoken narratives and the amplified detritus of consumer society to journey further into the subsurface of a world in which matter is a web of dynamic energy patterns rather than a comfortingly solid, static, three dimensional thing occupying only physical space.

In December 2004 I shared a bill in Athens with Lee Patterson, at that time an emerging sound artist from Manchester. As I watched Lee burning amplified peanuts and sparklers, or gingerly placing upturned wet bottles on amplified metal plates, I thought about Hugh. By that time, I knew he was terminally ill, and it brought a lump to my throat. After Lee's performance, I asked him if he knew about this man Davies, the doyen of circuit bending, hardware hacking, and contact microphones. Of course he did. He had sat behind him at a lecture at Middlesex University, where Hugh was researcher in sonic arts, but had been too overawed to speak to him. Now it was too late. There was a time, that brief period when the future looked shiny and bright, when Hugh's work was discounted as being too Heath Robinson, too shabby, and way out of touch with a world in which music would be created by the wave of a hand in the direction of a transparent screen. But just like hip-hop and its turntables, there is a new generation choosing a different sort of future, perhaps wired up to amplified egg slicers, playing power drills, recording the sound of bluebottles dancing on contact microphones, or hacking at the innards of a talking toy. How cruel that Hugh won't be here to give them the benefit of his knowledge and experience.

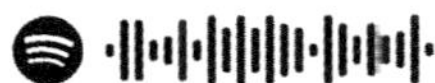

Filament Drums: the endless instrument, for Tomás Saraceno

Published in *Tomás Saraceno: Our Planetary Bodies*, Asia Cultural Centre, Korea, 2017

A door opens; loud sound (anthropomorphised) enters to occupy a bounded space. The human notes this sound, assumes an effect of the ears, reacts quickly to unwelcome audibility. The door closes and sound is pushed back yet still enters through other means, by vibration of walls, ceiling, floor. This sound is transformed by the solid mass of a visible medium, filtered of its higher frequency components, now perceptible as a different kind of intruder, less aggressive perhaps yet more insistently pervasive as a body effect with greater psychological depth, an undermining of the fallacy that the senses are distinct from each other, their sources only processed through the organs of reception, the ears, nose, mouth, eyes and fingers.

There is some familiarity with the fact that non-human species communicate and engage within their worlds by highly specialised means such as electrical impulses, colour and light, chemical systems, scent, ultrasonic and subsonic sounding and hearing, echolocation and, in the case of spiders, vibration. The so-called courtship dance of a spider is described as follows in *Animal Communication*.[1] "This usually involves specific movements of the legs, palpi, and body. In some lycosids, special hairs or colored areas on the legs are erected. In some salticids, the color of the eyes changes." All of these complex effects combine with vibratory signals generated by percussive, stridulatory and tremulatory actions.[2] The resultant displays, as can be seen on amateur video examples easily found on Youtube may seem strange and disturbing reminders of what biophilosopher Jakob von Uexküll defined as the *Umwelt*, the environment-world or unknowable world in which each animal lives.

Perhaps they are beyond description for human language, yet the unknowability of worlds interlocks, as Giorgio Agamben demonstrates in his description of the relationship between spider and fly: "The two perceptual worlds of the fly and the spider are absolutely uncommunicating, and yet so perfectly in tune that we might say that the original score of the fly, which we can also call its original image or archetype, acts on that of the spider in such a way that the web the spider weaves can be described as 'fly-like.' Though the spider can in no way see the Umwelt of the fly (Uexküll affirms – and thus formulates a principle that would have some success – that 'no animal can enter into relation with an object as such,' but only with its carriers of significance), the web expresses the paradoxical coincidence of this reciprocal blindness."[3]

If there is a gulf, perceived or otherwise, between human and non-human animals, how might it be crossed? In 1971 I came up with the idea  of Bi(s)onics, a means of working with sound inspired by natural environments and the animal world, particularly the phenomenon of bioacoustics. Bionics, the science of systems based on living things, was a talking point during that period, an embodiment of an imagined future in which humans would further extend their understanding and application of animal capacities (radar and sonar, for example) into realms of the superhuman. Bionics was not just futurology, however, since humans had been learning from non-human entities and biological processes from the beginning. Evidence of intimate connections between human and animal can be seen in rock art, more specifically in music through the Hohle Fels flute, made from the wing bone of a vulture and dated back to 35,000 years ago and the Divje Babe flute, made from a cave bear femur, more controversially dated back to c. 43,100 years ago.

This has been a rich, continuous history. Working in Nigeria in 1930, Professor Henry Balfour, first curator of the Pitt Rivers Museum in Oxford, became fascinated by a group of instruments used to disguise the voice for ritual purposes. Many of these simple instruments were speaking or singing tubes, but the attached material which added a buzzing timbre, the otherworldly quality that reinforced the impression that masked dancers

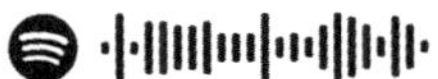

or hidden singers were voicing ancestral spirits, was often taken from spiders' egg sacs or webs. Balfour wrote, for example, of the Katab male secret society cult of the *Obwai*, in northern Nigeria's Zaria Province: "The *Obwai* is not seen but his shrill voice is heard and I gather the quality of his voice is due to the vibratory interference of a membranophone ... During the festival, the *Obwai*, who is apparently concealed in the roof, speaks to the assembly in a voice disguised by the use of a filament of spider's web."[4]

In ancient China the various methods of touching and plucking the silk strings of the Ch'in, the classical seven string lute or psaltery, were informed both by listening to the sounds of animals and through observation of their movements. Various touches of the left hand – pressure, movement and vibrato – called for mimesis of cicadas or the cry of a dove announcing rain; others were designed to evoke subtle natural phenomena, such as rain on bamboo, fallen blossoms floating down with a stream, floating clouds, a swimming fish moving its tail or the dim resonance of water heard in a mountain gorge. These touches were described in handbooks both with directive explanations and symbolic illustration. In Ming dynasty handbooks, vibrato was illustrated by a drawing of a cicada creeping up a tree, the plucking of one string by two fingers of the right hand by drawings of a wild goose carrying a reed stalk in its bill.

In his book, *The Lore of the Chinese Lute*, Robert Hans van Gulik gives the example of a rapid movement on one string, described for the student as "a purple crab walking sideways" (as van Gulik writes, "One should think of the rapid movement of the legs of small crabs when they scurry over the sand.").[5] The most refined of all these techniques established a threshold of perception accessible only to the most attuned scholar: "Remarkable is the *ting-yin* – the vacillating movement of the finger should be so subtle as to be hardly noticeable. Some handbooks say that one should not move the finger at all, but let the timbre be influenced by the pulsation of the blood in the fingertip, pressing the string down on the board a little more fully and heavily than usual."[6]

This image of stillness at the far reaches of abstinence exemplifies a Taoist ideal state, translated from the *Laotzi* by François Jullien as:

The great square has no corners
... the great tone makes only a tiny sound
the great image has no form[7]

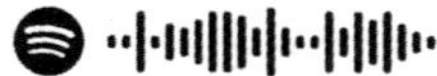

Action of inaction, the inner pulse of the body becoming the music, leads towards an intense listening practice, a state described by Jullien as "energetic capacity gathering itself up ... Sight aggressively projects attention outward whereas listening *gathers it up* within."[8] Another image is also irresistible, the fingertip resting on silk strings, slightest of vibrations registering and producing sounds so fugitive that only the most sensitive spirits are attuned to their presence.

This is the spider, who seems to float in empty space, in contemplation and patience, waiting for vibration to signal a gathering in. Webs and nets are very powerful ancient symbols in human culture, strangely contradictory metaphors for entrapment, catching (both the fishing net and the safety net), spatial extendedness and complex interconnection. Of entrapment, there is the story related by Plutarch, who wrote of an ingenious military device employed by Brutus during his siege of Xanthus: nets laid deep under the river that ran past the city. When Xanthians tried to swim to freedom they were entangled, their presence betrayed by the sound of small bells attached to the uppermost nets. And for extendedness and interconnectedness we have the image of ourselves in our present-future, tapping on keyboards, tablets and smartphones, listening to vibration, developing tactile skills and gathering up from invisible lines of information within the online environment we describe variously as the (inter)net or (world wide) web.

Spiders, we now understand, have given us a model of which the present is a simulacrum, though not just the technocratic, seemingly intangible future-present of life online but also the real-world urgency of environmental relationships and their fragility. Jakob von Uexküll's pioneering work in biology was popularised by another pioneer, zoosemiotician Thomas Sebeok. As Dorion Sagan writes in his introduction to Uexküll's *A Foray Into the Worlds of Animals and Humans*, Sebeok spoke of Uexküll's conception as "a 'semiotic web' – our understanding of our world being not just instinctive, or made up, but an intriguing mix, a spiderlike web partially of our own social and personal construction, whose strands, like those of a spider, while they may be invisible, can have real-world effects."[9]

For the spider, the little drummer, its web is an evolving instrument without distinct boundaries, a near-invisible extension of its own body, infinite in the interconnectedness of its architecture, an endless yet temporary instrument whose purpose is not so much a percussive sounding-out drum as its reverse – an ear-drum receptor for listening-in, a gathering in of impulses or signals that we think of as sounds, even though many of them are inaudible to the unaided human ear. In the greatly expanded world of environmental sound recording, electronic music and sounding arts perhaps we can accept the spider as its most rarefied practitioner?

Notes

1 Hubert and Mable Frings, "Other Invertebrates," published in Thomas A. Sebeok (ed.), *Animal Communication*, Indiana University Press, 1968, p. 258.

2 S. Sivalinghem and A. C. Mason, *Sensory Communication In a Black Widow Spider (Araneae: Theridiidae): From signals production to reception*, XIV International Conference On Invertebrate Sound and Vibration, University of Strathclyde, UK, 2013, p. 54.

3 Giorgio Agamben, *The Open: Man and Animal*, trans. by Kevin Attell, Stanford University Press, 2004, p. 42.

4 Henry Balfour, "Ritual and Secular Uses of Vibrating Membranes as Voice Disguisers," Royal Anthropological Institute Journal, 1948, p. 51.

5 Robert Hans van Gulik, *The Lore of the Chinese Lute*, Sophia University, Tokyo, 1968, p. 129.

6 ibid, p. 132.

7 François Jullien, *The Great Image Has No Form, or On the Nonobject through Painting*, trans. by Jame Marie Todd, University of Chicago Press, 2009, p. 47.

8 Ibid, p. 171.

9 Dorion Sagan, introduction to Jakob Uexküll, *A Foray Into the Worlds of Animals and Humans*, University of Minnesota Press, 2010, p. 4.

Tania Chen plays John Cage and David Tudor – Electronic Music for Piano

Published as liner note for Tania Chen with Thurston Moore, David Toop and Jon Leidecker: *John Cage: Electronic Music for Piano*, Omnivore CD, 2018

Vague scores – what's the point? For improvising musicians the thought is not uncommon. A vague score maps out some sort of terrain, absolves the composer from any great responsibility yet retains the hierarchical status quo of a particular world. John Cage's *Electronic Music For Piano* is a case in point, or maybe not. Dated September 2, 1964, it was handwritten on notepaper from the Hotel Malmen in Stockholm, then performed by David Tudor at Flykingen, Sweden's experimental music centre, on the 10th September.

The score, such as it is, is really a score that burrows inside, encircles and expands a score or attaches to it, limpet fashion, since it explicitly refers to an earlier piece – *Music For Piano 4–84*. Based on expediency and contingency, *Music For Piano* was a series of eight scores, eighty-four short compositions, written between 1952–56. At the beginning Cage was looking for a way to speed up the process of composing by using indeterminate methods, particularly since dancers such as Jo Anne Melcher and Louise Lippold were asking for new pieces and using the *I Ching* took more time than their performance imperatives allowed.

A detailed description of his method, first written for *Die Reihe* No. 3 in 1957, was included in *Silence: lectures & writings*. If you're not a pianist or Cage scholar then these details get arguably less interesting as time passes but the approach that he initially followed came from the physical characteristics of a basic material. "I looked at my paper," Cage told Daniel

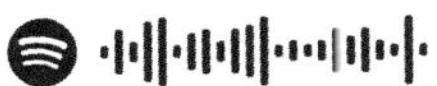 

Charles, "Suddenly I saw that the music, all the music, was already there." The so-called imperfections of paper, in other words those micro-variants in texture, colour and hidden pattern that make paper so sensual, provided existing marks of transformative potential. Draw a stave on a sheet of paper and pitches would automatically appear within them. As David Revill noted in *The Roaring Silence*: "It had symbolic as well as practical value; it made the unwanted features of the paper its most significant ones – there is not even visual silence."

As Tania Chen says, it's an artist's approach, more concerned with the process and materials than any teleological purpose. A score could become another score because the sense of music as a final object, a vase to be perused by the audience, never really exists. Perhaps there was further expediency in the composition of *Electronic Music for Piano*, with its hurried instructions for "feedback and changing sounds (microphones, amplifiers. loudspeakers – separate system for each piano)" and then, more cryptic, "without measurement of time," "Consideration of imperfections in the silence in which the music is played" (not unlike the paper), a corrected "ossciloscope" and the single word "Friction."

The close and long-lasting working relationship between Cage and David Tudor was certainly fruitful but may also have entailed some degree of productive friction. Current thinking leans toward the idea that Cage blithely and with a big smile took a lot of credit for Tudor's input and Tudor happily acquiesced. This may have been an unwitting consequence of the way in which exploitative mechanisms in many social roles and professional/personal identities were made to seem "natural" and fixed at that time; it may also have suited the dynamics of their relationship. "[Tudor] didn't find what he was writing interesting," Cage told Joan Retallack in 1992. "Later I think he ... when he left the piano and became involved with electronics ... then he began to think of himself as a composer. Not immediately, but some years thereafter. And he does that now, so that he's not always a performer. He is himself a composer. (pause) But how he composes is unknown, because he loves keeping secrets. He doesn't want people to know what he's doing. He said once – even as a performer – I want to have an instrument that no one else knows how to play."

At this point in time an unspoken strictness in these self-imposed roles feels rather odd, so much so that the difficult question of who was

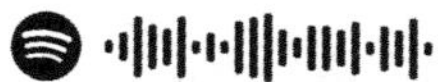

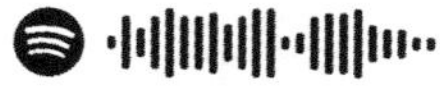

responsible for what in the Cage/Tudor relationship can be understood as a significant turning point in music history. In the 21st century we are free to be more fluid so take it for granted yet it was hard won. Equally, in many of these works there is the suggestion of crisis in relation to notation. After all, Cage was only jotting down on a sheet of hotel notepaper the broadest outlines of what Tudor already knew how to do, which was to make an instrument – those forests of cables – that nobody else (Cage included) knew how to play. In Yasuhiro Yoshioka's famous photograph of Cage and Tudor in Japan, Cage's head is inside the temple bell but Tudor is about to strike the bell with its wooden clapper.

While visiting Japan with Tudor in 1962 Cage composed *0'00,"* another piece handwritten on a single sheet of letter paper. This was dedicated to Yoko Ono and Toshi Ichiyanagi, its instrumentation described as: solo to be performed in any way by anyone. The performance itself was the score, or vice versa, its main instruction being: In a situation provided with maximum amplification (no feedback), perform a disciplined action. In among all these ambiguities there is the question of why such actions needed scores?

Three days after the Swedish performance of *Electronic Music for Piano* in 1964, Tudor devised a piece called *Fluorescent Sound* for Robert Rauschenberg's *Elgin Tie*, using contact microphones to transform the fluorescent lighting of Stockholm's Moderna Museet into what Nina Sundell described as a "bell sounding instrument." Tudor composed a timing score for the seventy-five switches activating the lighting system, in part because Rauschenberg's entrance with a Brahma cow required a gap in the music. An interview with Tudor (by Teddy Hultberg in 1988) has him self-deprecatingly laughing about this: "That was my first composition."

With hindsight it's easy to see how timings (what in other contexts, like theatre or television would be called cues) were among the last traces of compositional intent to vanish. Tudor's initial response to the museum's fluorescent lights coming on – "... the most beautiful music" – was not only first thought-best thought (no composition necessary because as a phenomenon it's spectacular) but also a solo to be performed in any way by anyone (until the cow enters). As *0'00"* makes explicit, open compositions on the edge of being not compositions at all have the purpose of creating focus, whether Fluxus type scores which indicate a distinct field of possibilities and its restrictions – from George Brecht, "turn on a radio/ at the first sound turn it off" – or *Fluorescent Sound* which (long before Martin Creed) is all about the lights being switched on and off.

Cornelius Cardew came away from working as Karlheinz Stockhausen's assistant in the 1960s wondering why there was a need for all that fanatically detailed Darmstadt school notation when an improvisation could sound much the same. Such provocations customarily initiated savagely polarised debates about the pros and cons of improvisation versus composition – an adversarial impasse obscuring the emergence of an approach that was both and neither. Informed by all the complex entanglements of twentieth-century soundwork at all points on the spectrum of technologies, histories, geographies, popular/unpopular and human/non-human, this less partisan approach proposed a way of making-with-sound no longer dependent on the authority of the score yet not tied to any particular school or ideology of improvisation. You could describe it as a way of switching the lights off based on intuition and experience.

Performing the electronics part of *Electronic Music for Piano* is quite possible without being overly conscious of performing a score, though the piano part will always act as a reminder that somebody must. "But the magic of this piece is where the piano writing and the piano takes on a kind of invisibility," Tania writes, "and becomes a translucent vessel like a sound portal of all possibilities." What is tacit (and Cage with his silences and Tudor with his secrets both liked tacit) is the main point: as Tania says, "make the piano into a giant amplifier." So the work is about resonance on a number of levels. The piano frame, its strings and air are sounded by means that adhere to various 20th- and pre-twentieth-century conventions yet attempt to go beyond that into a realm of vibrant materiality. At the same time the piece is resonant with implications that arise out of the open work and its deep, convoluted history. Without John Cage these possibilities would almost certainly be stuck in a far less accessible place.

Trembling Between States: Rie Nakajima

Combined texts from sleeve note to Rie Nakajima's *Four Forms* LP, Consumer Waste 17, 2014, and catalogue essay for Rie Nakajima's *Cyclic* exhibition, Ikon Gallery, Birmingham, 2018

Where we listen; how we listen. These two elements form a sculptural reality in relation to the object of sounding. The object of sounding is indifferent. There is its place in a room but when listening begins there is no object, no room – only the sculptural reality created by the listener. But then there is presence – the sources of sound that have the lives of small creatures, maybe small creatures that hibernate in darkness but then come to life when exposed to the light. Within their limitations they have their own minds; in the same way that certain invertebrates are stimulated into making sound by signals such as airborne chemicals, vibration or sound made by their own species, these creatures of which I speak are activated to perform their own cycles of drumming or scraping, all working together as if moving inexorably toward the sudden miraculous synchronicity of flashing light that a few fortunate observers have seen in firefly displays. But then there is proximity, to be close and to see, or to be closer still, almost inside but seeing nothing, hearing the microaudial detail of what might be a fish drumming on its swim bladder or a spider spinning a web. If we were speaking about Japan then I might refer to the different cycles of a shishi-odoshi deer scarer in a garden, bamboo filling with water, tipping over to strike another bamboo with a reverberant dok, then returning to rest before filling again, or a pachinko parlour in which many thousands of tiny steel balls ricochet around their enclosures with utter indeterminate

indifference to the wins or losses of their operators. There is the feeling of dipping the head into a pond filled with dry water or of descending by bathosphere to the depths of the ocean to observe those strange animals that burrow just under the abyssal floor or traverse its soft surface on stilts. If we are thinking about speculative music then another sculptural reality can come into being whereby every combination of these creatures is an ensemble of musical actors whose place in the spectrum of life is not dissimilar to that of a sponge, a porous living creature that lacks most, if not all, of the attributes that we associate with organic life yet can live, some surmise, to more than five-thousand years of age. What I am saying, should it be unclear, is that this is a kind of intensely rhythmic music performed by an extra-human ensemble not susceptible to the orthodoxies of human culture, in a sense a step into another dimension.

Through repetition and familiarity we become habituated to an idea of objects as subjugated entities, disposables enfolded within the human domain, alongside animals, plants, even slaves. This is why the revolt of objects in films by the Marx Brothers and Jacques Tati is so compelling. Artists dwell on and with materials, their nature, temporality and energy, living as a material, and so they become the materials they work with. Uncommon attunement is how it might be described. So much of human life is governed now by a limited, tyrannical perception of time; in this case it's humans who are subjugated entities, even slaves. But time is one of those materials with which artists develop uncommon attunement, bending and stretching it, foraging far ahead or gathering memory, working outside its knell by marking according to small deaths, renewals, emptiness, those cycles of personal ritual that become shared.

Things I have learned from Rie Nakajima – it could be a long list. The slow, apparently soundless impact of a sponge against a wall contains an infinitude of learning. This is about sense and sensibility, an adjustment, waiting for nothing. But ... allowing, this is one of those things. To allow objects or materials or time to be themselves flips the coin of subjugation. Note that the coin falls on its edge, trembling between states. Silent questions arise, maybe, like, "Who is busy here? Artist or materials?" Rie often tells me she is lazy but this word – lazy – takes on a transformed meaning, not some slothful individual who wants to avoid responsibility but a person who allows things to happen. This is not passivity. It takes an iron will. Discipline, also, to set aside all that is unimportant. I think of what Byung-Chul Han says in *The Scent of Time* about modes of being – hesitancy, releasement, shyness, waiting, restraint – resting on an experience of duration. What lasts and is slow, he says, " ... evades being used up and consumed." Instead, it allows.

List of tracks

16) p. 45b "Plan 3 from Outer Space" – Andrea Centazzo, Tania Chen, Henry Kaiser
https://open.spotify.com/track/6La1v1FTA6x0D3VV5MpPTu

17) p. 48 "In the Face of Overwhelming Forces" – Joachim Koester
https://camdenartscentre.org/whats-on/view/koester

18) p. 50 "Sāmaveda – Rathamtara" – Ekamra Sastri, Sivarama Sastri
https://open.spotify.com/track/4mKWdNY0iBY9NCbBLPUcQu

19) p. 51 "Dhikr (Remembrance)" – Fayoum Oasis, Egypt
https://open.spotify.com/track/2duD6BIH8nY3Epf9zsNOok

20) p. 52 "Gion Shoja" – Akira Sakata
https://open.spotify.com/track/4D3ZiidSPYa4W6kYWJsMsa

21) p. 54 "Department of Abandoned Futures" – Joachim Koester and Stefan Pedersen
https://soundcloud.com/abandoned-futures/department-of-abandoned-futures

22) p. 59 "Lyst" – Camille Norment
https://open.spotify.com/track/2z1BtBZeuUqPxZ3M6J68vl

23) p. 67 "An Eavesdropper with a Woman Scolding" – Nicolaes Maes
https://artuk.org/discover/artworks/an-eavesdropper-with-a-woman-scolding-52160

24) p. 69 "Als Jakob erwachte (The Dream of Jacob)" – Krzysztof Penderecki
https://open.spotify.com/track/2u8YsEkeOVKaiRQ691LMLU

25) p. 70 "The Singing Sculpture" – Gilbert & George
https://bbc.co.uk/programmes/articles/22ywTtt7WhmgzJt5Z4N942h/
living-sculptures-at-home-with-gilbert-george

26) p. 74 "Underneath the Arches" – Flanagan & Allen
https://open.spotify.com/track/088cCeh4k3cGxElX0syBhK

27) p. 78 "Jesus' Blood Never Failed Me Yet" – Gavin Bryars
https://open.spotify.com/track/3LhUvATbveDYzSp9q86Cjf

28) p. 81 "Solid State Survivor" – Yellow Magic Orchestra
https://open.spotify.com/track/3rAluZ29n7zj10Ukg1E7Ic

29) p. 81b "Bend It" – Dave Dee, Dozy, Beaky, Mick & Tich
https://open.spotify.com/track/4LZIfS27F8B3CEc9BdaV7g

30) p. 83 "Seymour Writes Back: Disc A" – Seymour Wright
https://seymourwright.bandcamp.com/album/disc-a

31) p. 89 ""4" 33 RPM" – Haroon Mirza
https://soundcloud.com/icalondon/haroon-mirza-4-33-rpm

32) p. 90 "50 Locked Grooves" – Haroon Mirza
https://discogs.com/Haroon-Mirza-50-Locked-Grooves/release/8007705

33) p. 90b "Solid Vibration of the Surface of Concrete Ground" – Toshiya Tsunoda
https://intransitive.bandcamp.com/track/solid-vibration-of-the-surface-of-concrete-
ground-pts-1-2-maguchi-bay-miura-city-kanagawa-prefecture

34) p. 91 Fred Sandback
https://fredsandbackarchive.org/

35) p. 96 "Washing Machine" – Mr. Fingers
https://open.spotify.com/track/5RNOPiFhFVyTpoPmm0XNCa

36) p. 96b "Phone System" – Ricky Smith
https://open.spotify.com/track/5DcRLDoL98ILBpgD3brjxE

37) p. 99 "Cry" – Johnnie Ray
https://open.spotify.com/track/4wnFpRgQu2gLN8d76dYOGr

38) p. 103 "Piano Film" – Annabel Nicolson
https://luxonline.org.uk/artists/annabel_nicolson/piano_film.html

39) p. 105 "Whirled Music" – Max Eastley, Steve Beresford, Paul Burwell, David Toop
https://discogs.com/Max-Eastley-Steve-Beresford-Paul-Burwell-David-Toop-Whirled-Music/master/1313632

40) p. 108 "I Am Sitting in a Room" – Alvin Lucier
https://open.spotify.com/track/09pHNT4RIvhLui6eIi4HKO

41) p. 110 "In Memoriam Stuart Marshall" – Alvin Lucier
https://open.spotify.com/track/7E1ihmbF8zKrXR2ElMsImS

42) p. 111 "A Celestial Tune by 108 Glass Marbles" – Mieko Shiomi
https://ubusound.memoryoftheworld.org/fluxus_box/Fluxus-Anthology-30th_14-05_Mieko_Siomi.mp3

43) p. 111b "Requiem for George Maciunas" – Mieko Shiomi
https://discogs.com/Mieko-Shiomi-Requiem-For-George-Maciunas/release/1037361

44) p. 112 "Song from the Fast" – Rolf Julius
https://open.spotify.com/track/0C8KrJkTdbQuCcvAG4ETOJ

45) p. 124 "John Henry" – John Jacob Niles
https://open.spotify.com/track/4IXx63swh2lWJlYwWIpp6M

46) p. 125 "Blue Blazes Blues" – Emery Glen
https://open.spotify.com/track/4iDg5bhjDchKVSPE0QCLK5

47) p. 125b "Xango" – Roland Hayes
https://open.spotify.com/track/5lM8id19S4xW8zlGkGX4IZ

48) p. 128 "Night Flights" – Christina Kubisch
https://open.spotify.com/track/4d5EhKmJ9iYA053j0qO11D

49) p. 132 "Rayon Hula" – Mike Cooper
https://open.spotify.com/track/15YB90qedaidaBBtw1eZ9Q

50) p. 132b "Mele Manu – Ho'okani Pila" – Mike Cooper
https://open.spotify.com/track/6qjKF3O74l1bYygN73nw2A

51) p. 136 "Analapos" – Akio-Suzuki
https://discogs.com/Akio-Suzuki-Analapos/release/1672148

52) p. 138 Lav Diaz
https://en.m.wikipedia.org/wiki/Lav_Diaz

53) p. 140 "Memory Maps" – Loré Lixenberg
https://lorelixenberg.wordpress.com/discography/

54) p. 140b "Star-shaped Biscuit: Scene 3 – Invocations" – David Toop
https://soundcloud.com/david-toop/star-shaped-biscuit-scene-3

55) p. 142 "Koku-Reibo (Bell Ringing in the Empty Sky)" – Goro Yamaguchi
https://open.spotify.com/track/4fUHZZVHasty2MECsq7hp2

56) p. 144 "Oni" – O YAMA O
https://open.spotify.com/track/39p1n4WtWFULwSkZGzzsqG

57) p. 151 "Neon" – Hans Berg
https://open.spotify.com/track/7MTv9qFqGrDQ6eRfnx5CBG

58) p. 158 "Anna" – Bob Cobbing
https://open.spotify.com/track/0l8yz1BYznEUiOsLVf90oU

59) p. 160 "Rayday" – Jeff Keen
https://open.spotify.com/track/5y3rn7I5Xeouc64VsDKIxd

60) p. 164 "Alphabet of Fishes" – Bob Cobbing
https://open.spotify.com/track/0oTyTQR0QaNh82rIlZoiqQ

61) p. 165 "The Congo" – Vachel Lindsay
https://open.spotify.com/track/1vxHbwxojagOERmbwkq3nO

62) p. 169 "When Faith Moves Mountains" – Francis Alÿs
http://ubu.com/film/alys_faith.html

63) p. 171 "PoorManMusic" – Philip Corner
https://open.spotify.com/track/4v3Mytww4gdU9kAD6Dvu54

64) p. 177 "Martyrdom of Saint Matthew" – Caravaggio
https://caravaggio.org/martyrdom-of-saint-matthew.jsp

65) p. 180 "Idiophonics" – Stuart Marshall
https://luxonline.org.uk/artists/stuart_marshall/idiophonics.html

66) p. 184 "Navajo Film Themselves"
https://penn.museum/sites/navajofilmthemselves/

67) p. 189 "Into the Tranquility" – Ryoji Ikeda
https://open.spotify.com/track/2ahb5VetdGAQ1EfQpc9ZtD

68) p. 191 "Do the Bathosphere" – David Toop
https://open.spotify.com/track/3uRlSiu9T7TOaZBc2K32QZ

69) p. 196 "Projection Esemplastic for White Noise" – Joji Yuasa
https://open.spotify.com/track/5Il3zM3TLQzXw29vgUqCLC

70) p. 199 Ad Reinhardt
https://moma.org/artists/4856

71) p. 203 "We the People Who Are Darker Than Blue" – Curtis Mayfield
https://open.spotify.com/track/19EcD3vKOGLi5T7Xi57Ik7

72) p. 207 "Dansa del Maestrat" – Joan Blasco
https://open.spotify.com/track/000DT03as8Pao3Bi7UyKpl

73) p. 208 "8 petites pièces romantiques: VIII. Valse romantique" – Déodat de Séverac
https://open.spotify.com/track/0tfdYHGRkbdFH4USs9NXx5

74) p. 209 Shirazeh Houshiary
https://lissongallery.com/artists/shirazeh-houshiary

75) p. 213 "Why Patterns?" – Morton Feldman
https://open.spotify.com/track/Ca7kqJ5P1utfJigUvyyIQ5

76) p. 216 John Latham
http://flattimeho.org.uk/

77) p. 218 "Roaratorio: An Irish Circus on Finnegans Wake, Pt. 1" – John Cage
https://open.spotify.com/track/1xhpjkSqgTl2YzpJvQ8Fxd

78) p. 219 "Yeibichei (Night Chant)" – Navajo
https://open.spotify.com/track/3YhagEcOIeoSvJc1ejxRHo

79) p. 220 "Prelude for Meditation" – John Cage
https://open.spotify.com/track/2BsmLPW1DYhKXJXuE47D4X

80) p. 228 "Fold" – Lucie Stepankova
https://luciestepankova.com/fold

81) p. 229 "Rosie" – C. B. Cook & unidentified prisoners
https://open.spotify.com/track/6LH94JKGmGqakhshtsGnNz

82) p. 233 "Cross Hands Boogie" – Winifred Atwell
https://open.spotify.com/track/0JSsSANk93G71t7FIYtviY

83) p. 233b "Boogie Woogie"
https://en.m.wikipedia.org/wiki/Boogie_Woogie_(album)

84) p. 236 "The Clock" – Christian Marclay
https://tate.org.uk/whats-on/tate-modern/exhibition/christian-marclay-clock

85) p. 236b "Moving Parts" – Christian Marclay, Otomo Yoshihide
https://discogs.com/Christian-Marclay-Otomo-Yoshihide-Moving-Parts/release/262857

86) p. 237 "Guitar Drag" – Christian Marclay
https://discogs.com/Christian-Marclay-Guitar-Drag/release/684902

87) p. 238 "Sidewalk and Shopfront, New Orleans" – Walker Evans
https://artic.edu/artworks/118143/sidewalk-and-shopfront-new-orleans

88) p. 238b "Tea for Two" – Thelonious Monk
https://open.spotify.com/track/3SSDhknexqI9YrrLj3sof6

89) p. 240 "Sensyo" – John Zorn's Cobra
https://open.spotify.com/track/6EzpiFJeUrOnqnQJRi2ETV

90) p. 241 "Archery" – John Zorn
https://discogs.com/John-Zorn-Archery/release/644736

91) p. 246 "The Etchings" – Robert Ashley
https://open.spotify.com/track/13A8FhDgi0sIIHOUWbdeBM

92) p. 246b "Vespers" – Alvin Lucier
https://open.spotify.com/track/1woKbhunp3nUUNbtBuzMhF

93) p. 247 "Funny How Time Slips Away" – Joe Hinton
https://open.spotify.com/track/5lZ3sezVWHTpmdVbHyMDBT

94) p. 248 "Stones – Sonic Boat Journey: Hailuoto-Pyhäjoki" – Ryoko Akama
https://open.spotify.com/track/4sK2EwLiwxyNf0MVvv4mQJ

95) p. 248b "Automatic Writing" – Robert Ashley
https://discogs.com/Robert-Ashley-Automatic-Writing/master/67752

96) p. 254 "Balfour Papers" – Henry Balfour
https://prm.ox.ac.uk/balfour-papers

97) p. 255 "Pression" – Helmut Lachenmann
https://open.spotify.com/track/3k4jn3PbCeLnnYXavucP4d

98) p. 256 "Ride On Josephine" – Bo Diddley
https://open.spotify.com/track/2peFiSzlqcTZ32ZiBhipL0

99) p. 256b "Dōtaku: Ancient Japanese Bronze Bells from Yayoi Period"
https://discogs.com/土取利行-銅鐸黄金の鼓動-Dōtaku-Ancient-Japanese-Bronze-
Bells-From-Yayoi-Period/release/1822417

100) p. 258 "Quintet" – Hugh Davies
https://open.spotify.com/track/79aI7E81r5m9ByLsQ4Za0I

101) p. 259 "Gesang der Jünlinge" – Karlheinz Stockhausen
https://open.spotify.com/track/5pqgzs5NQ8f1CX5NO45R0V

102) p. 260 "London, 1969" – The Scratch Orchestra
https://discogs.com/The-Scratch-Orchestra-London-1969/master/1425710

103) p. 262 "Dragon Path" – Derek Bailey, Evan Parker, Hugh Davies, Jamie Muir
https://open.spotify.com/track/63EkOXdMHjXVauuRw9bdSs

104) p. 262b "Shozyg Music for Invented Instruments" – Hugh Davies
https://discogs.com/Hugh-Davies-Shozyg-Music-For-Invented-Instruments/master/
865792

105) p. 262c "Eden" – Talk Talk
https://open.spotify.com/track/5kSAwHSqqr05MqWJ1o1KaG

106) p. 263 "Work on What Has Been Spoiled" – Borbetomagus
https://discogs.com/Borbetomagus-Work-On-What-Has-Been-Spoiled/release/
462769

107) p. 264 "Melodic Group Shapes iii" – Daphne Oram
https://open.spotify.com/track/1CQa0vRTpL47Fc179A82OX

108) p. 264b Three Pullovers
https://discogs.com/Three-Pullovers-Three-Pullovers/release/3741040

109) p. 264c "Interplay" – Hugh Davies
https://discogs.com/Hugh-Davies-Interplay/release/1043252

110) p. 265 "The Divination of the Bowhead Whale – David Toop
https://open.spotify.com/track/70y3mjUV3RrhV60l4NY3Dv

111) p. 267 "Móine Mhór Wires" – Lee Patterson
https://open.spotify.com/track/2cL4BfBHeCW7k3wkxwsvW6

112) p. 268 Tomás Saraceno
https://studiotomassaraceno.org

113) p. 269 "Bi(s)onics: the science of (sound) systems based on living things" – David Toop
https://tape-mag.com/David_Toop___Bisonics_the_Science_of_sound_
systems_based_on_living_things+Mirliton_Publications_Quartz_Mirliton_
Publications+PRESS_und_ORG_LITERATURE-1-1-2560-7.html

114) p. 269b "Improvisation (Hohle Fels 1)" – Anna Friederike Potengowski
https://open.spotify.com/track/69OBzYEUb9evvDqy5G6jEf

115) p. 270 "Moon Over Guan-Shan Mountain" – Hong Ting
https://open.spotify.com/track/7wShJWtHf4XQZoFAalPVQl

116) p. 273 "John Cage: Electronic Music For Piano" – John Cage, Tania Chen,
Thurston Moore
https://open.spotify.com/track/3RIH14dIptxc4Fw35r0pSU

117) p. 273b "Music for Piano 1" – John Cage
https://open.spotify.com/track/767PW0Ebh2xbojMGTbm5Yo

118) p. 274 "Music of Changes" – John Cage
https://open.spotify.com/track/4JrA9gY2LWOoexRGEZbIFX

119) p. 274b "Indeterminacy 1" – John Cage and David Tudor
https://open.spotify.com/track/491M3LNUj1y6xq0YNFZOJ8

120) p. 275 "Ryoanji" – John Cage
http://johncagetrust.blogspot.com/2013/07/john-cage-ryoanji-catalog-raisonne.html

121) p. 275b "Fluorescent Sound" – David Tudor
https://davidtudor.org/About/fsound.html

122) p. 277 Rie Nakajima
https://rienakajima.com

Index

Letter from the Editor...

Welcome to Mind & Body, a collective look at 360 wellness. This book offers practices for you to explore, a way to be able to uplift and enable yourself to flourish and thrive in life.

Wherever you are in your wellness journey, this is a beautiful guide of some of the practices from ancient Chinese methods, to some of the newer wellness trends.

I think many of us feel that we need to be 'woo woo', look a certain way, eat particular food or, on a deeper level, understand what it is we are looking for, to feel that connection to ourselves and others and to understand wellness. The truth is to 'BE WELL' is for all of us.

As the founder of a wellness sanctuary in the heart of Kent,

I have spent several years meeting experts in their chosen fields, to understand what they do, learn the in-depth benefits of the experiences they offer and how integrating these practices into our day, week or month can allow people to feel release, less anxiety, progress in letting go of passed trauma, improve their health, build strength, develop happiness and ultimately feel a joy to be alive.

In this book we look at the reasons why moving our bodies increases our mental and emotional wellbeing as much as the physical. We shake off the myths of what hypnosis is; explore the benefits of open water swimming and why we should be including infrared as part of our weekly routine. We also look more closely at how a coach can really help you make that change. My aim for this title is that you take

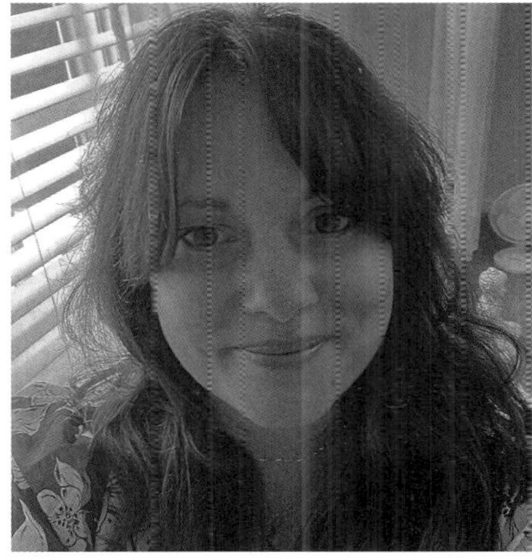

something from it for yourself, to access whatever it is that you truly desire to feel. That you understand wellness is for everyone AND is available to you.

Flourish and thrive all,
Hannah

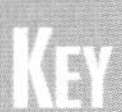

ISBN: 978 1 80282 845 0
Editor: Hannah Sander
Photography: Sam Tom @samtomphoto
Thanks: The Nest and Health Group
Senior editor, specials:
Roger Mortimer
Email: roger.mortimer@keypublishing.com
Design: Panda Media
Advertising Sales Manager:
Brodie Baxter
Email: brodie.baxter@keypublishing.com
Tel: 01780 755131

Advertising Production:
Becky Antoniades
Email: rebecca.antoniades@keypublishing.com

SUBSCRIPTION/MAIL ORDER
Key Publishing Ltd, PO Box 300, Stamford, Lincs, PE9 1NA
Tel: 01780 480404
Subscriptions email: subs@keypublishing.com
Mail Order email: orders@keypublishing.com
Website: www.keypublishing.com/shop

PUBLISHING
Group CEO & Publisher: Adrian Cox

Published by
Key Publishing Ltd,
PO Box 100, Stamford, Lincs, PE9 1XQ
Tel: 01780 755131
Website: www.keypublishing.com

PRINTING
Precision Colour Printing Ltd, Haldane,
Halesfield 1, Telford, Shropshire, TF7 4QQ

DISTRIBUTION
Seymour Distribution Ltd, 2 Poultry Avenue, London, EC1A 9PU
Enquiries Line: 0207 429400C.

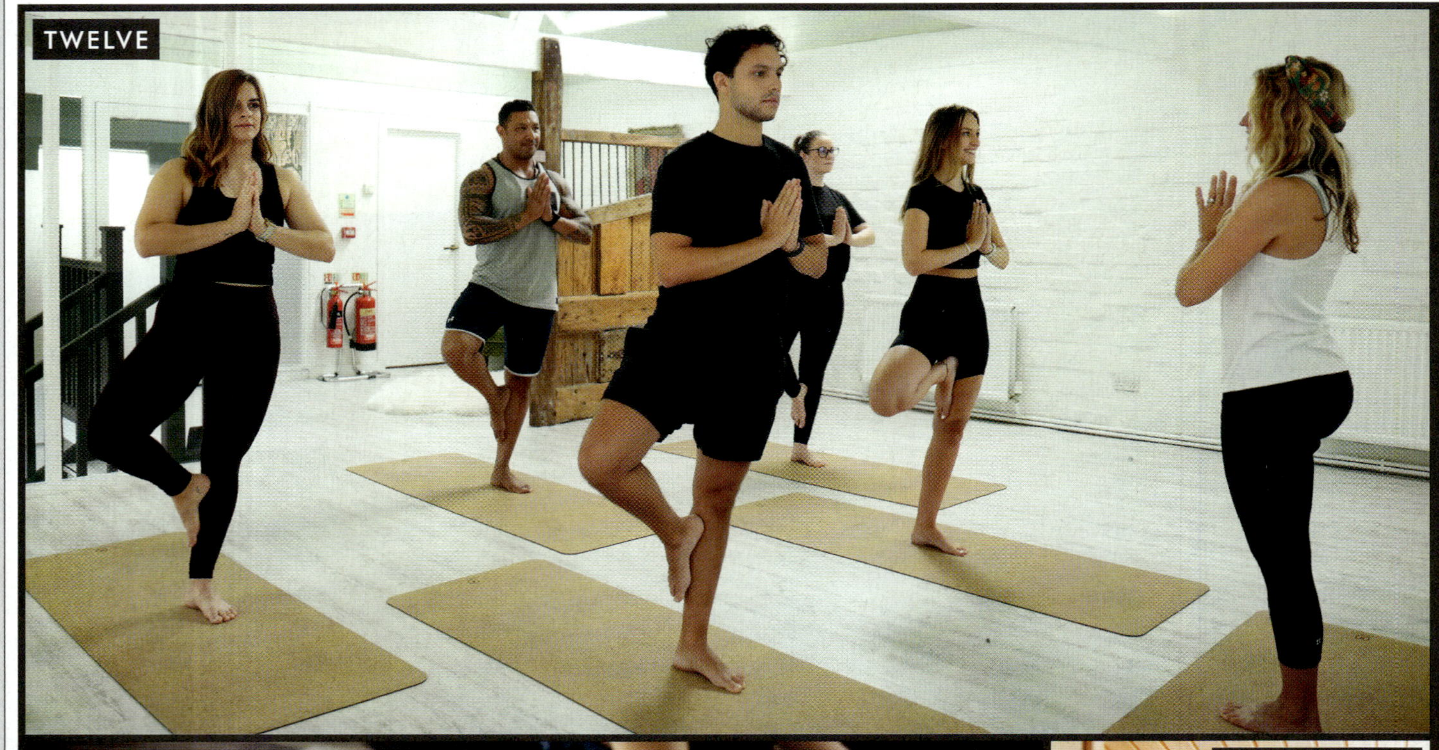

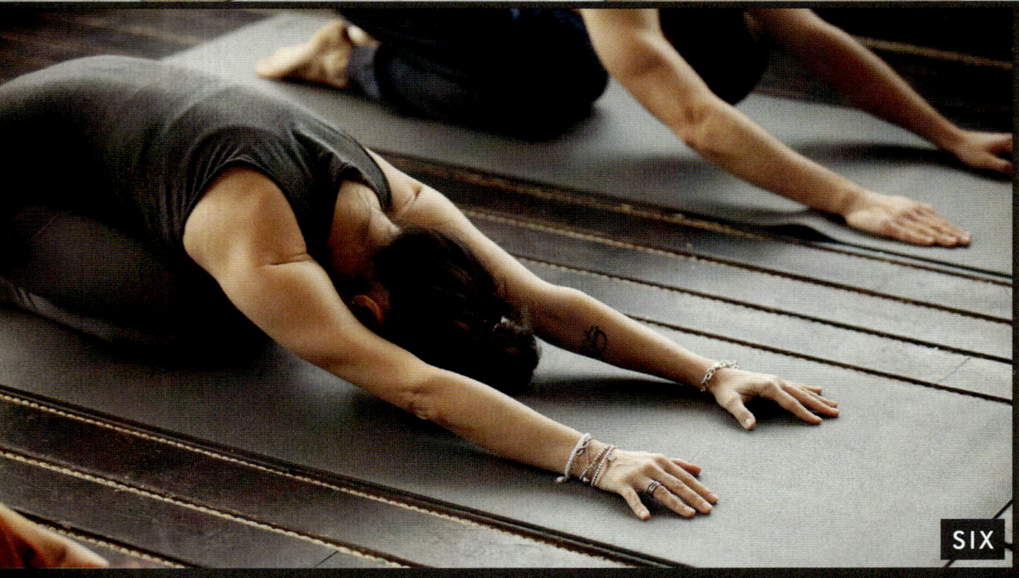

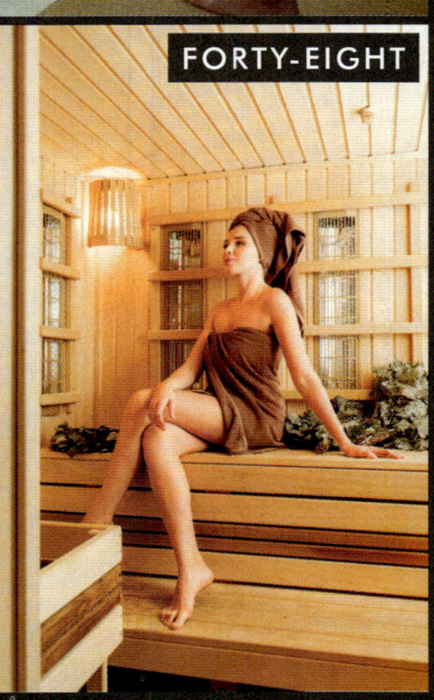

Mind & Body contents

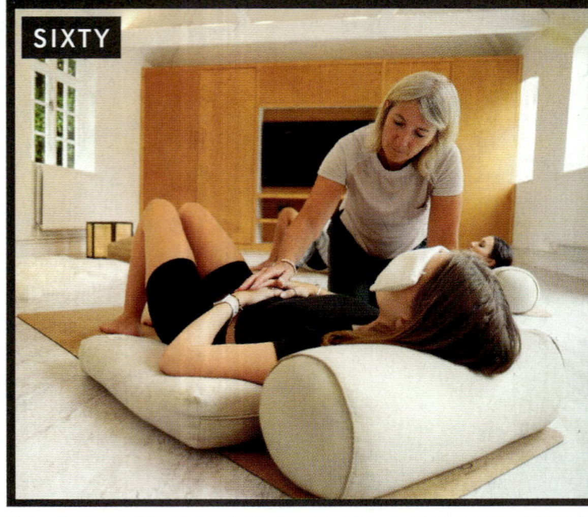

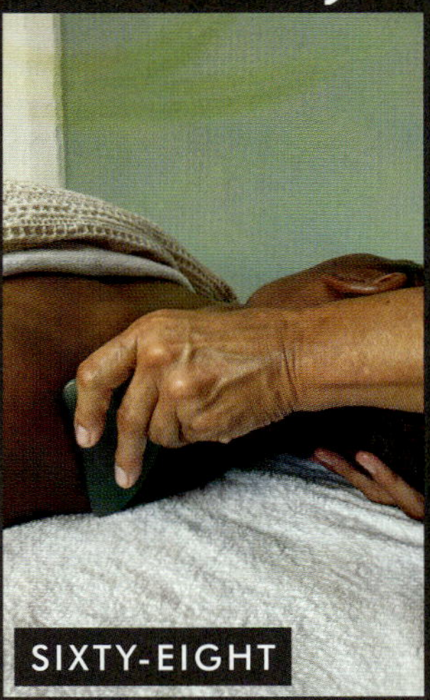

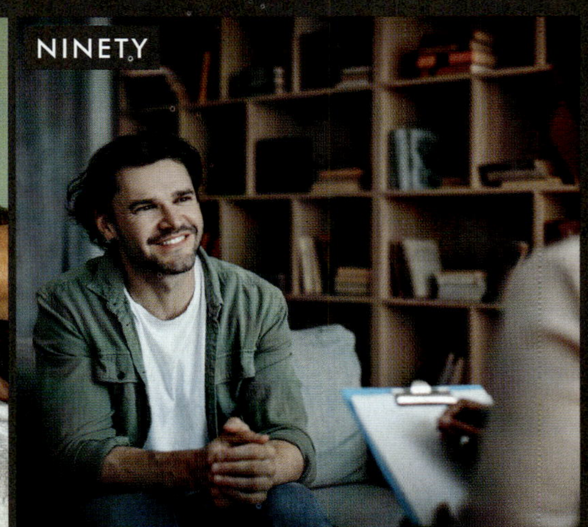

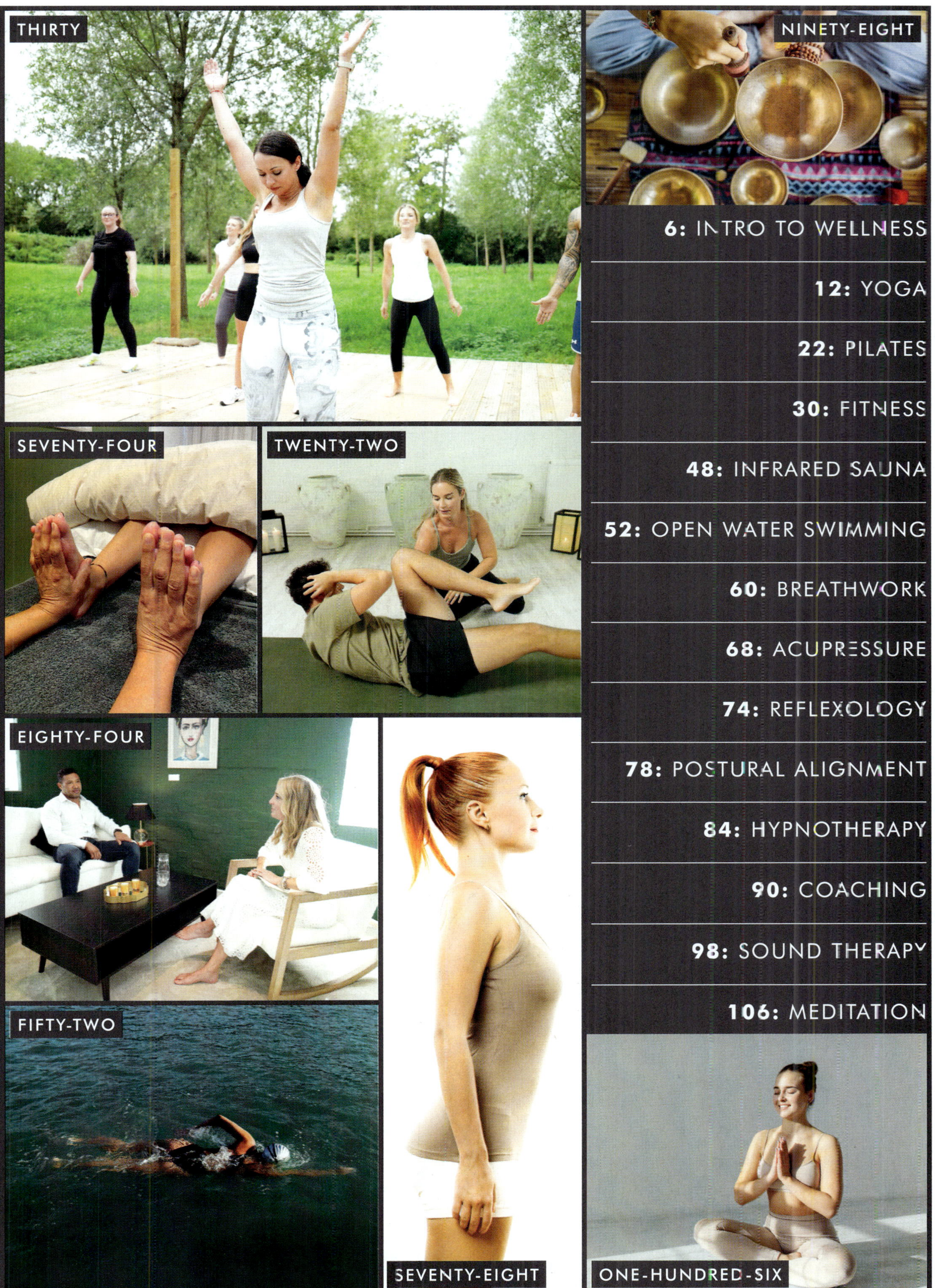
THIRTY
NINETY-EIGHT
SEVENTY-FOUR
TWENTY-TWO
EIGHTY-FOUR
FIFTY-TWO
SEVENTY-EIGHT
ONE-HUNDRED-SIX

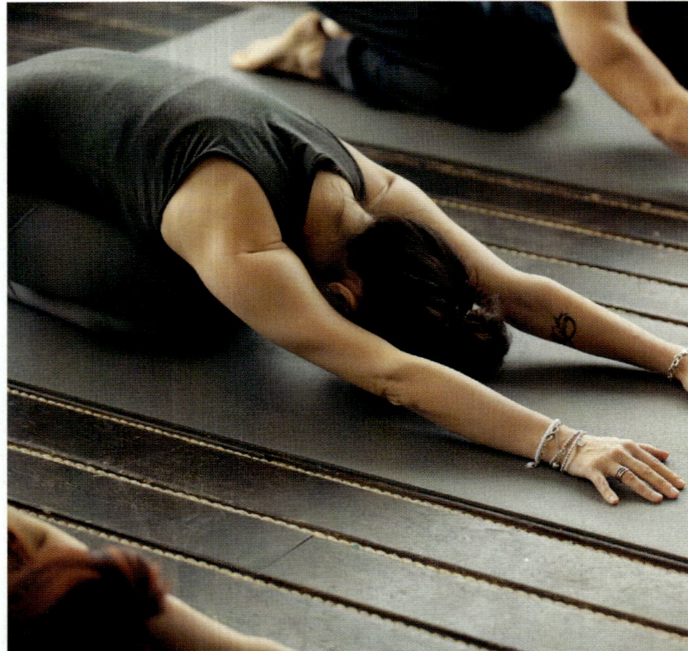

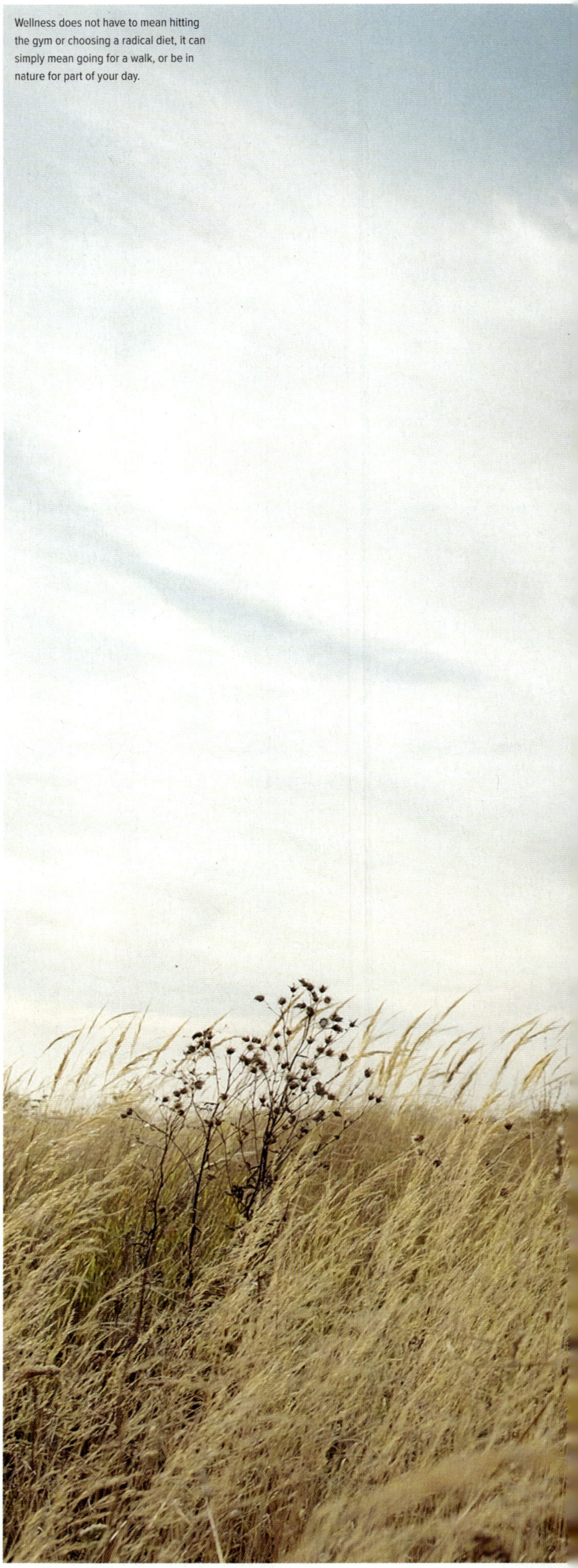

Wellness does not have to mean hitting the gym or choosing a radical diet, it can simply mean going for a walk, or be in nature for part of your day.

What is Wellness?

Hannah Sander, founder of The Nest and Health Group, explores what the meaning of wellness is and why she is so passionate about it being available for all. A champion for social prescribing, she wants to see more people take their health into their own hands.

For me, introducing someone to wellness is like lighting a fire on a dark winter's night, or seeing that sunrise on a long-awaited holiday, the sun warming up your body. Being 'well' is like the awakening many are searching for, and not knowing how to achieve this feels like being permanently on hold. We may go to gym, or listen to an inspirational pod cast, which is all amazing, but what does it mean to truly optimise all areas of your health and life?

Sadly, many of us do not look into things to make us feel well until we have a reason to, for example, an illness, a 'dis'-ease, or even we feel age creeping up on us. We spend years relying on resources outside of our bodies to be healthy. We rely on a health service which, for all the amazing things it does and the people within it doing their very best to keep us well, it is more of a 'sick service' than a fully-functioning organisation, with budgetary, time and staff restraints pushing it into its own point of crisis.

Now, with the pandemic behind us, people, more than ever, are pausing for a moment of realisation,

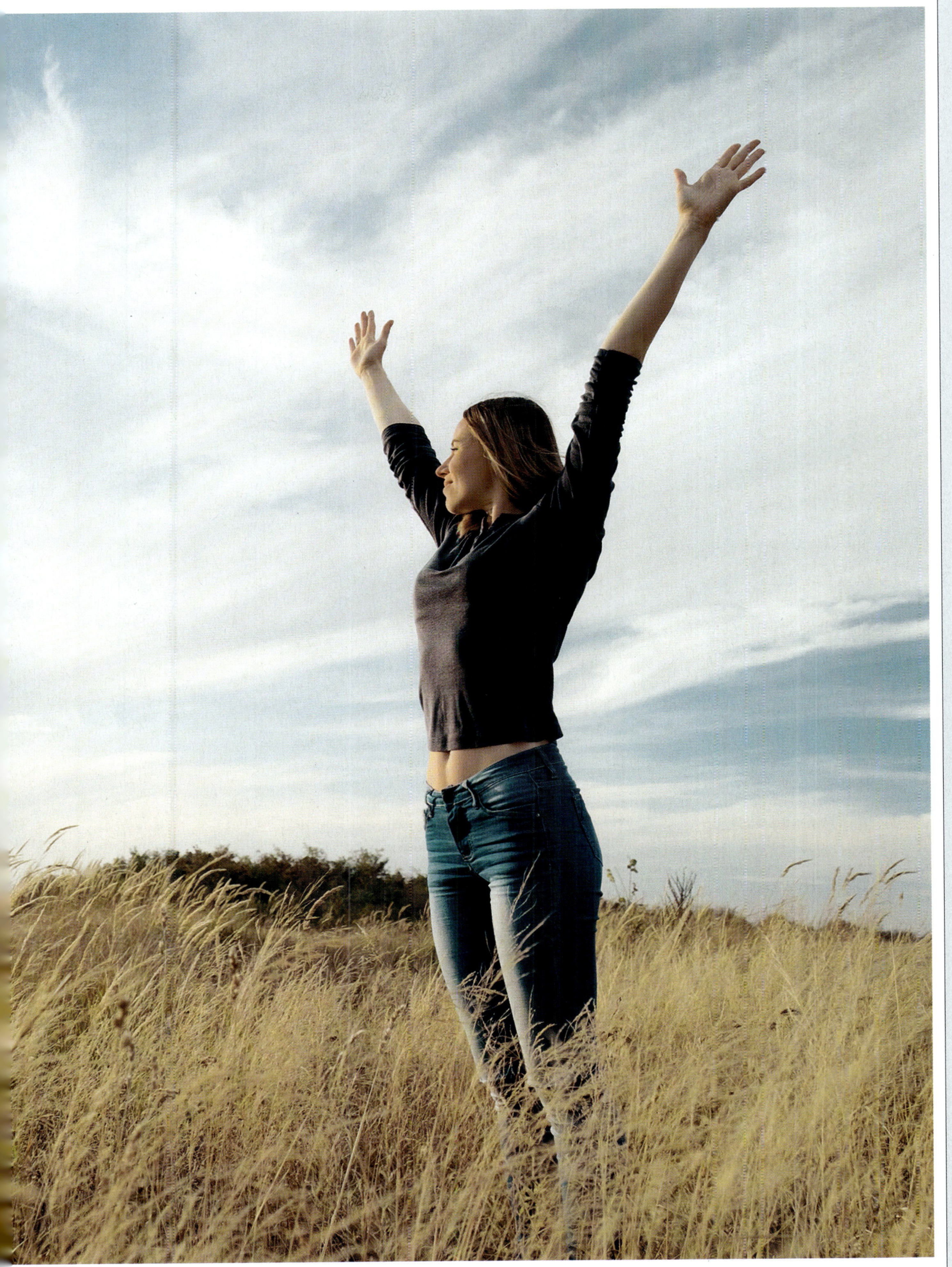

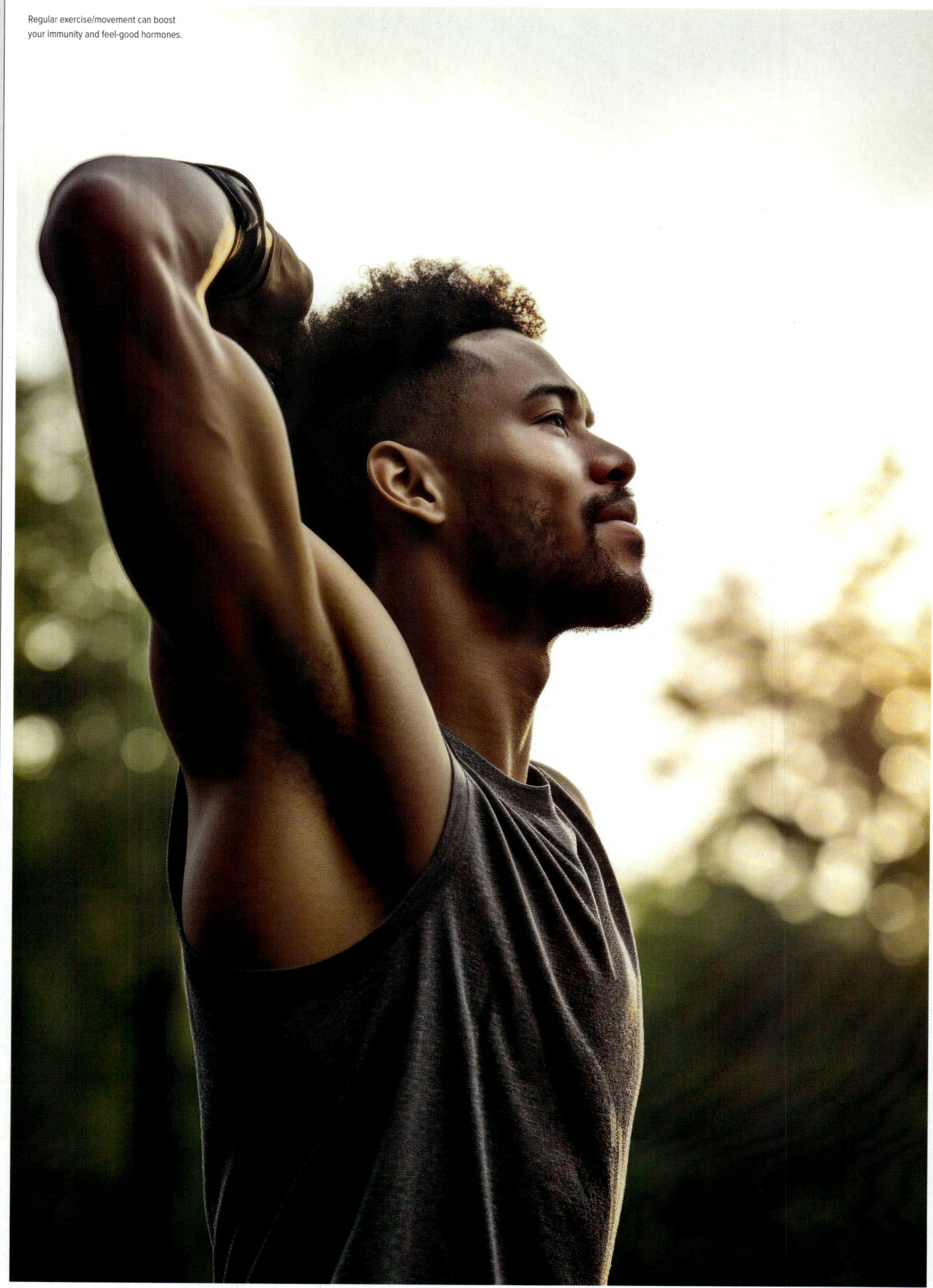

Regular exercise/movement can boost your immunity and feel-good hormones.

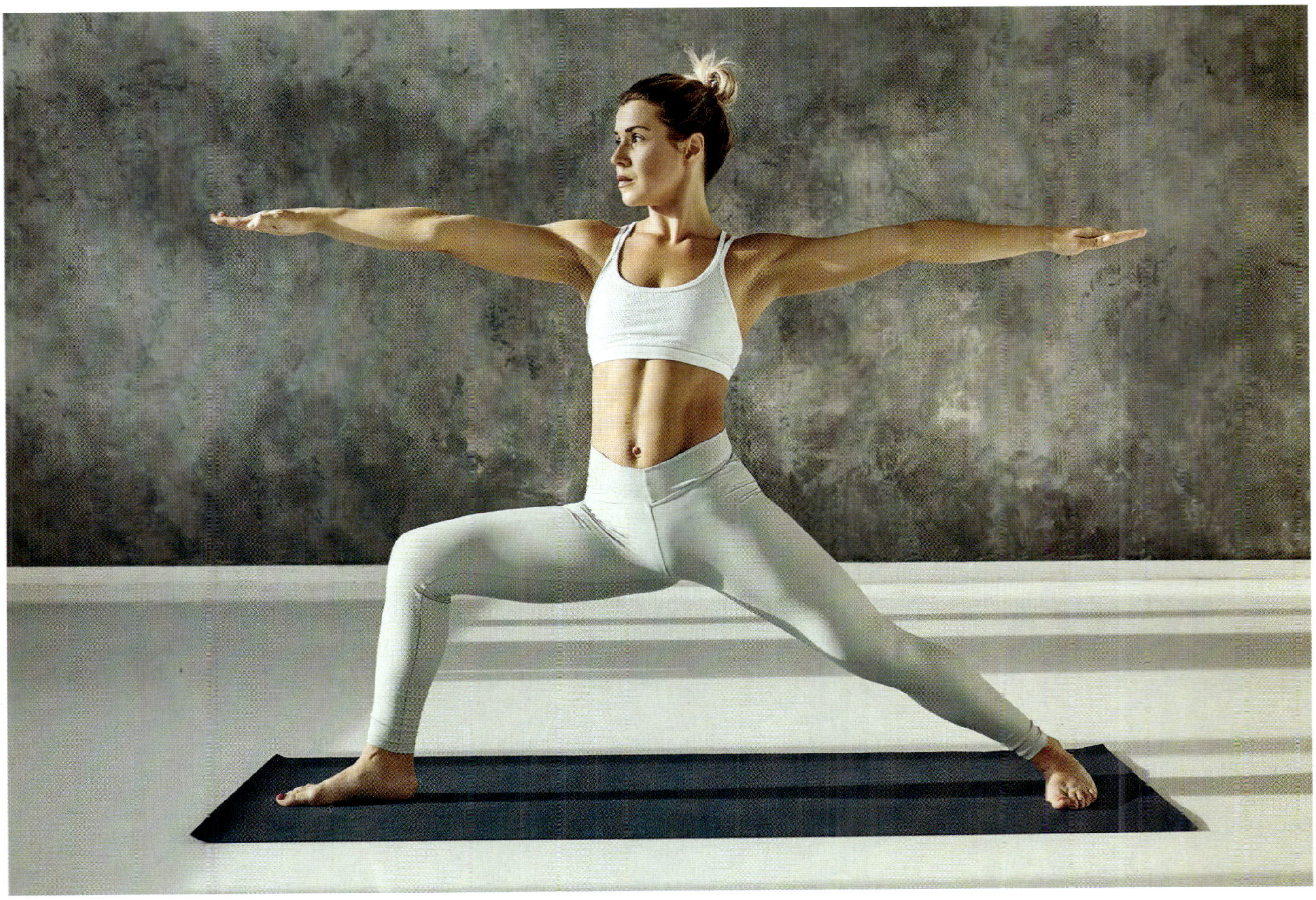

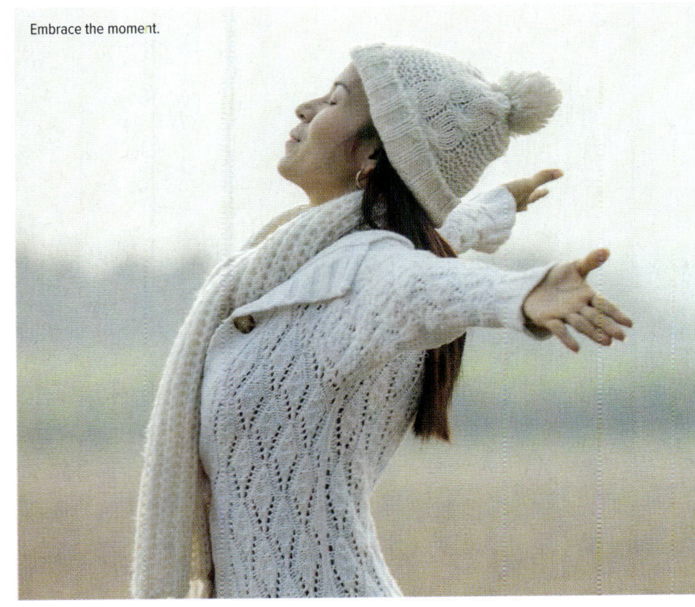
Embrace the moment.

that they can take ownership of their own health and do something different. Like Einstein said "if you keep doing the same things, you'll end up getting the same results". Large swaths of society are feeling fatigued with the same same.

However, knowing where to start can feel equally confusing and overwhelming. I hope this book submerges you into wanting to take control of your own health and future. Take control, for everything we need is within us to change and live better than we are.

Deciding to make that change and trying to focus on your mind and body can seem like a big challenge, especially when you feel like maybe you are starting from scratch. The importance is realising it's a journey and not to have it all sorted in one go. Integrate small changes, for it's the small changes that collate into big ones.

Social commentators talk about starting with a routine, making small changes bit by bit. This can be implementing waking up 15 minutes before you do currently and meditating and setting your intention for the day. This could be joining a local class that you have been meaning to do, reading or listening to something/ someone as this will help you feel growth, for if we are not growing we are dying.

Take a walk, meet up with friends. Wellness does not mean keeping up with all the trends, quality sleep and social wellness (meeting with friends and loved ones) is just as important.

Wellness is making a conscious change to grow and develop yourself. It is simply the tools to help you be

Being 'well' is like the awakening many are searching for, and not knowing how to achieve it.

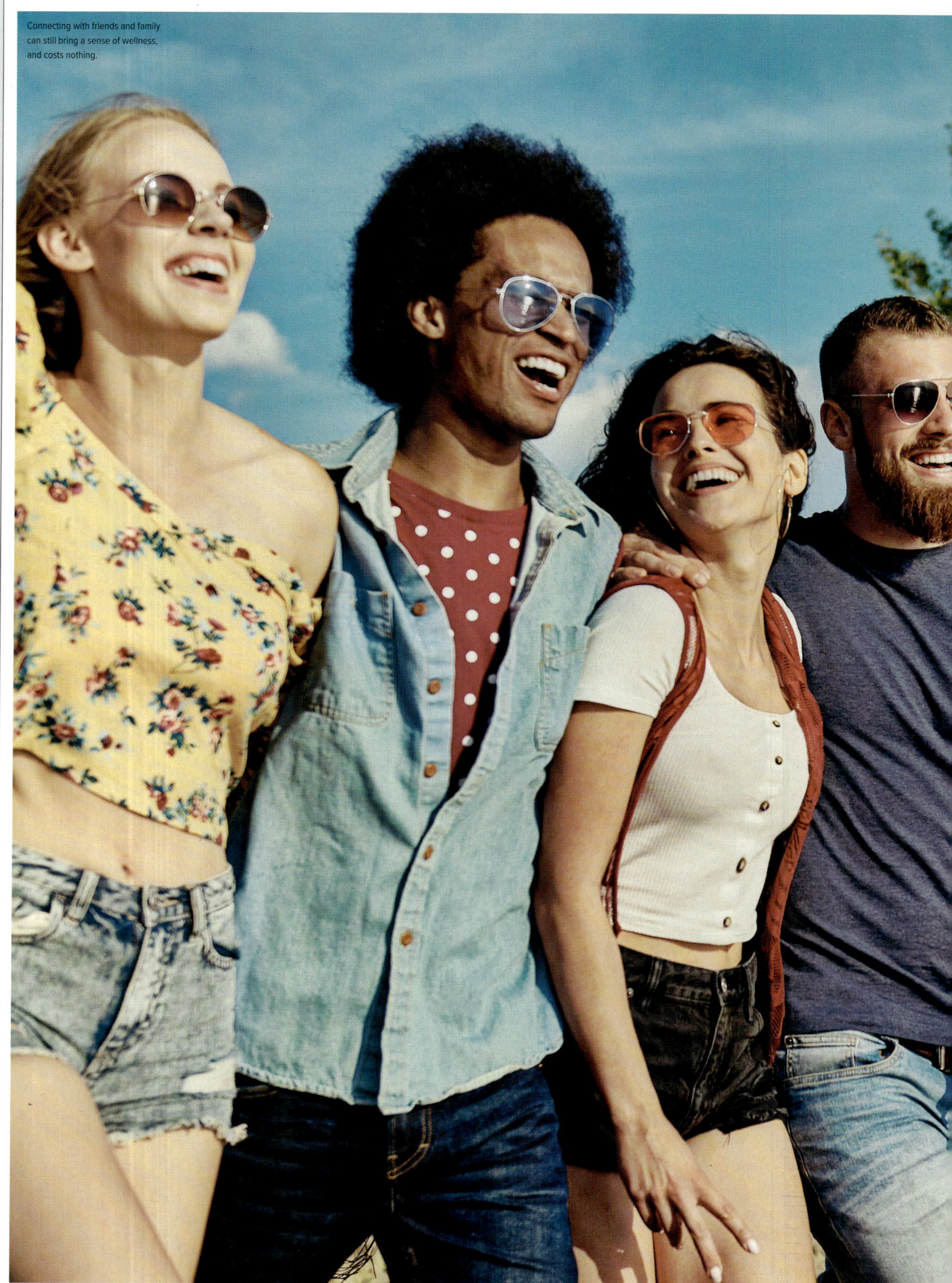

Connecting with friends and family can still bring a sense of wellness, and costs nothing.

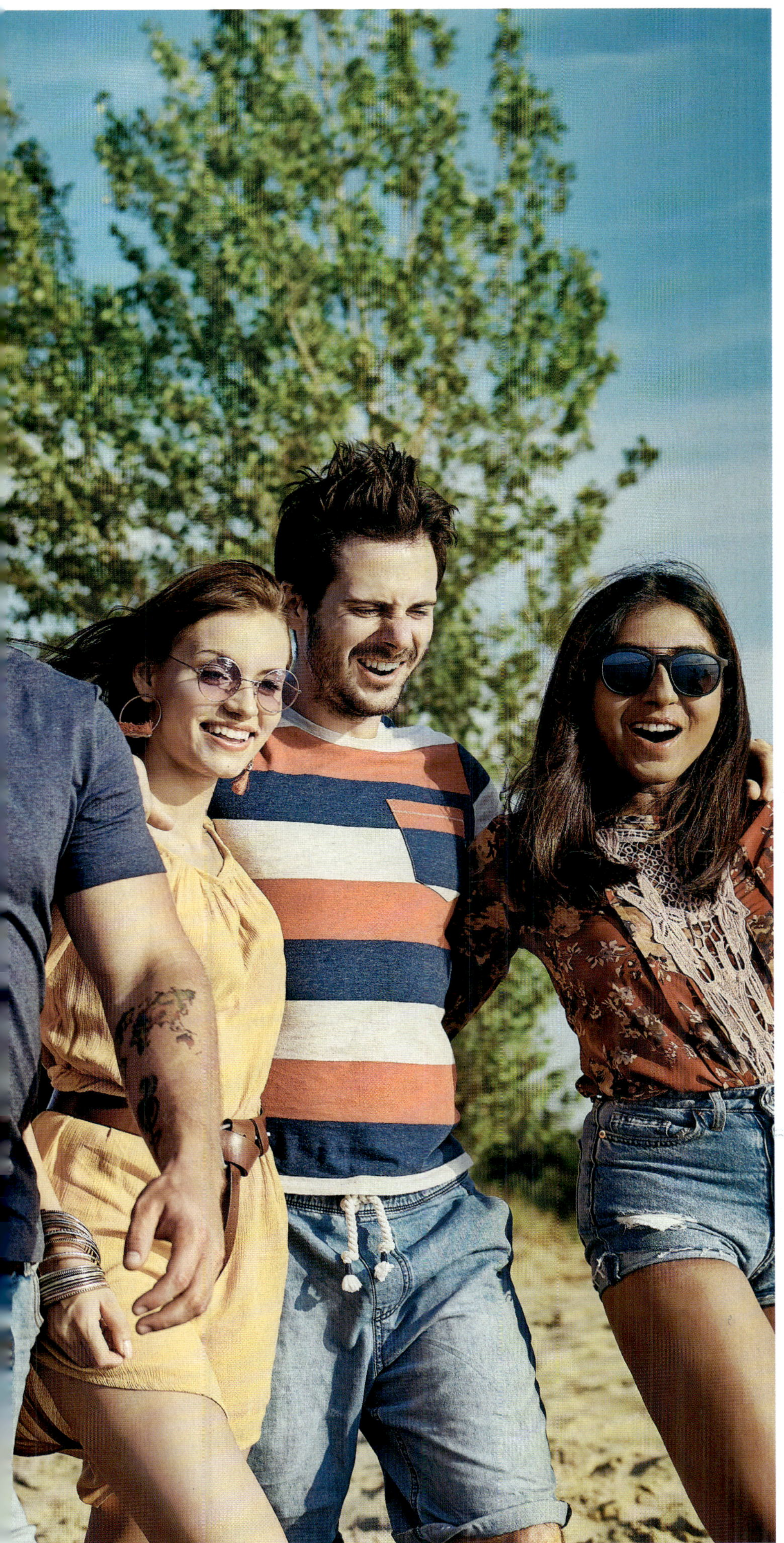

a happier and healthier human being. Wellness is described as one of the most powerful words in the English dictionary, with many thinking it is a new 'it' word thrown around corporates and hippy communes. In truth, it is a word that has been used since the 1950s and by definition, in medical terms, means to be healthy of both mind and body as a result of deliberate effort.

Millions of people around the world have already taken control of their own health and optimising their wellbeing. On a recent trip to a conference in mainland Europe, I was blown away with the advances of such things as biohacking, while also listening to the benefits of micro dosing. DNA nutrition and future therapies. Science is evolving so quickly and we are learning so much about the links to how our mind really is the power over our body, even by the way our cells rejuvenate

Although in the UK we are a little slow of the mark with this, practices like these are coming. With medical backing, an understanding is developing that there is power in going back to what the land has provided us and that natural remedies, tonics and therapies could mean us turning to these instead of relying on prescription medicines, that in turn can lead to their own problems.

A big part of my own wellness journey was letting go of judgement that I had to be a certain type of person to be well. All aspects of wellness look different for everyone. Our bodies are a physical story of our past. Our social wellbeing are the moments we have already shared with others, and our mental resilience is a makeup of past both good and bad.

There is no end point to wellness, no trophy for being the best at it, just a journey of showing up for yourself in all areas, to achieve the healthiest and happiest version of you. *M&B*

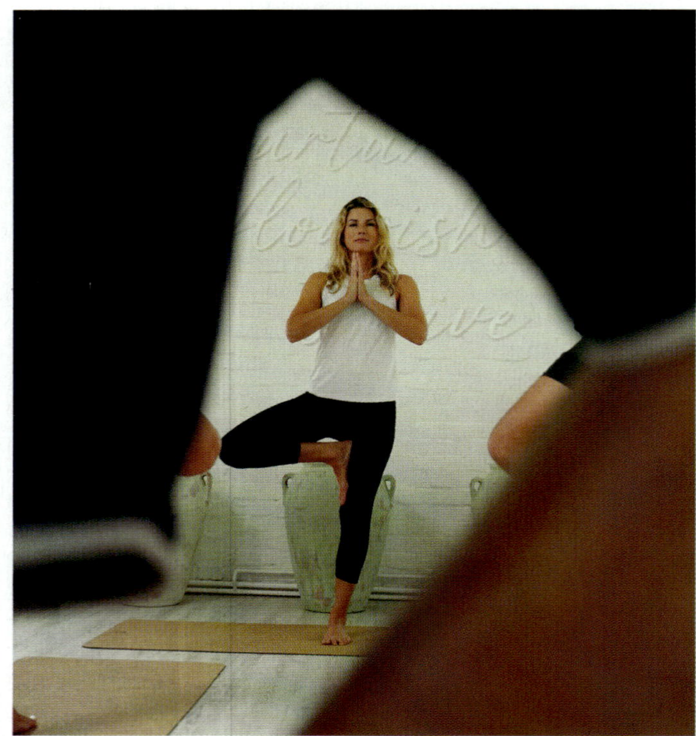

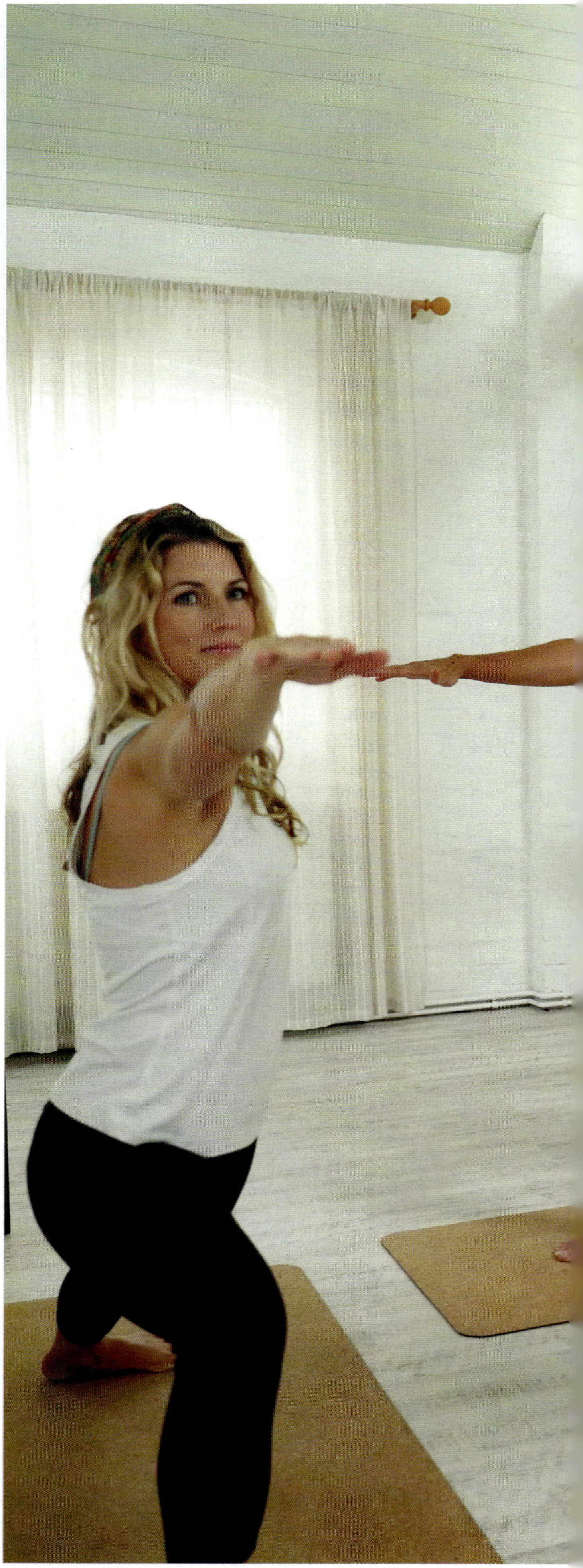

Calm Mind and Body

Practiced for thousands of years, Yoga is now studied and enjoyed throughout the world with many adaptations and versions.

Yoga is the epitome of our three elements mind, body and soul. Through physical movement we become aware of ourselves and the universe around us. The word yoga comes from the route word 'YUJ' which means 'to yoke' or to bind. This is all about connection of both mind, body and spirit.

A beautiful and ancient art to practice, a lot of us are not sure which is the right type of yoga to get involved with. Here we unpick the different branches so you can decide which one is right for you.

➤ Ashtanga yoga is known for its regimented style. It is a full body workout that consists of a sequence of asanas (postures) that develop flexibility, muscle strength and toning. It is a disciplined practice that can also support clarity of mind, focus and better concentration.

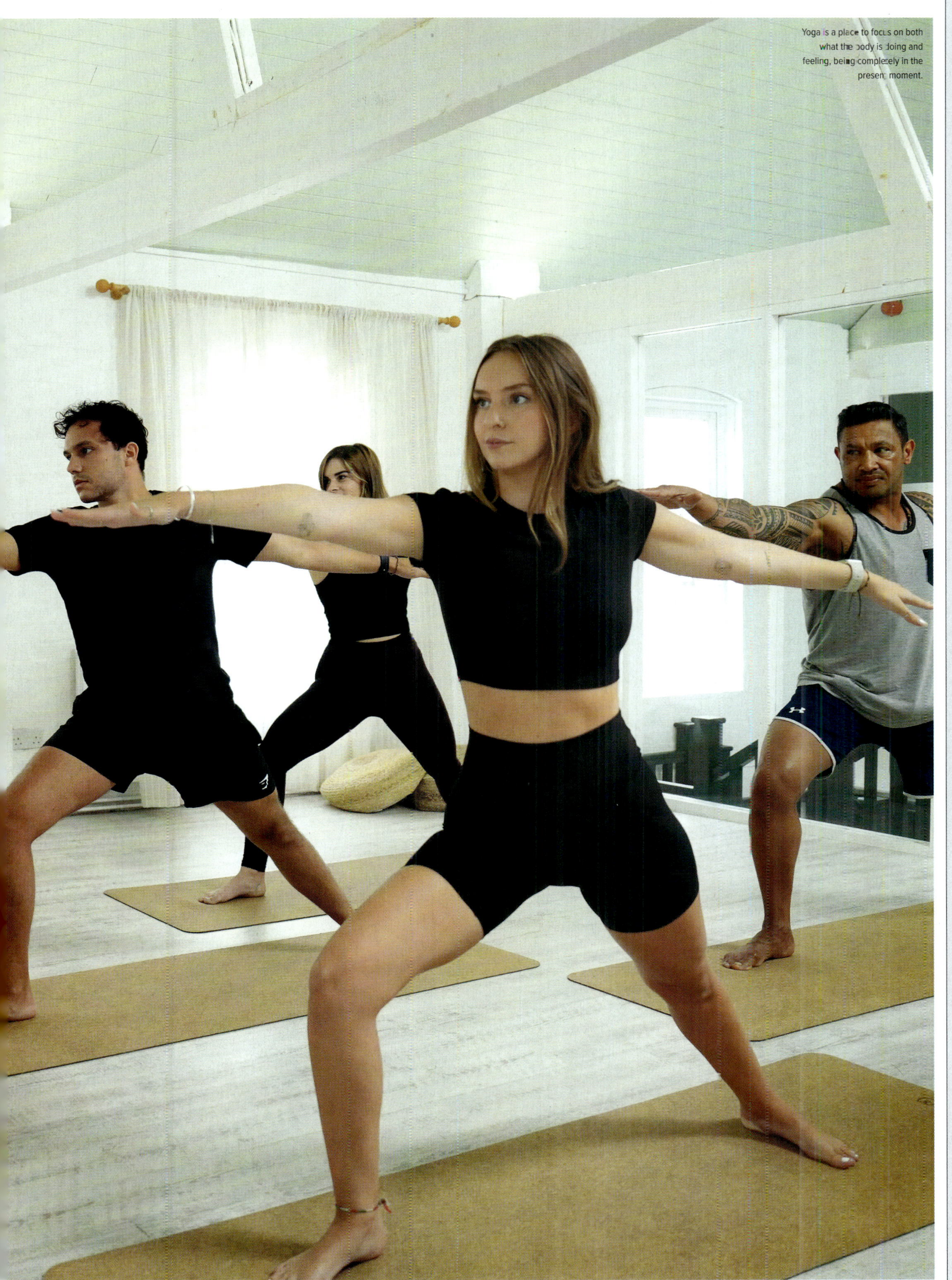
Yoga is a place to focus on both
what the body is doing and
feeling, being completely in the
present moment.

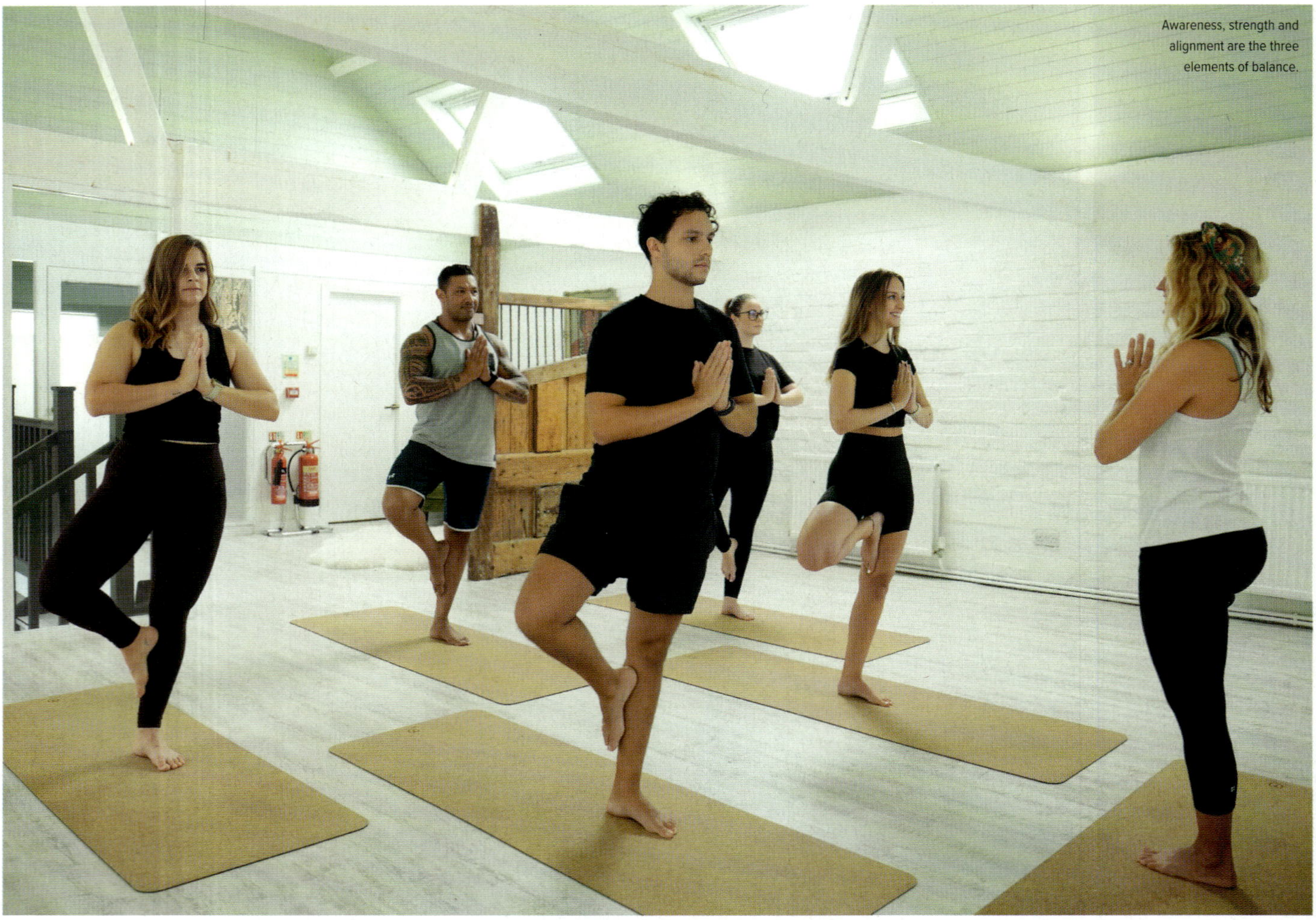

Through this style of yoga, the body is able to release suppressed emotions and excrete toxins.

➤ Vinyasa flow is derived from Ashtanga. Its variety of postures, synchronised with the breath, create an easy flow of movement. Throughout the poses your organs benefit from a gentle massage, your cells regenerate from increased oxygen flow and you can expect to leave a session feeling a sense of calm and relaxation.

➤ Yin yoga is a much slower paced practice. It involves holding poses for a longer duration, which promotes the unfolding of deeply connective fascia tissues and lubrication of joints. Through this style of yoga, the body is able to release suppressed emotions, excrete toxins, increase flexibility and optimise a healthy flow of energy. In the stillness emerges discomfort, both bodily sensations and challenging thoughts. This is an opportunity to practice calming the mind and body, which better prepares us for the day-to-day challenges that face us in our lives.

➤ Bikram yoga, also known as hot yoga, is a series of postures practiced in high temperatures at approximately 40°C. In a hot environment your pours open, allowing the body to detox itself deeply and

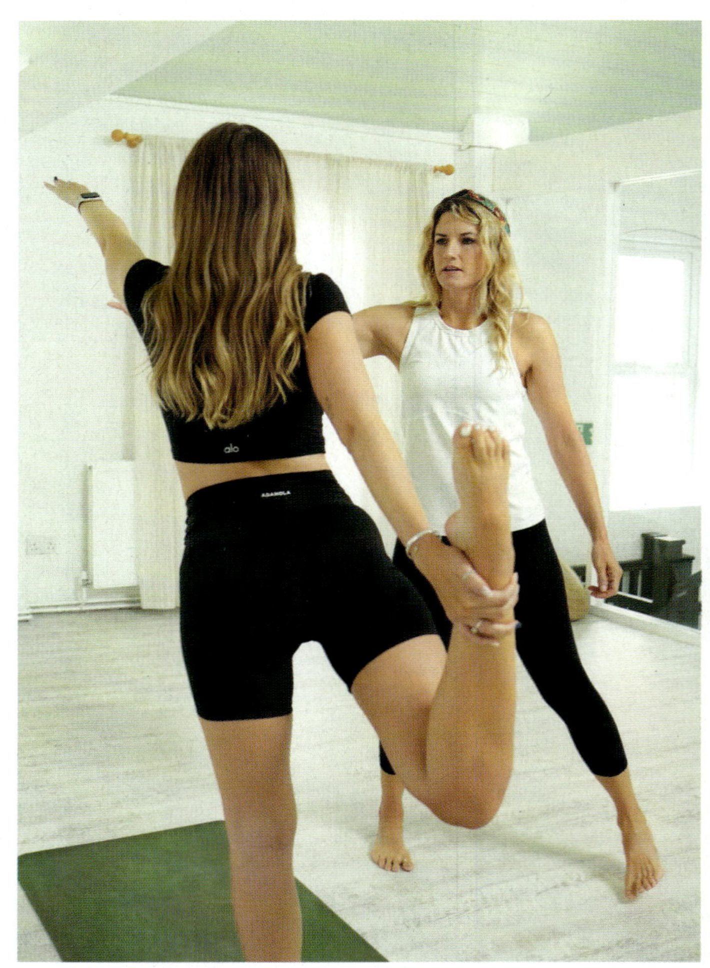

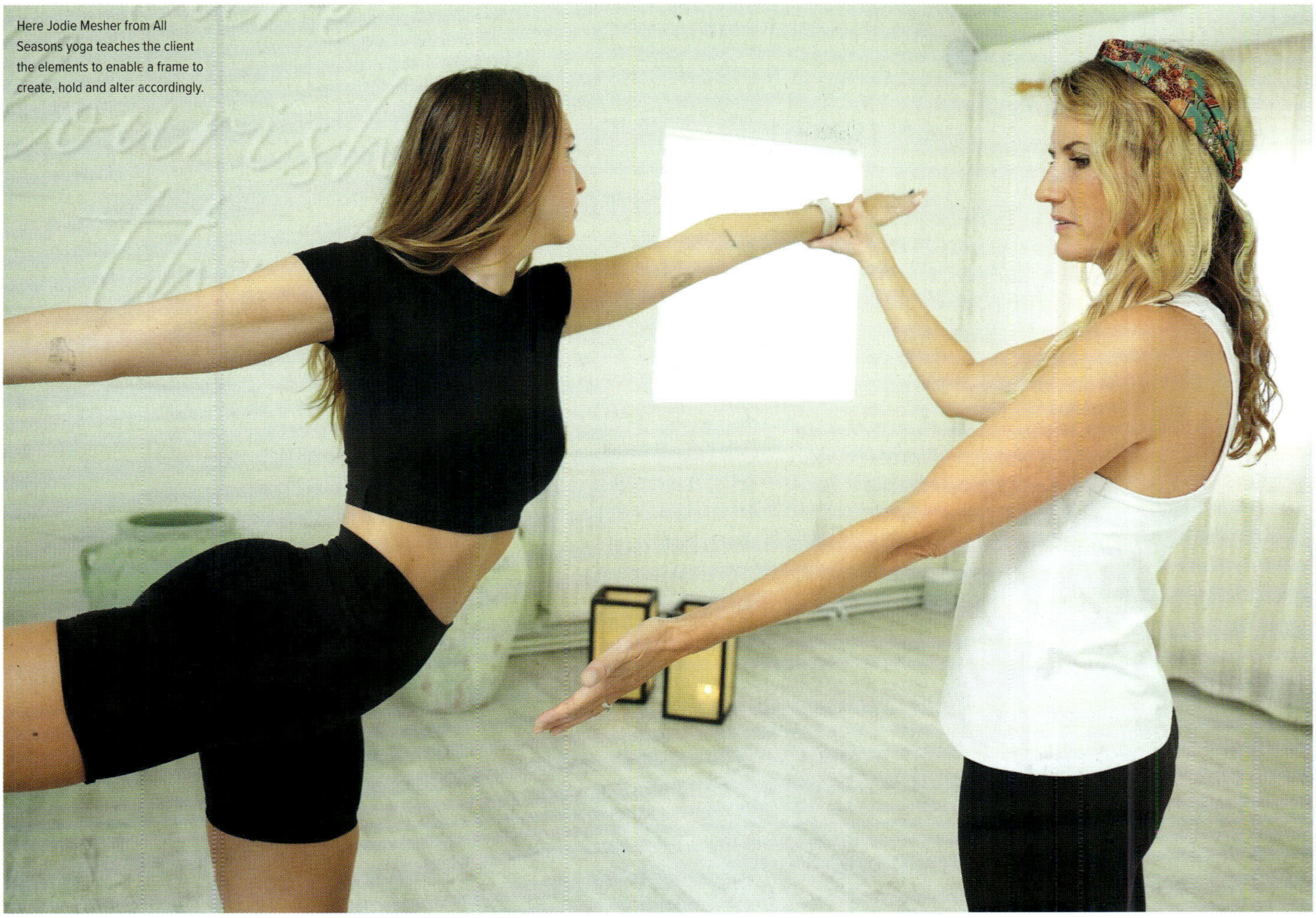

Here Jodie Mesher from All Seasons yoga teaches the client the elements to enable a frame to create, hold and alter accordingly.

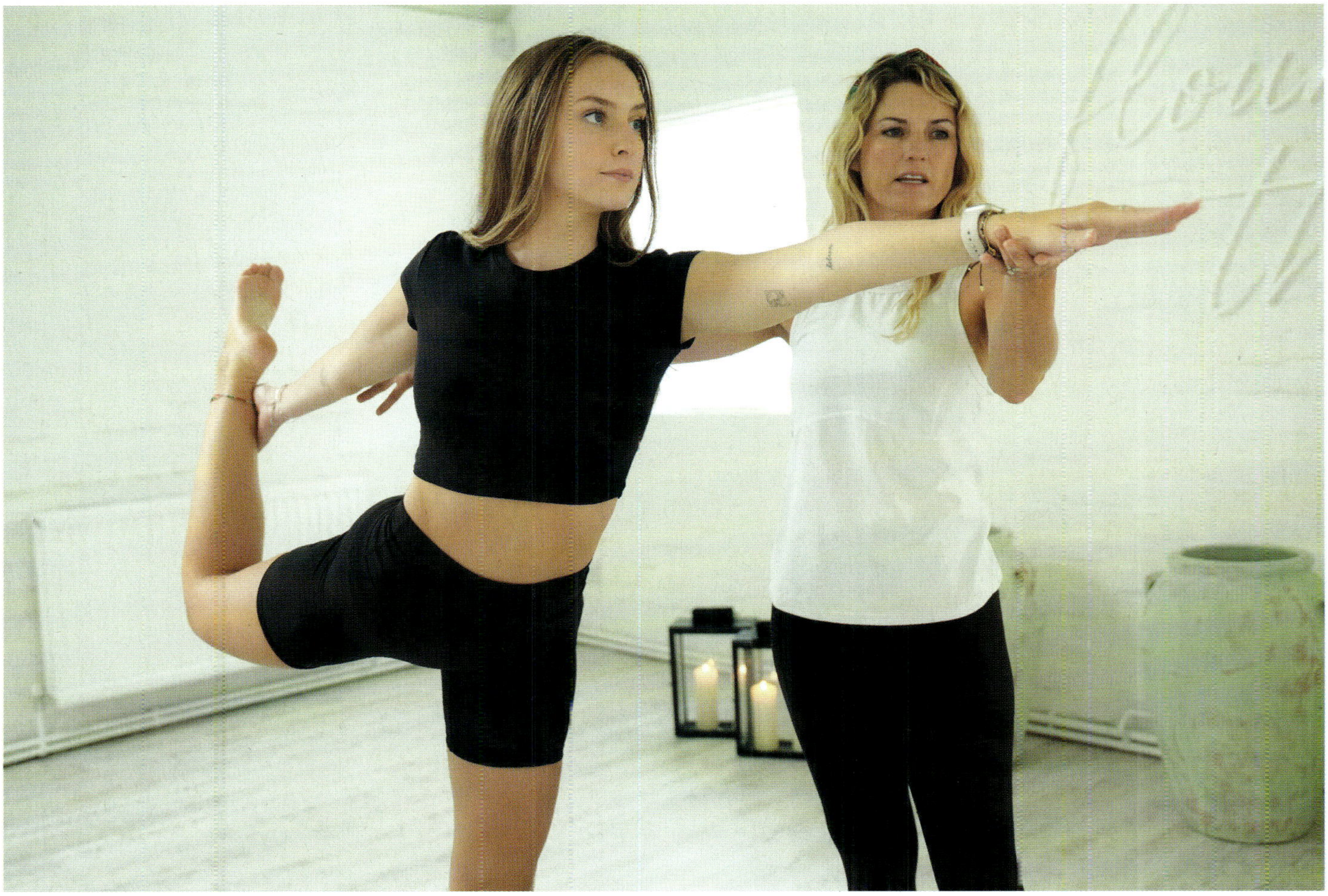

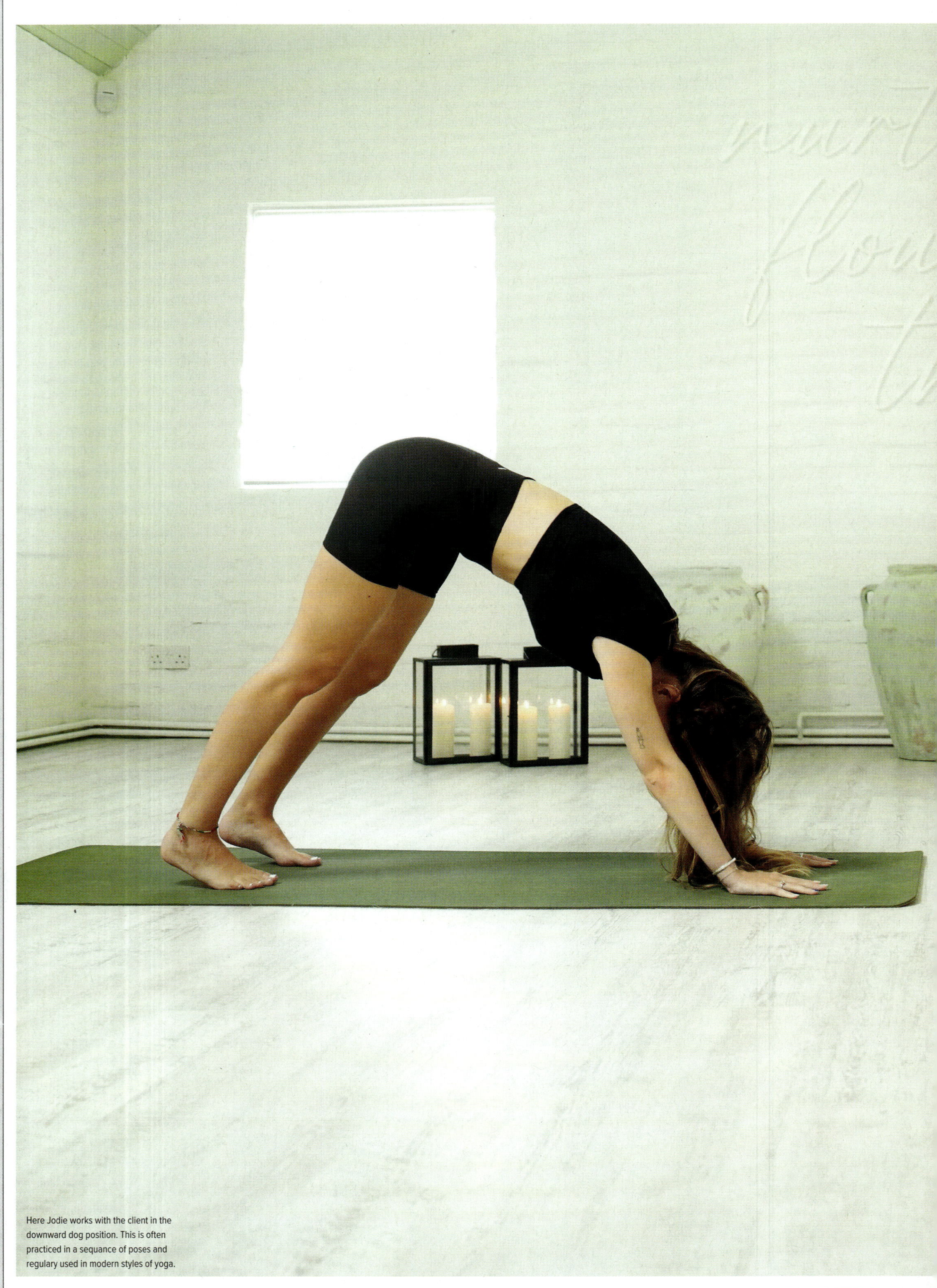

Here Jodie works with the client in the downward dog position. This is often practiced in a sequence of poses and regulary used in modern styles of yoga.

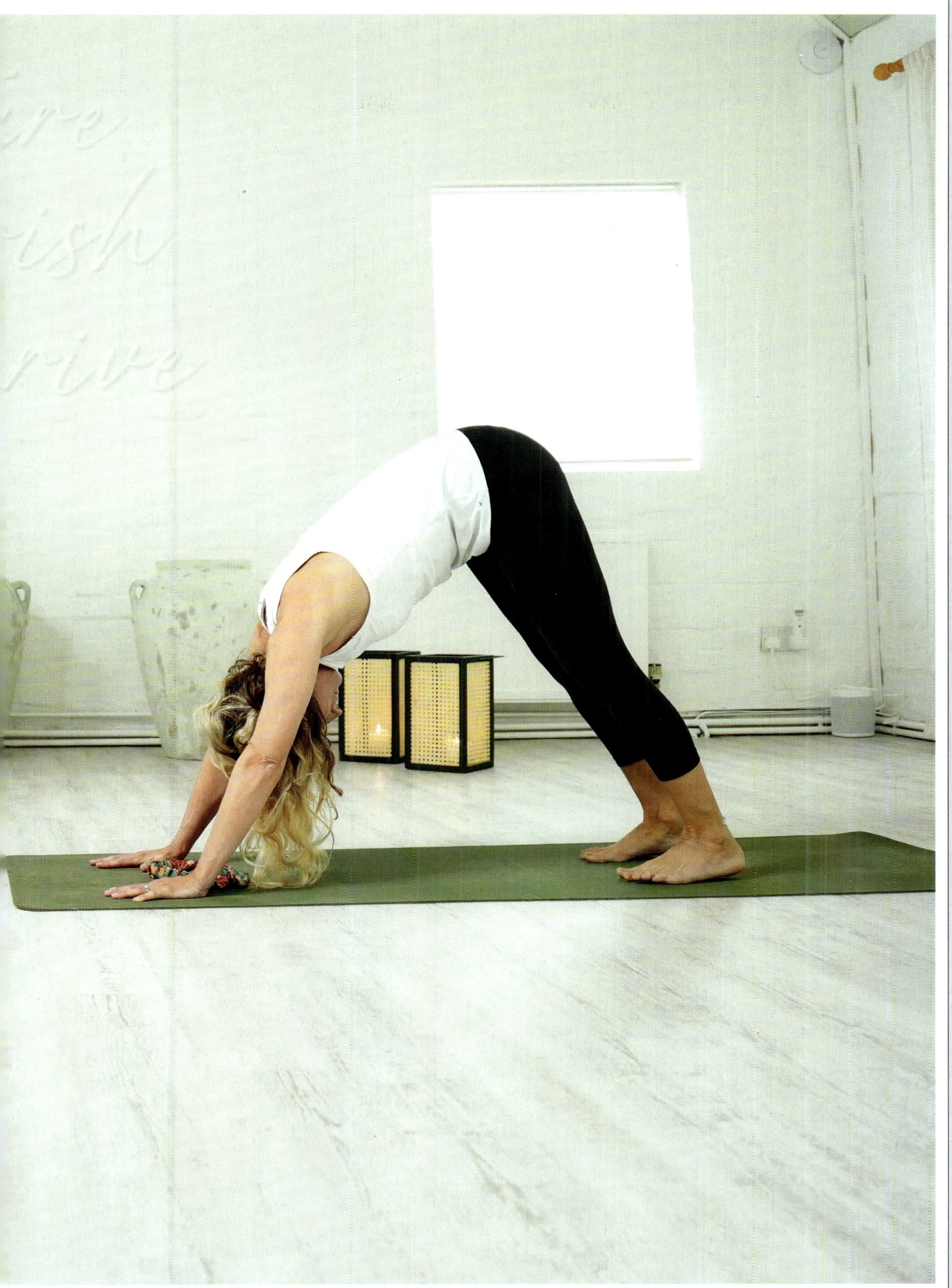

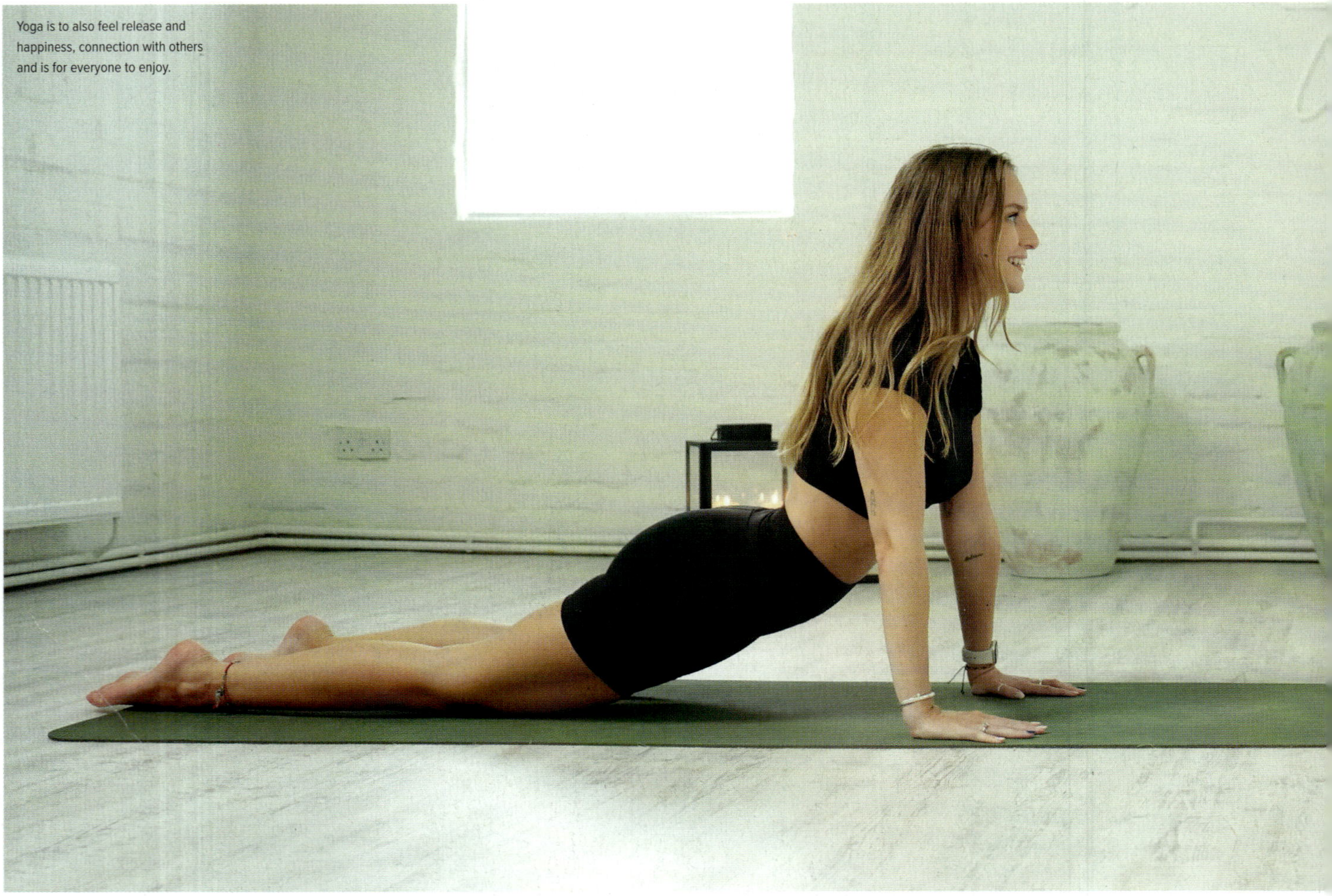

Physical benefits include improving your cardiovascular and circulatory health.

easily. Your metabolic rate also increases and warm muscles mean joints can stretch more easily. Many have experienced recovery and healing from old injuries during Bikram classes.

➤ Hatha means force. This is a gentle form of yoga that aims to restore balance in the body by working with our energy. 'Ha' means sun which is our masculine energy, while 'tha' means moon, our feminine energy. When the body reaches a state of inner equilibrium, it can function optimally which means improved health. You can expect to practice asanas (postures), pranayama (breathing techniques), mantra (chanting), shakti (cleansing), mudra (hand gestures) and various forms of visualisation.

➤ Kundalini yoga is a spiritual practice that seeks to awaken a dormant energy in the base of the spine. In this practice you start from the root chakra (one of the seven energy centres of the body) working the way up the spine finally reaching the crown chakra. This class involves a sequence of components: chanting, pranayama, kriya (postures), relaxation and meditation.

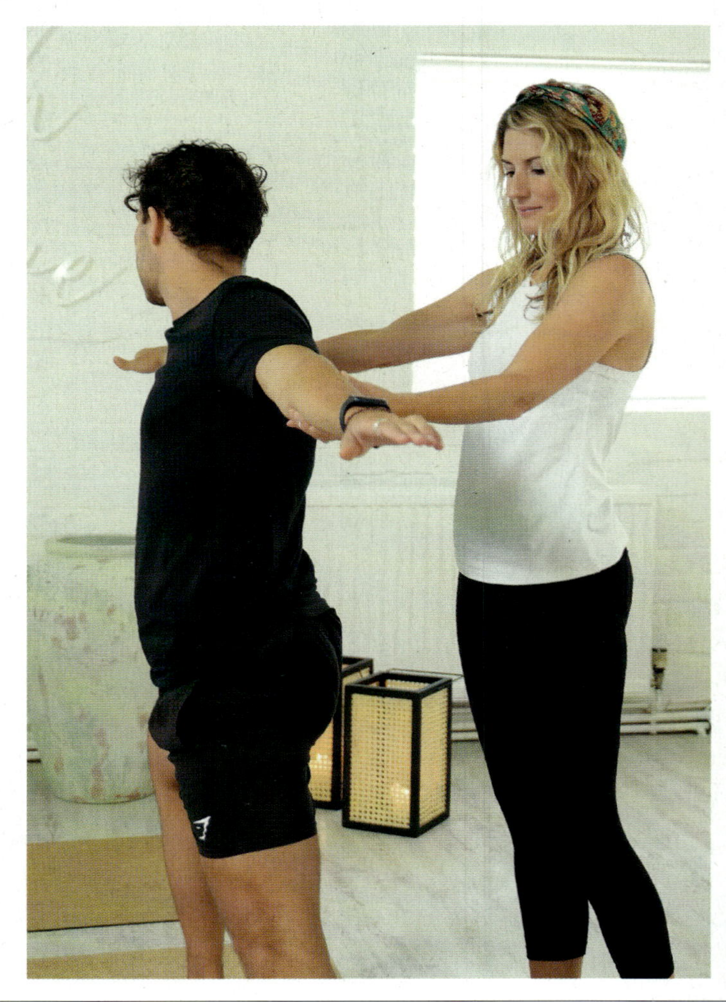

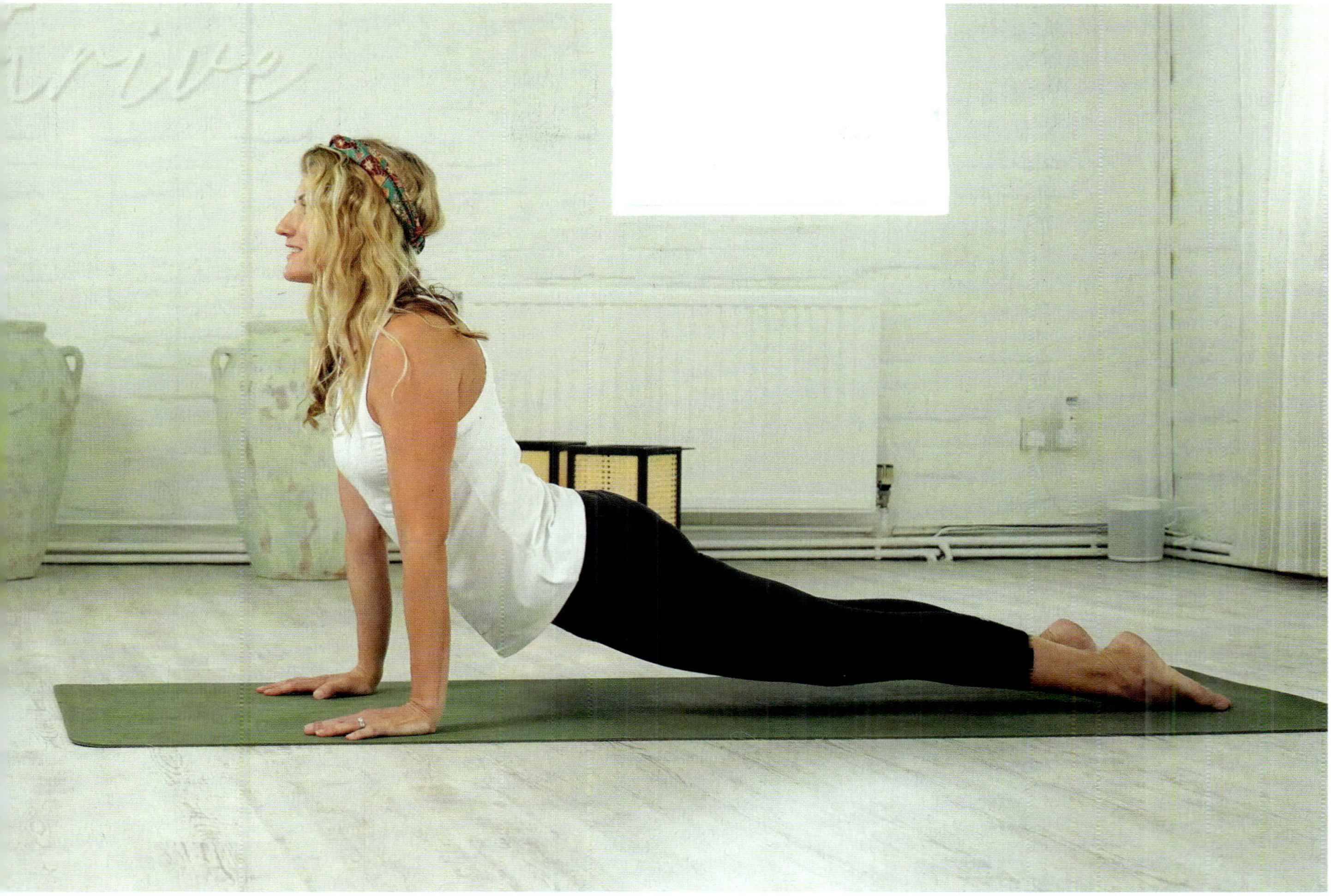

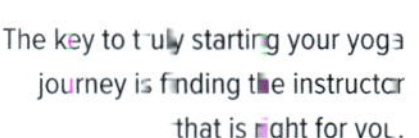

The key to truly starting your yoga journey is finding the instructor that is right for you.

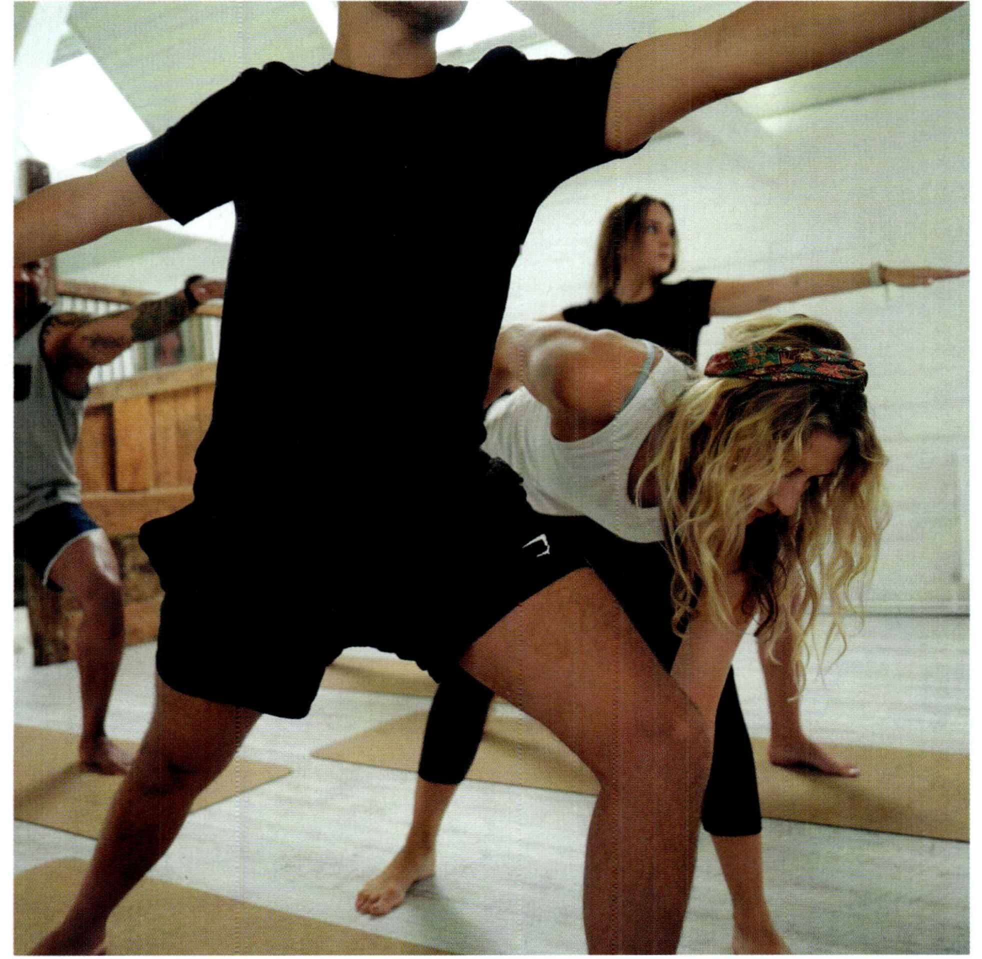

Take a moment to think about what your body and mind truly desire and explore different instructors until you find the one that you align with, as everyone has their own style and it may take time to discover which feels right for you.

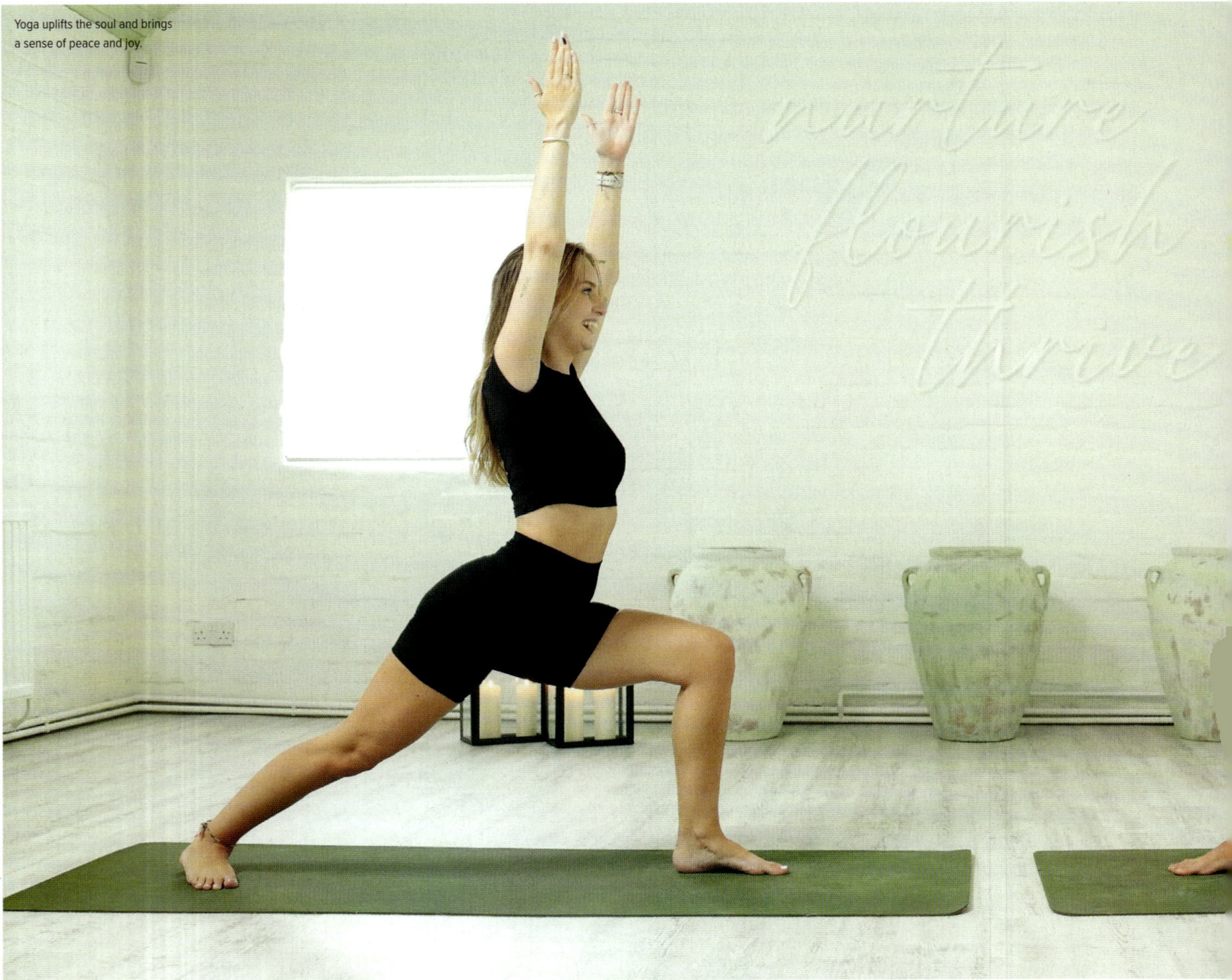

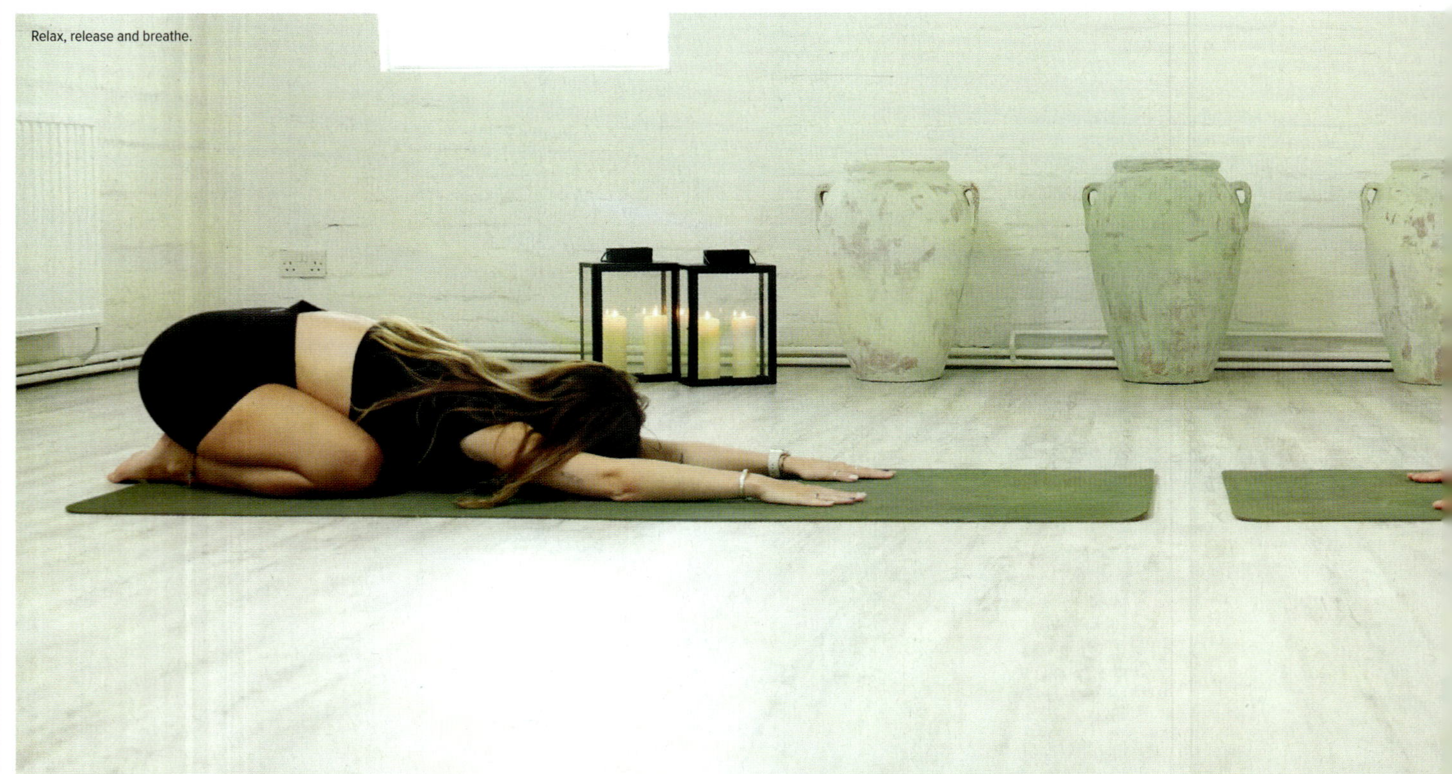

Yoga uplifts the soul and brings a sense of peace and joy.

Relax, release and breathe.

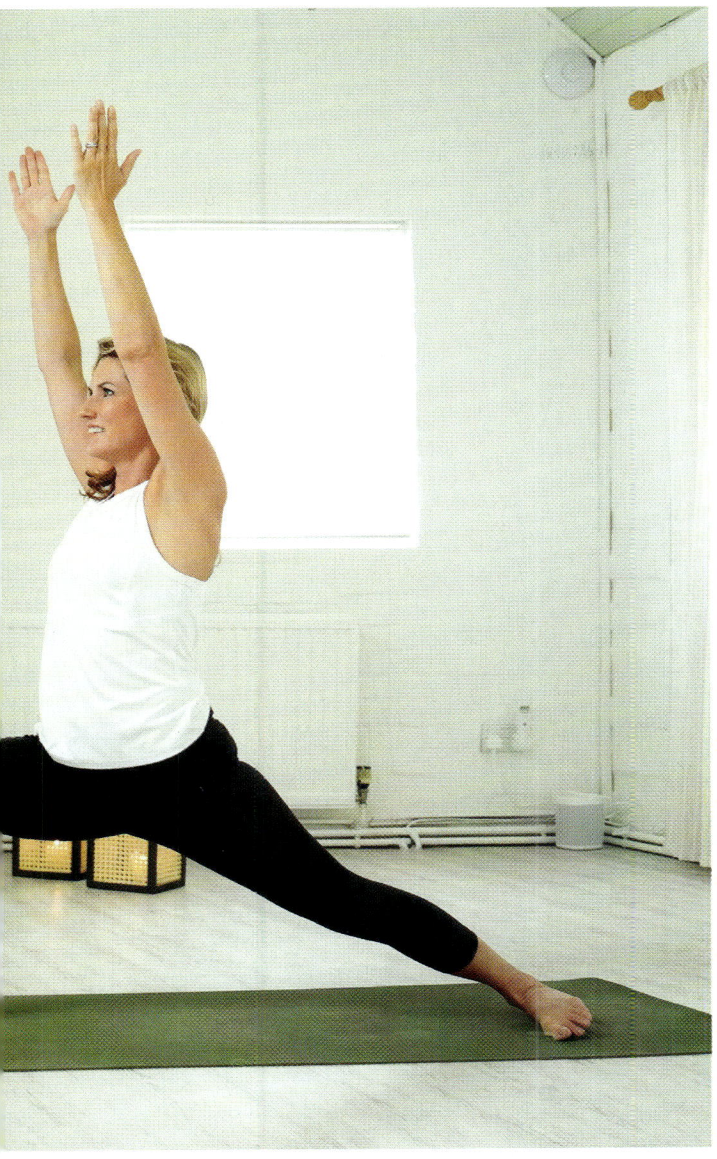

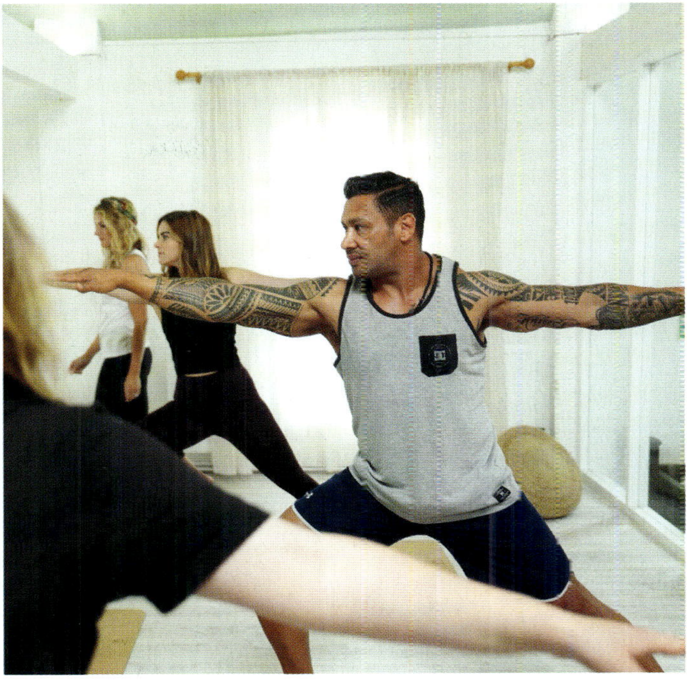

So, what are the benefits of yoga?

Yoga has multiple benefits to practice, with plenty of us thinking it's purely physical. Of course, there are a great number of physical benefits such as improving your cardiovascular and circulatory health, increasing flexibility, tone and muscle strength. And yes a great number of athletes use yoga for improving performance and preventing injury. As well as those, yoga can speed up metabolism and be great for weight reduction. It is also fantastic for reducing inflammation and helping with balance.

That said, there are also a plethora of mental health attributes which include: helping reduce anxiety, increasing focus, gaining clarity and if practiced in a class setting, it is an amazing opportunity for human connection and community, that we all search for as a core need. *M&B*

Jodie Mesher
All Seasons Yoga

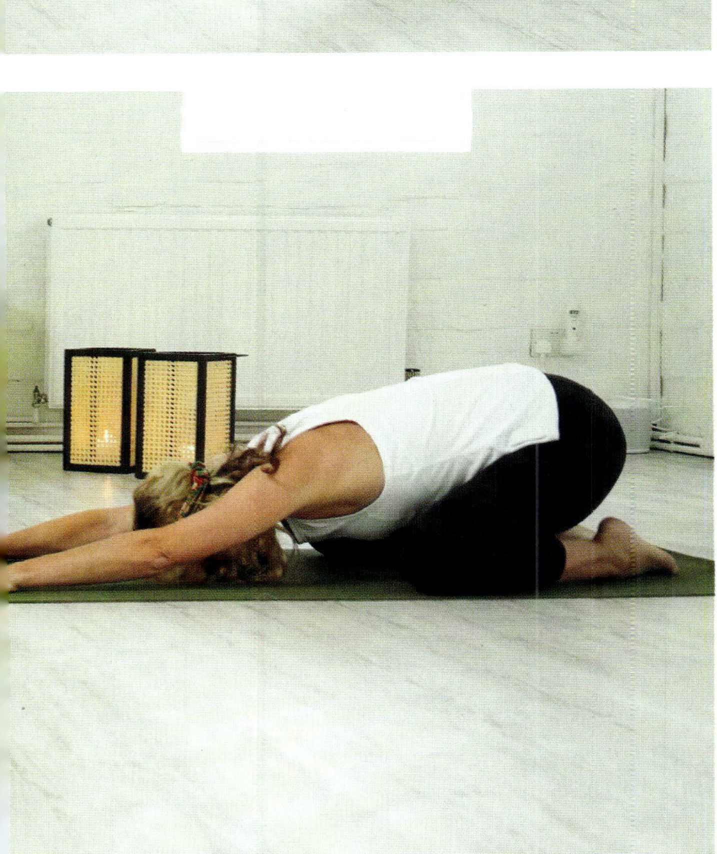

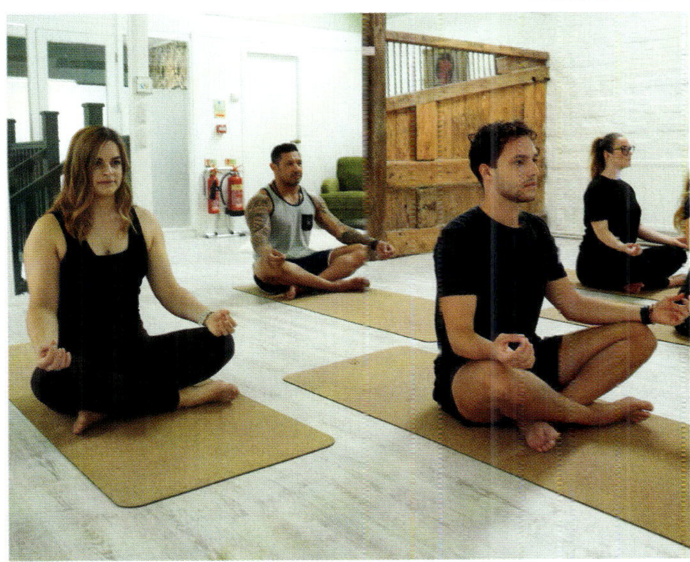

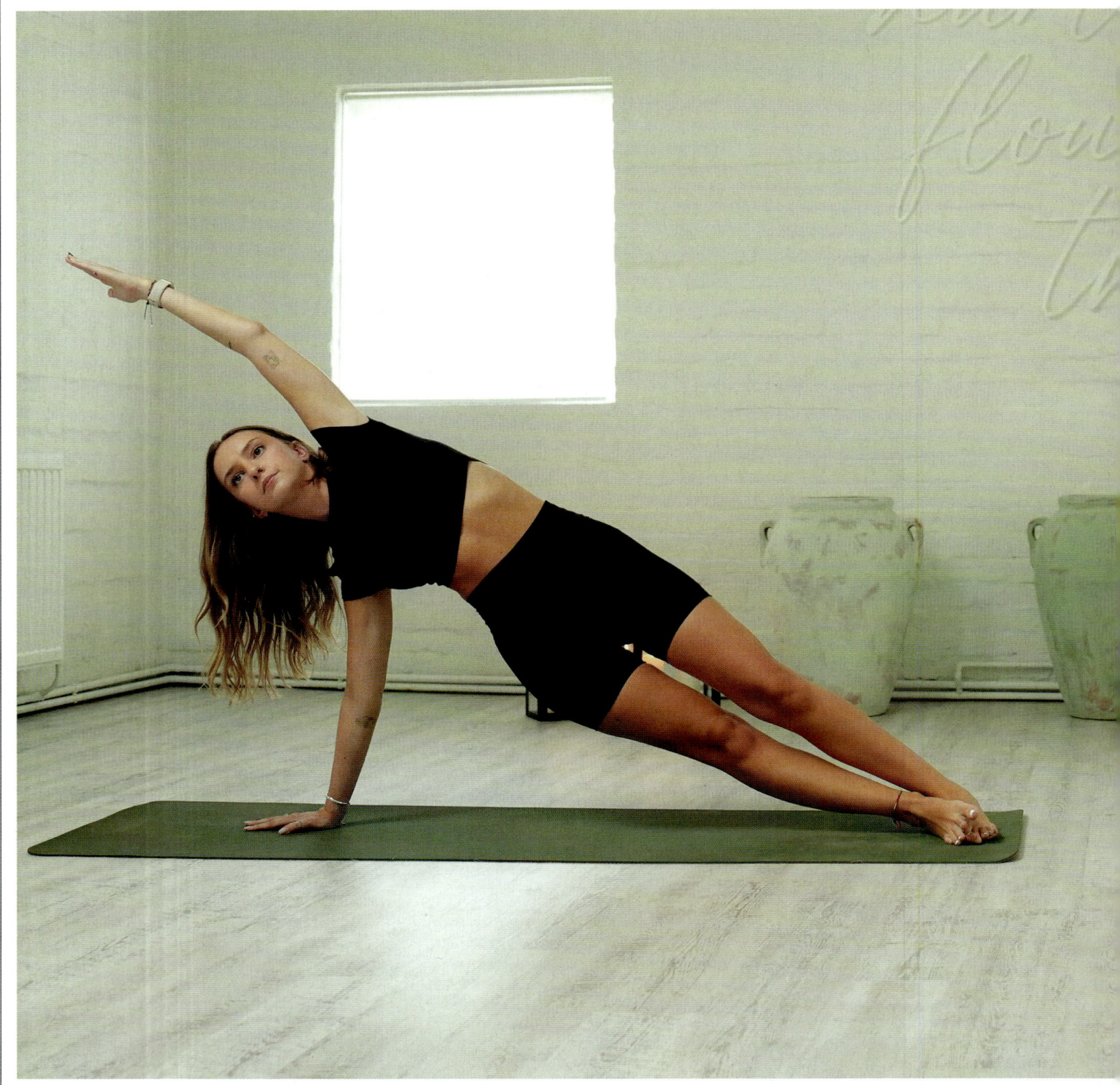

Controlled movement

Using his background in injury rehabilitation, BSc in Sports Science and MSc in Sports Therapy, led **Mike Georgiou** to specialise in Pilates, teaching in studios around the UK and Europe, including Kent's The Nest.

What is Pilates? Well, it depends on who you ask? Some would say it's a workout focusing on abs. Pilates purists may say it's a set of carefully crafted exercises done either on a mat or the various pieces of large equipment and they must be replicated precisely and/or in a specific order. Others say "Isn't it that thing women do on a mat? A bit like yoga?". I have my own idea of what it is, but before we get there, it would be good to visit the source, the methods' creator, a man named Joseph Pilates.

Joseph Humbertus Pilates was born in Germany in 1883 and was a physically weak child who overcame his various ailments through exercise and exposure to nature. Throughout his early years he experienced a

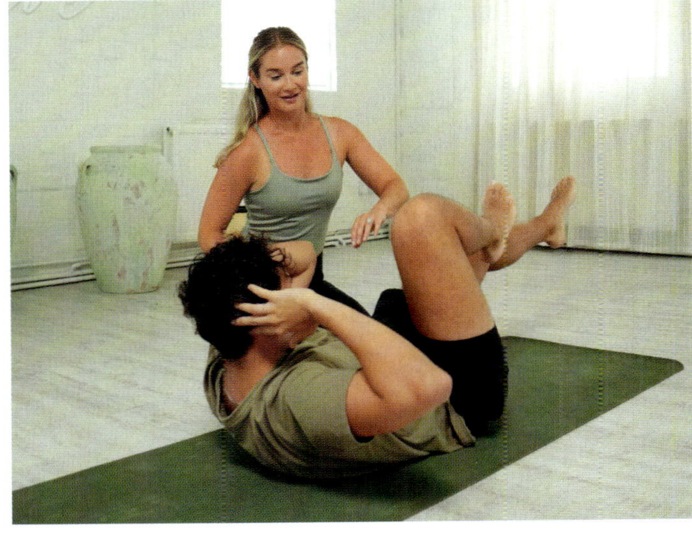

multitude of movement disciplines such as gymnastics, body-building, and martial arts. After moving to England in 1912 he earned a living as a self-defence trainer in police schools as well as a boxer and circus performer.

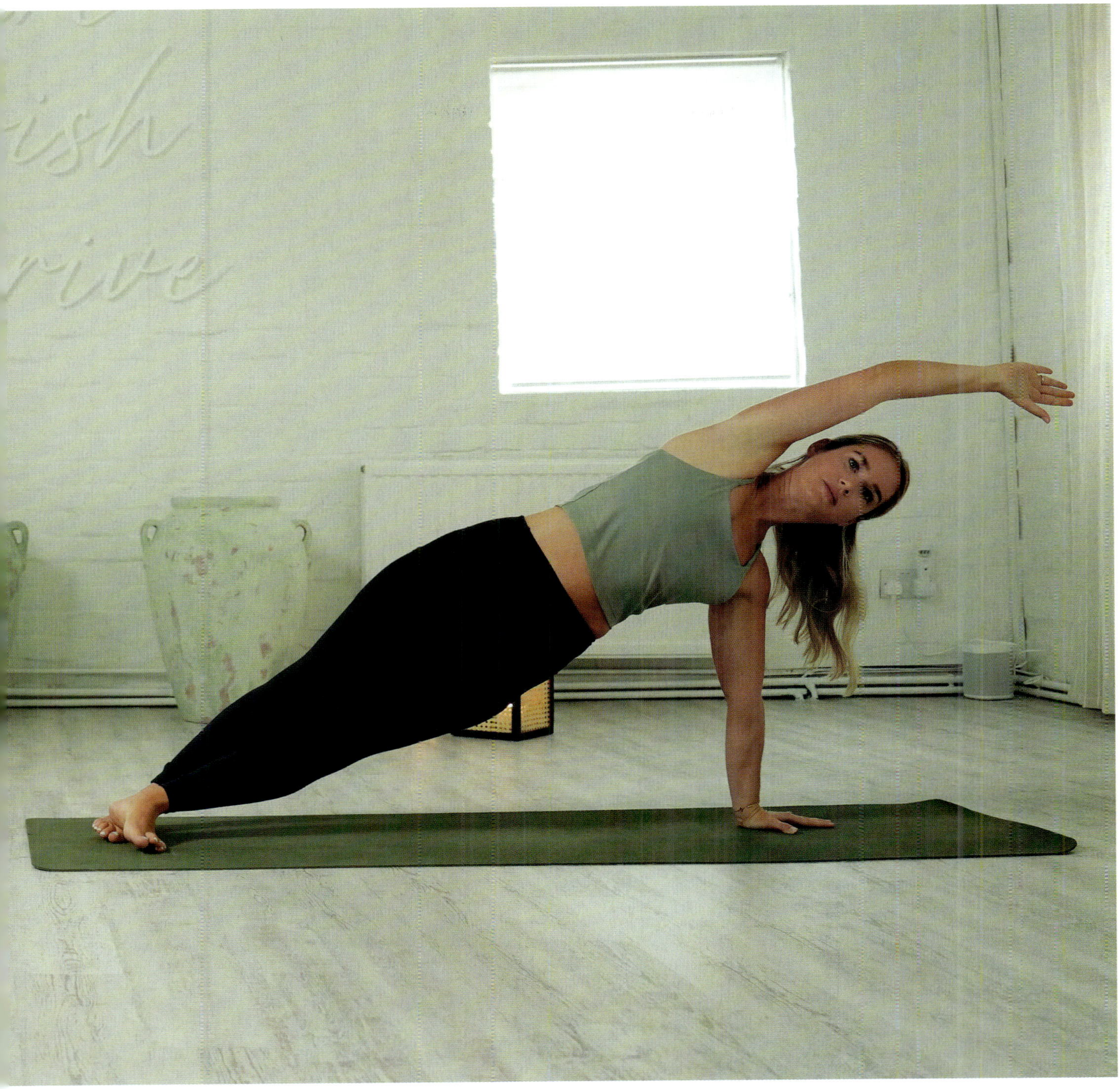

During World War I Joseph Pilates was interned by the British authorities to help train troops in wrestling and self-defence. As an avid scholar of human anatomy and movement, he never stopped learning but also

Joseph Pilates created method of body conditioning which he named 'Contrology'.

innovating, and started to consolidate his knowledge into creating a method of body conditioning which he named 'Contrology', as in control of the mind over the body. His method evolved further whilst working with injured soldiers who he helped rehabilitate through his ingenuity. For those who were bedridden, he thought of using the springs from underneath their beds to add resistance to their exercises to facilitate and speed up the rehabilitations process. This is believed to be the time where he started imagining the creation of the early resistance equipment which later evolved into the Reformer, the Cadillac, and the Wunda chair (among others) as the fundamental apparatus of the complete method. In his own words: "Contrology was conceived to limber and stretch muscles and ligaments so that your body will be as supple as that of a cat, not muscular like a brewery-truck horse or professional weightlifter you so much admire at the circus"…"Contrology develops the body uniformly, corrects wrong postures, restores physical vitality, invigorates the mind, and elevates the spirit."

A few years after the war ended, he moved to New York where he and his wife Clara opened the first Contrology studio. His method became famous within the professional dance community both for its training and rehabilitative components. In fact, some of his early students, some of whom then went on to carry his method forward and teach the next generations, were dancers. It is these dancers who helped evolve the method further to become more refined and with an added finesse and elegance to certain movements.

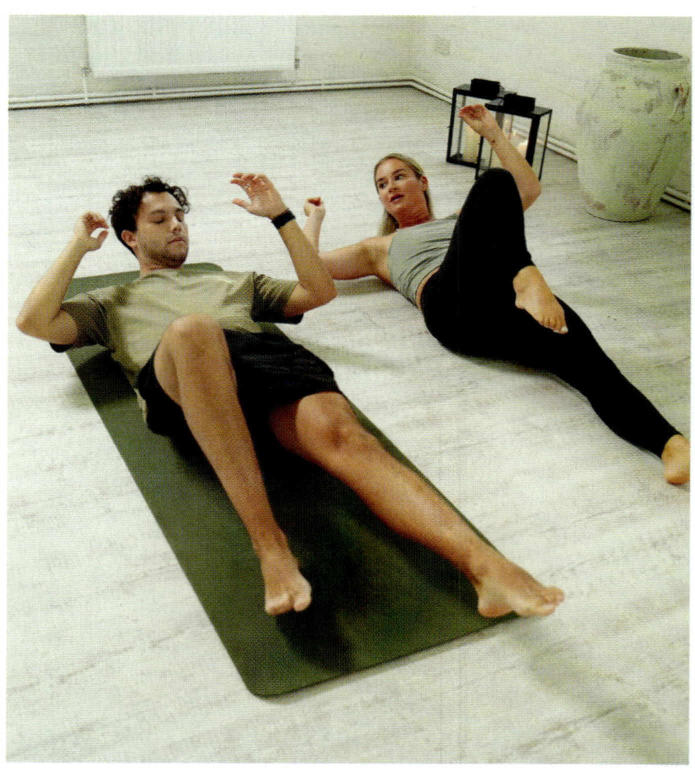

Contrology develops the body uniformly, corrects wrong postures and restores physical vitality.

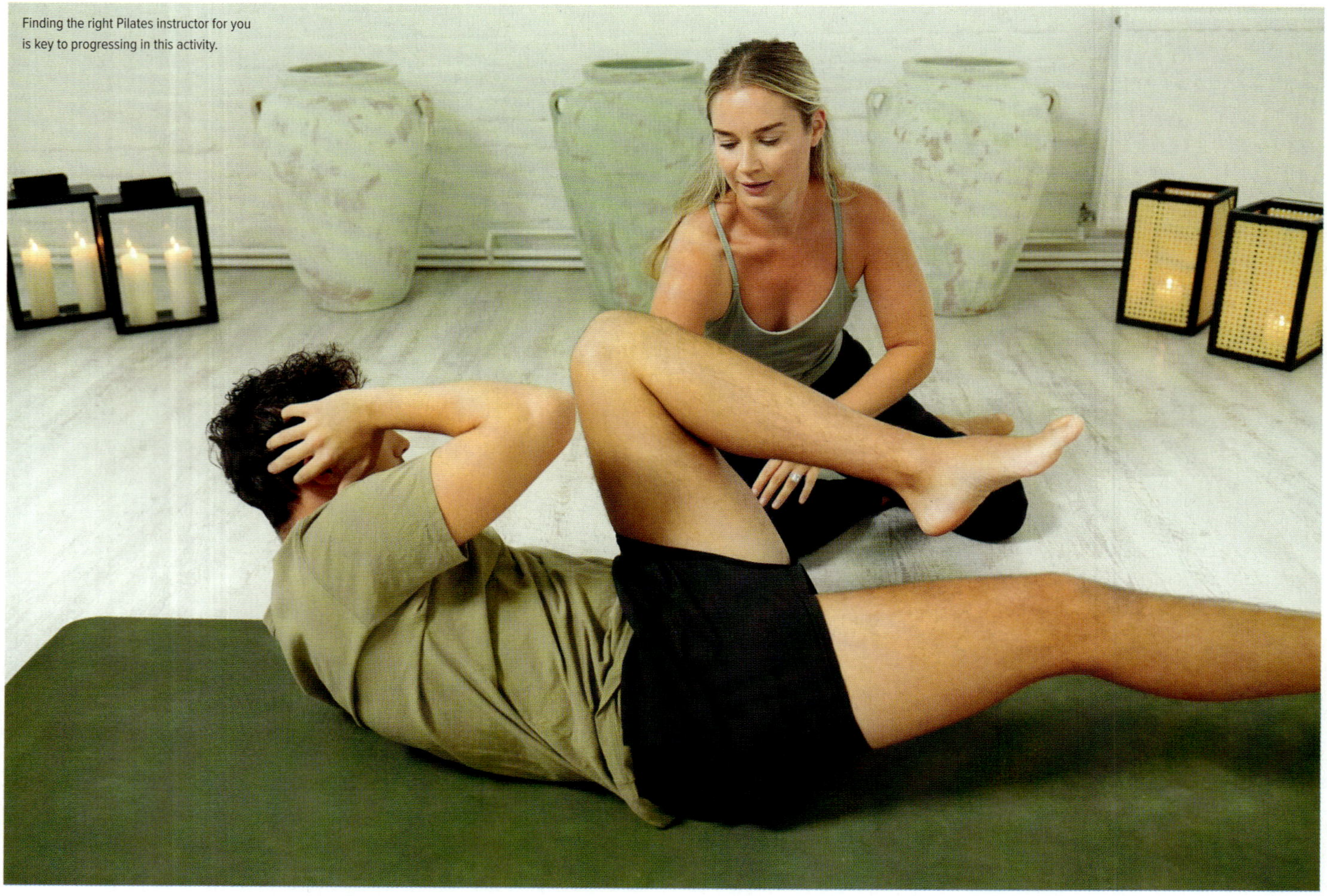

Finding the right Pilates instructor for you is key to progressing in this activity.

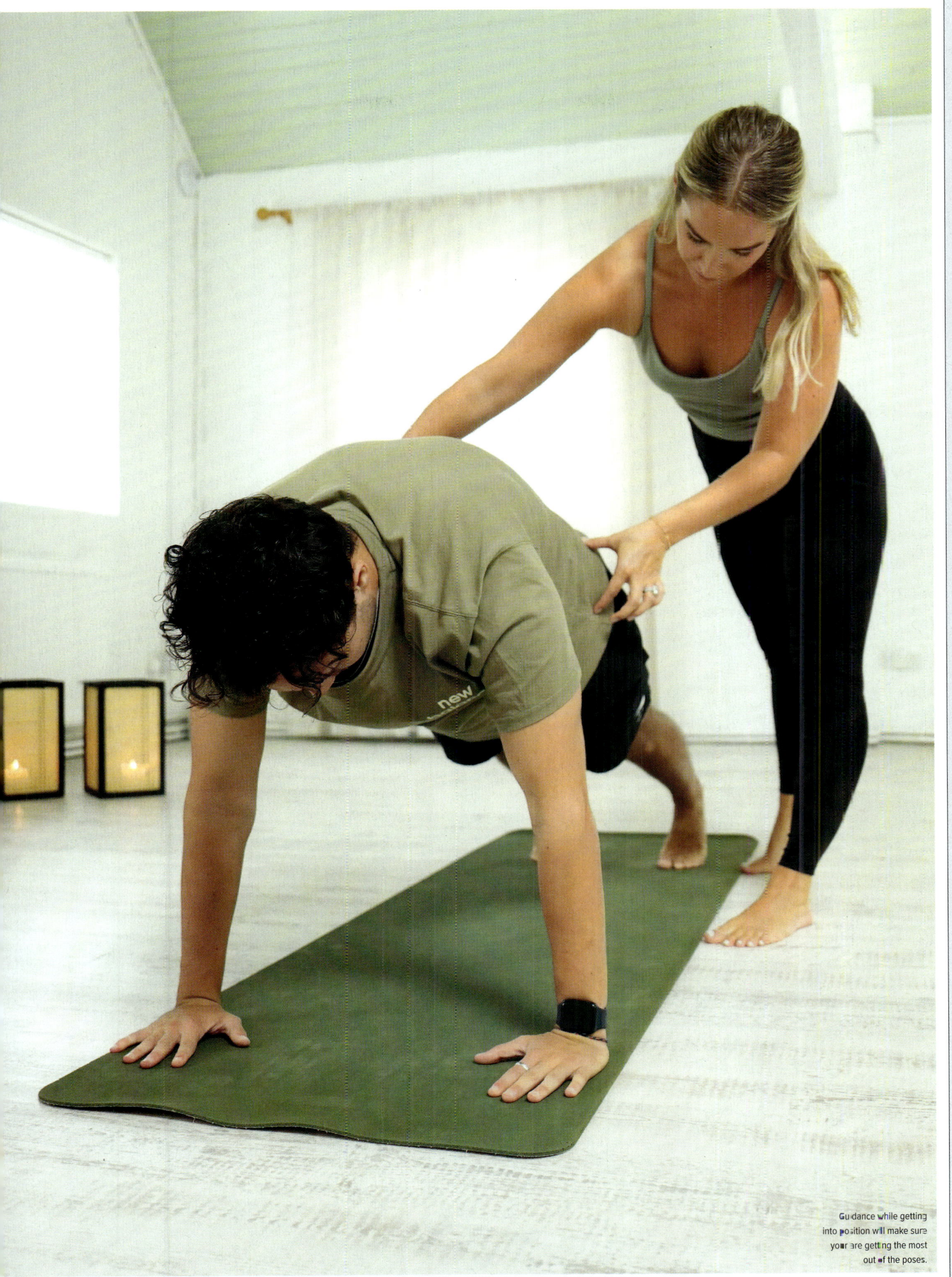

Guidance while getting
into position will make sure
your are getting the most
out of the poses.

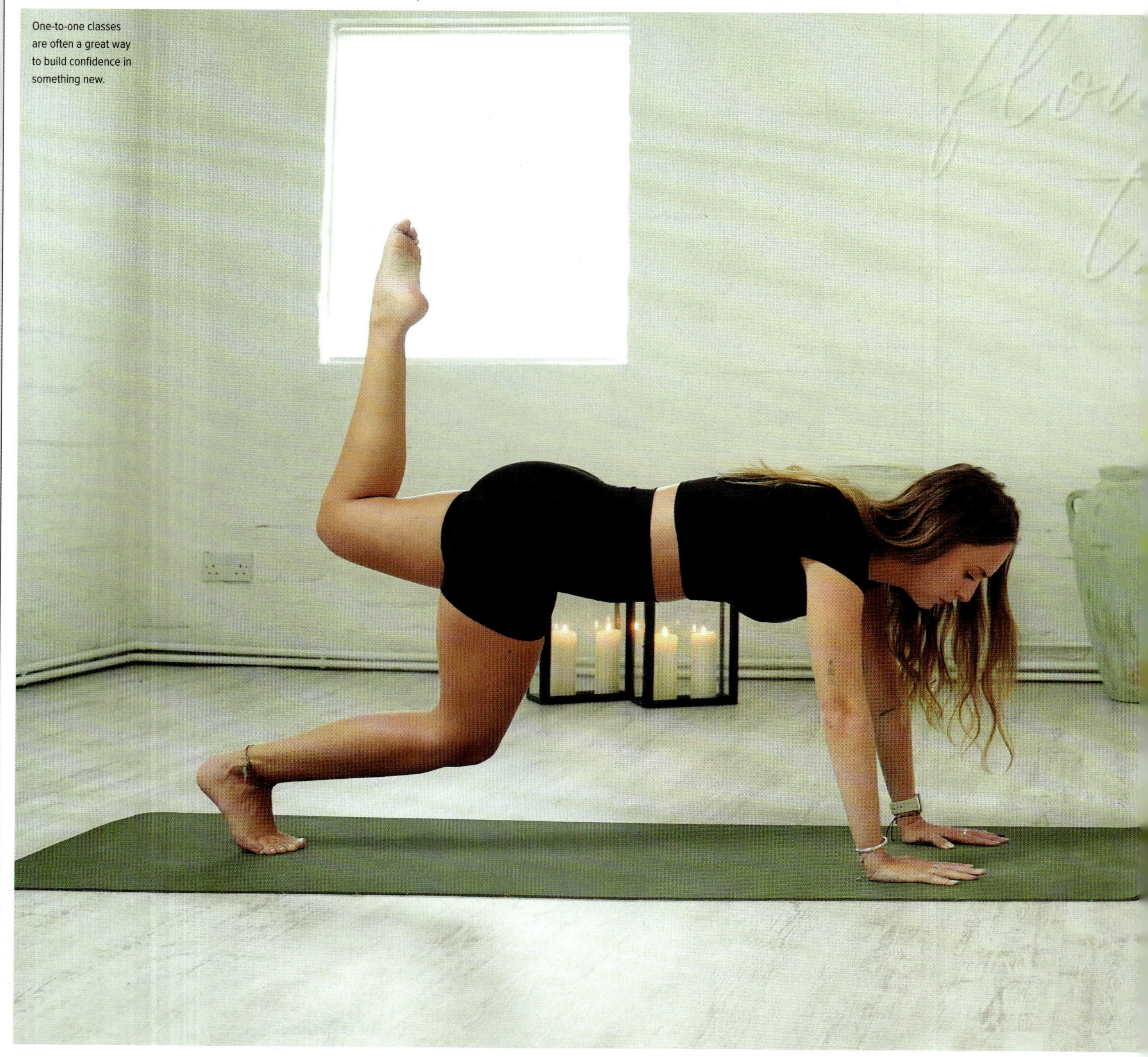

As the method was inherited by various first generation teachers, each went on to add their own personal touch whilst at the same time staying true to some core fundamental principles. It is certainly the case that none of the original exercises have been lost in time, but have, in some cases, been slightly modified. Certain additions would have also been made to the repertoire over time, which would bring us to the current reality of the classical and contemporary schools of thought. Classical being as true as possible to the source, and contemporary being explorative in terms of utilising the main concepts and Pilates philosophy to create exercises that are either a modification or a deconstruction of the original exercises, with the aim of making the method more accessible to a wider audience.

I am certainly no purist. Whilst in awe of the man and his work and whilst respecting how he would have wanted his exercises to be performed, I live and work

> These days Pilates is recommended to absolutely everybody, of any age, physical capacity, or disability.

in a time long after his, where things have changed. Gone are the days where one could stand on a client's torso whilst they were performing an abdominal exercise (yes, he really did that!). These days Pilates is recommended to absolutely everybody, of any age, physical capacity, or disability, and is recommended by doctors and physiotherapists to people suffering from all sorts of injuries. I have worked with people with such complicated disabilities they would never have possibly been able to attempt, let alone master, some of the original Pilates exercises.

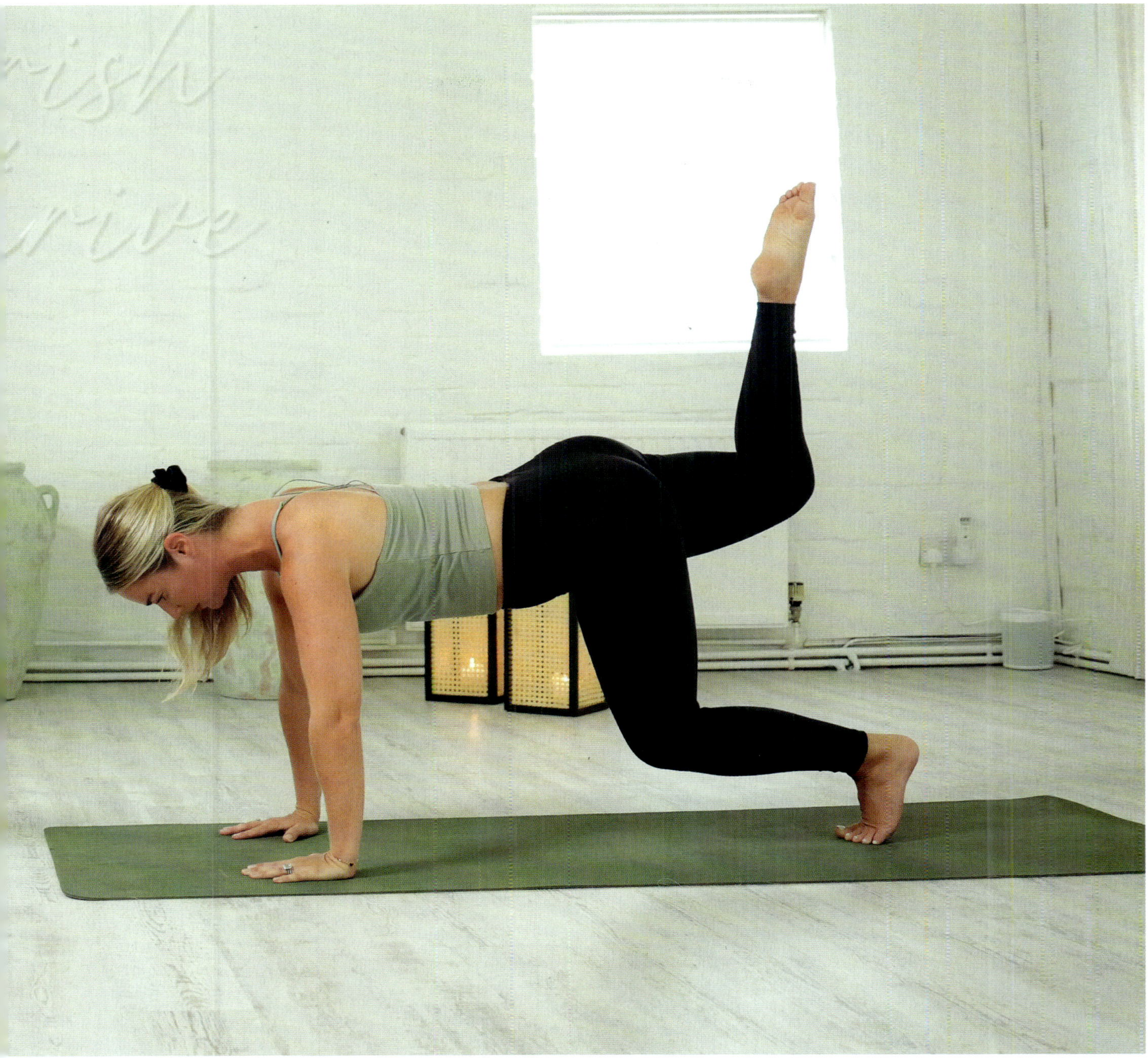

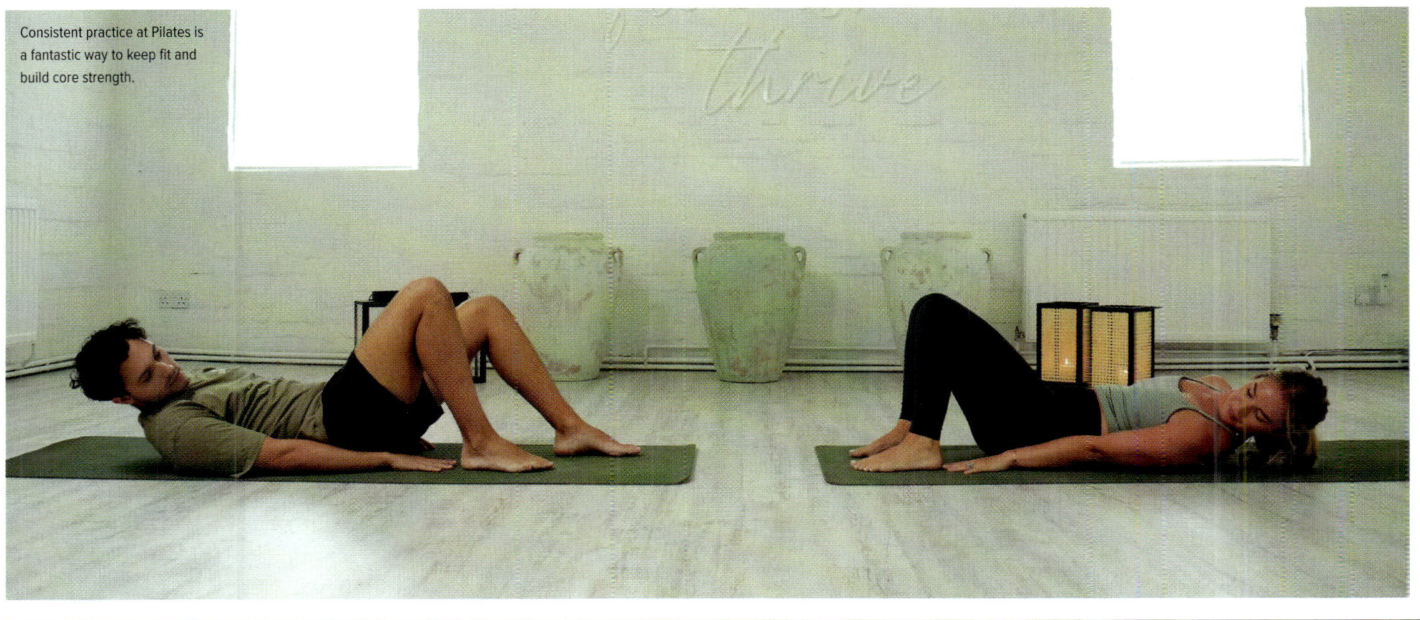

Consistent practice at Pilates is a fantastic way to keep fit and build core strength.

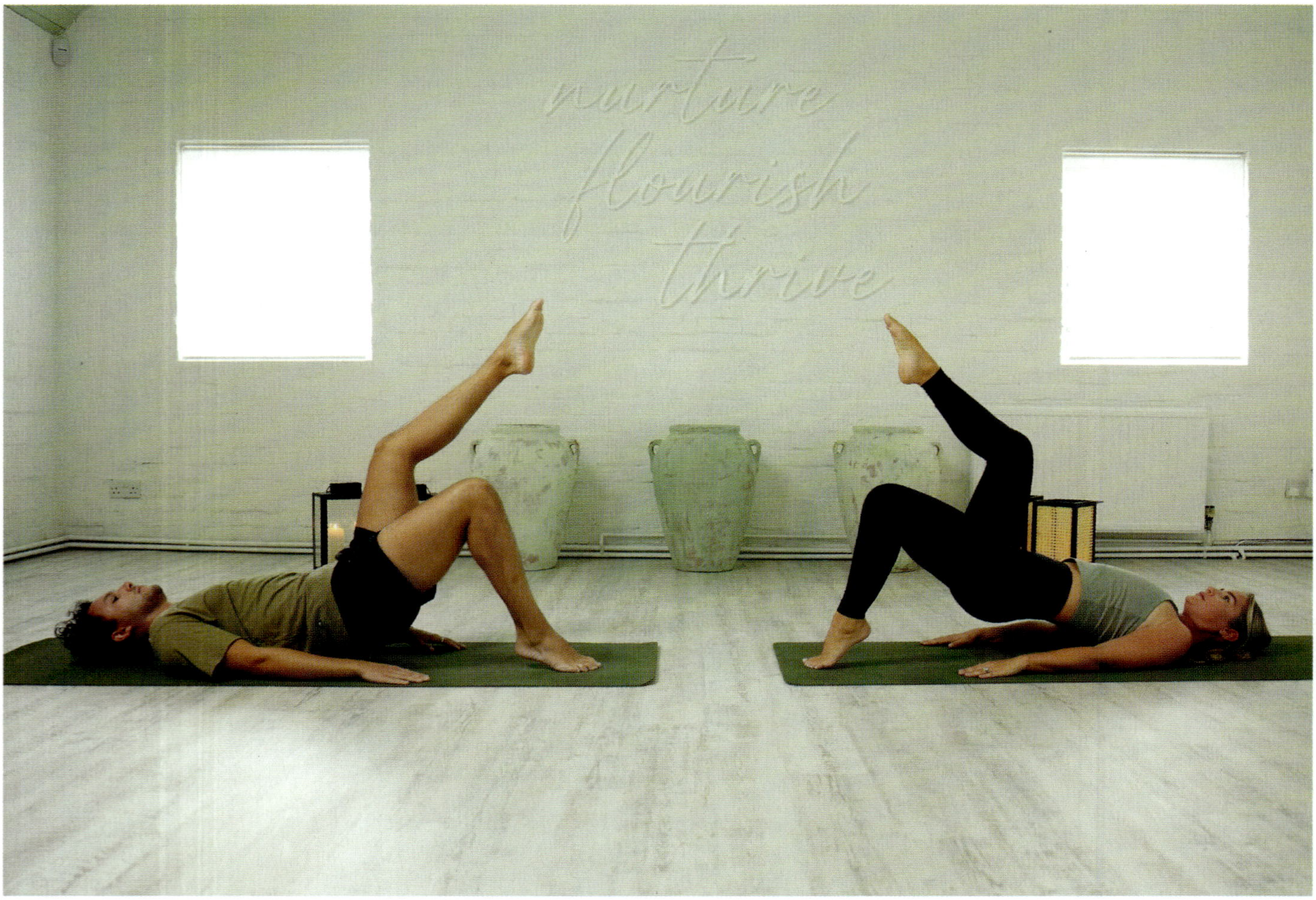

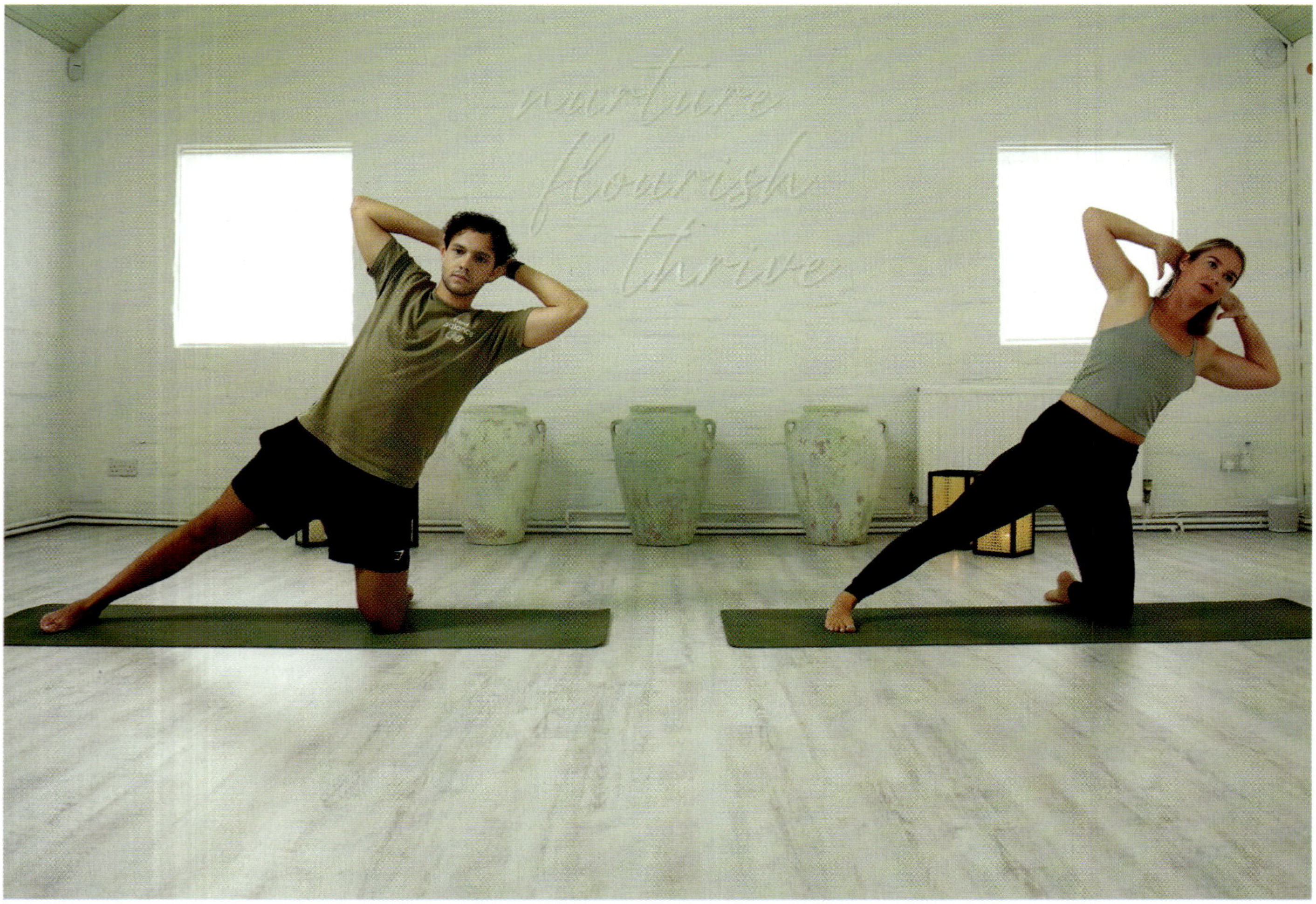

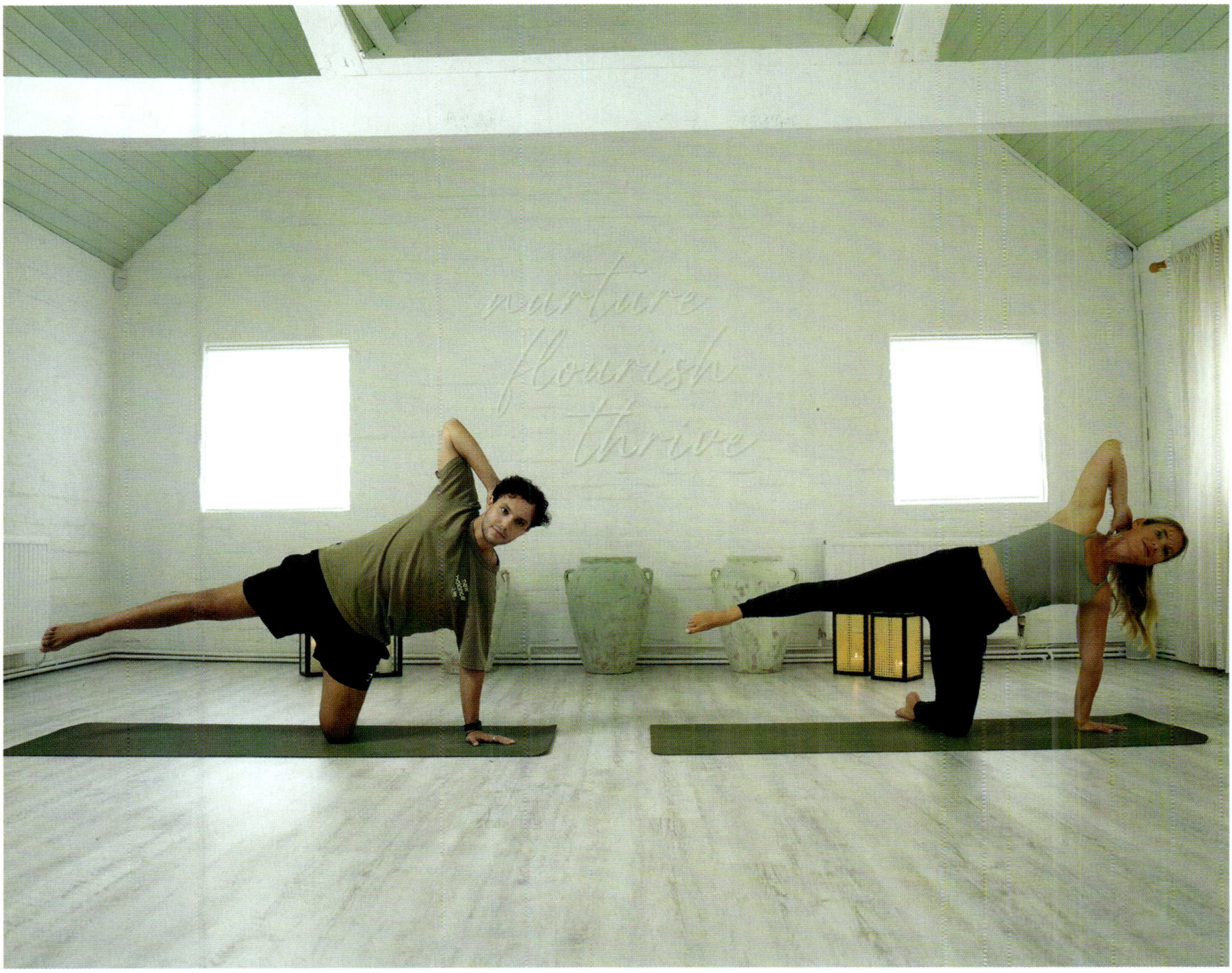

Which is why for me Pilates is a wonderful movement philosophy based on a clear and sound underlying technique. It is, for me, perhaps closer to the conceptualisation imagined by Joseph Pilates when he gave it the name 'Contrology'. It is a method through which the practitioner can gain better control and understanding of their body. It can help them overcome muscular imbalances and/or injuries and move in the most biomechanically correct and efficient way possible. Throughout each exercise, priority is given to the spine. To know when and how it must be stabilised or mobilised. In either case, how it can be supported by the appropriate core activation. How the breath can affect and be affected by the core activation, and how the two elements connect to slow and controlled movements that flow from the centre of the body to the extremities.

Pilates is truly for everyone. It can be a powerful addition to the conditioning repertoire of elite athletes. It can support the aging body in maintaining or improving posture and mobility. It can prepare the body for pregnancy and aid a swift recovery post-partum. It can be an immensely useful tool in the rehabilitation of musculoskeletal injuries. In general, it is a method that can help people move better and feel better in movement. *M&B*

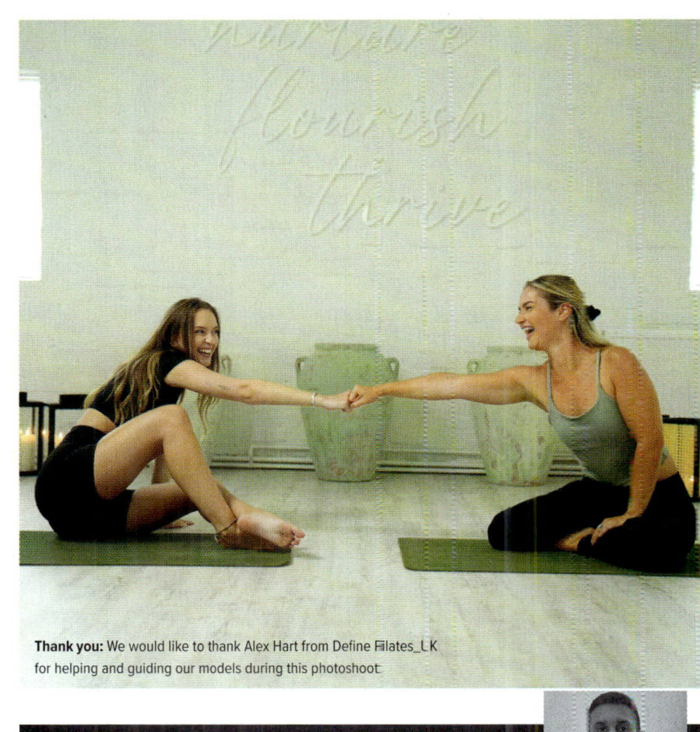

Thank you: We would like to thank Alex Hart from Define Pilates_LK for helping and guiding our models during this photoshoot.

Michael Georgiou
Pilates and Rehab Sevenoaks

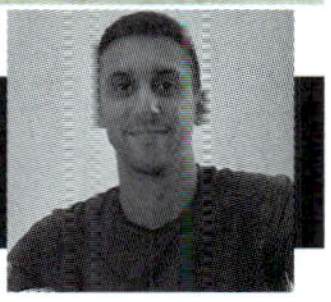

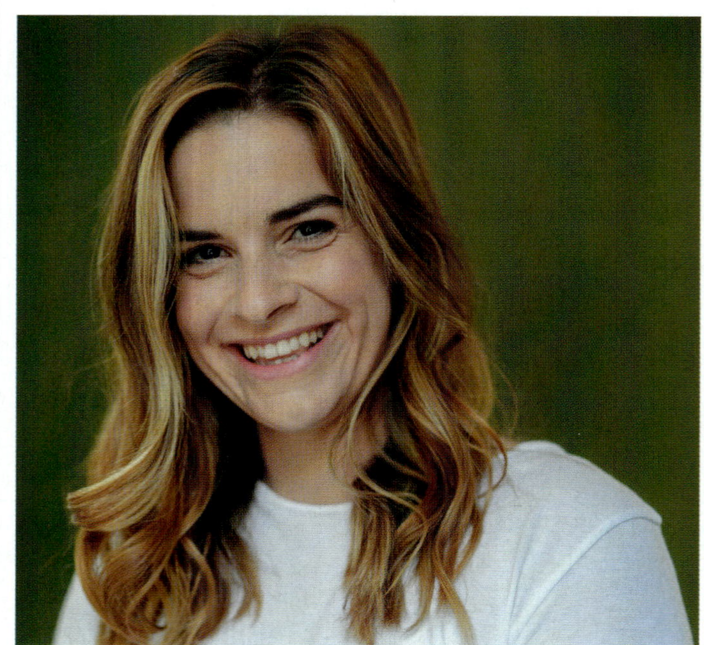

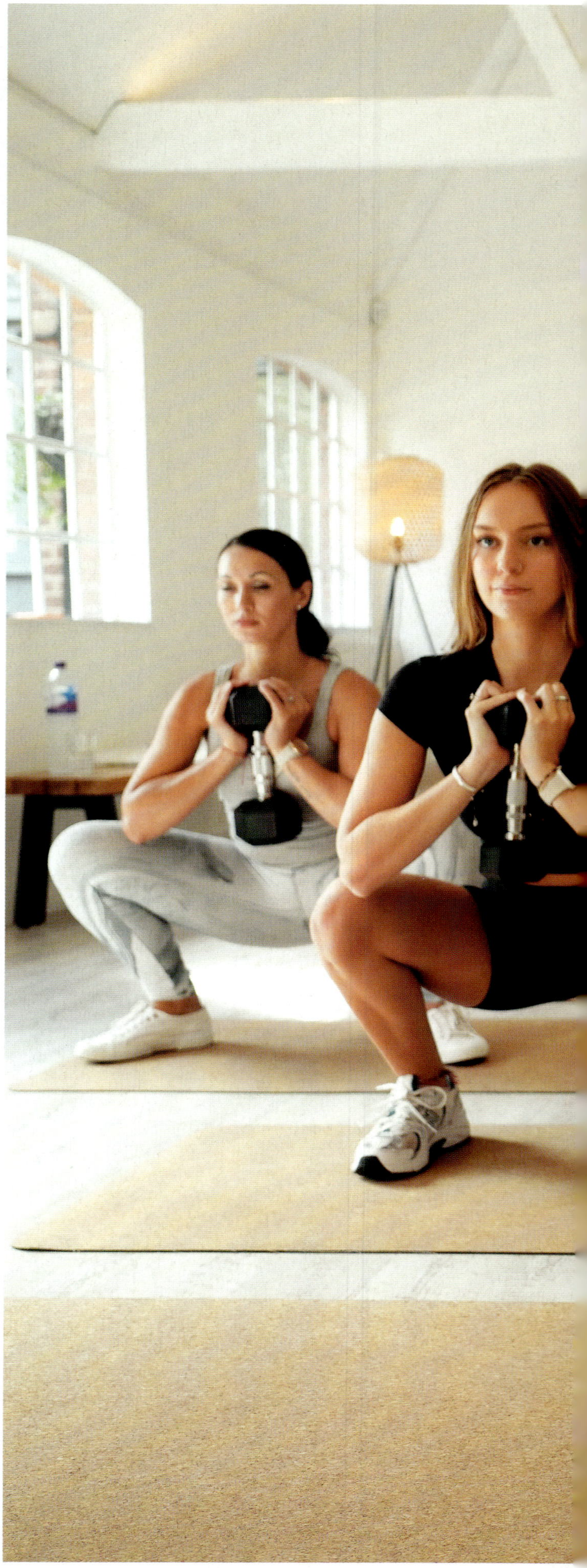

More than gyms and diets

Realising the impact good physical health has on both your mental health and mindset, is a key driver for **Aniela Szkoda**. Passionate about helping people achieve a fit and healthy lifestyle, she is a keen advocate that working on your mindset and learning to love and respect yourself alongside exercise, can lead to the results we strive for physically.

L ike many, my introduction to fitness was in no way motivated by being fitter or healthier but purely to change my body. I was only young, but I had already created the belief that my "big thighs" and "belly podge" did not conform to the body standards displayed everywhere I looked.

Being a personal trainer, people assume you have always loved exercise. The truth is, I hated it. I was that kid in school who would do anything to get out of PE lessons. In my late teens and early 20s, as I got even more self-conscious of my body, I would begrudgingly fling myself around the living room to home-exercise DVDs and signed up to a gym that I went to probably two-to-three times a month.

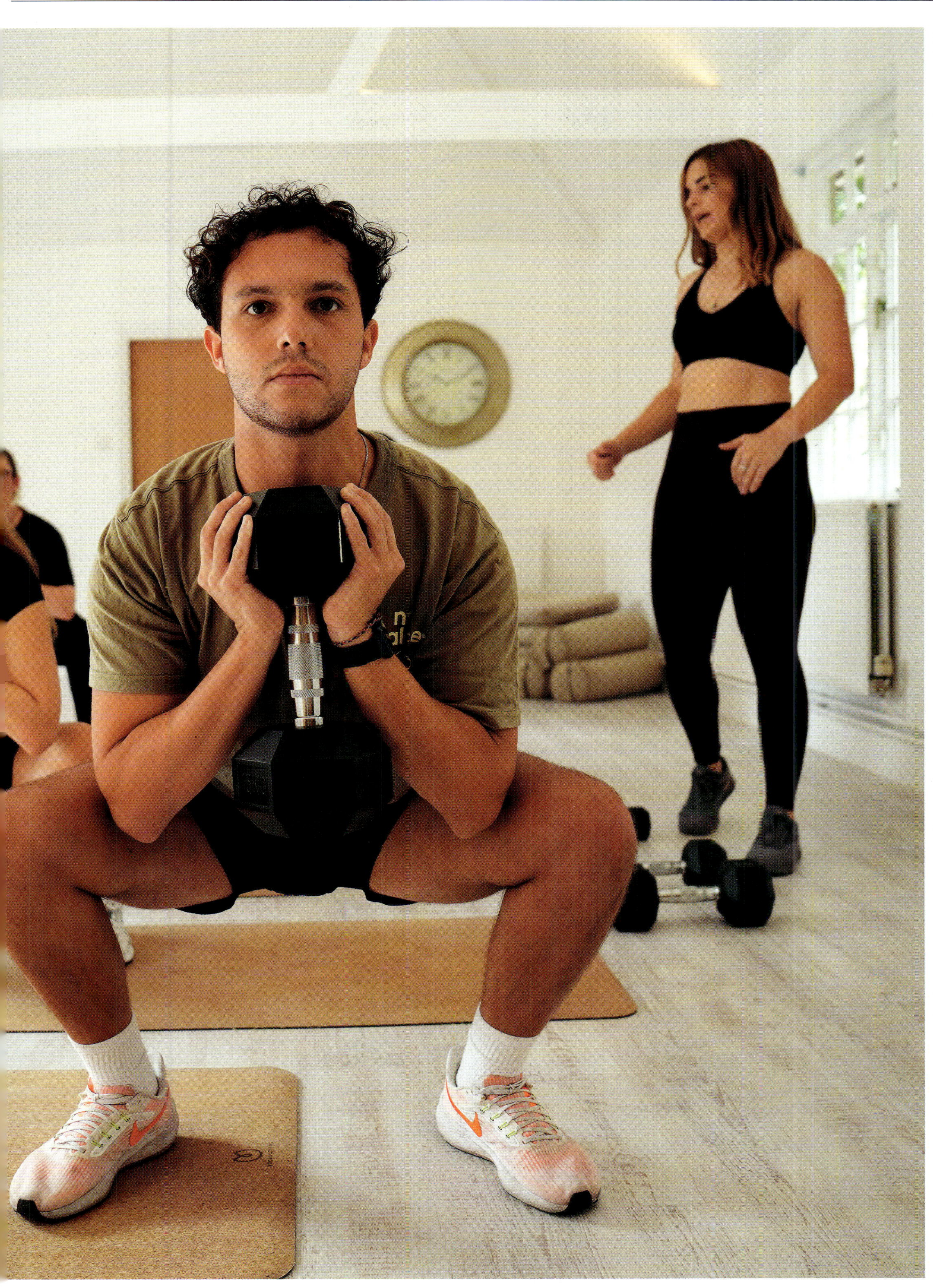

There are so many amazing physical, mental and emotional benefits of fitness.

Nothing stuck. I saw exercise as torture. As something I had to do to get slimmer and definitely not something I enjoyed even slightly. It wasn't until I started running with my best friend and eventually ran the London Marathon that I realised what my body was capable of; that the benefits of exercising are so much more than just losing weight or getting leaner.

Fast forward 10 years and I now get to help others improve their health, fitness and more importantly their strength both physically and mentally. I get to show them the incredible things their bodies are capable of and teach them to fall in love with movement, the way they look and the person they are becoming. I wish someone had shown me from a young age all the amazing physical, mental and emotional benefits of fitness. Let me be that person for you and let's look at how fitness can change your life.

Physical health
There are a magnitude of physical health benefits of exercise including reducing the risk of developing heart disease, diabetes, improving immunity and has

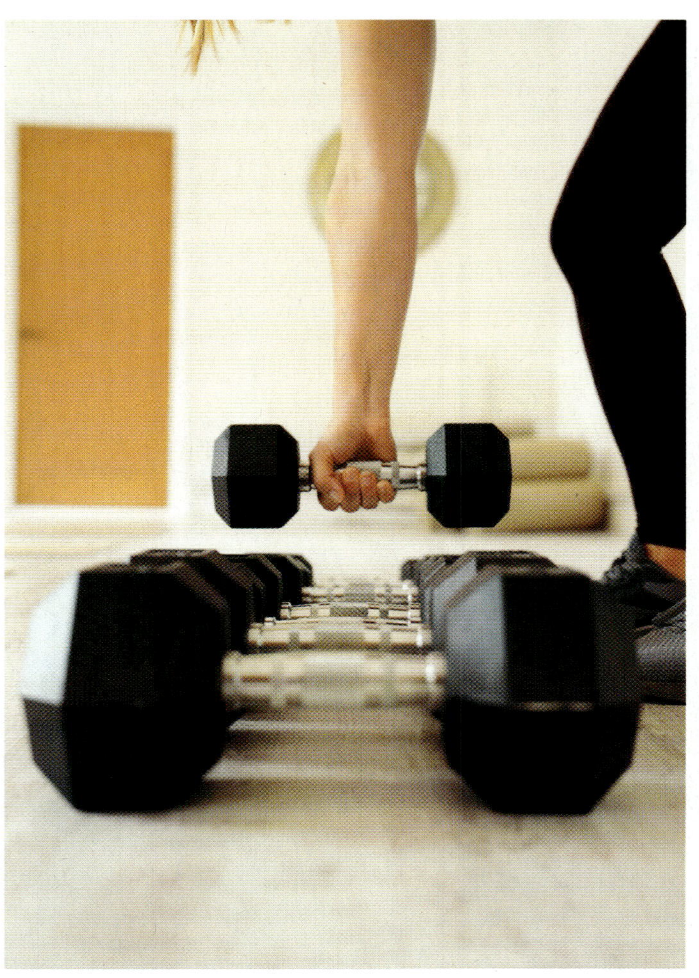

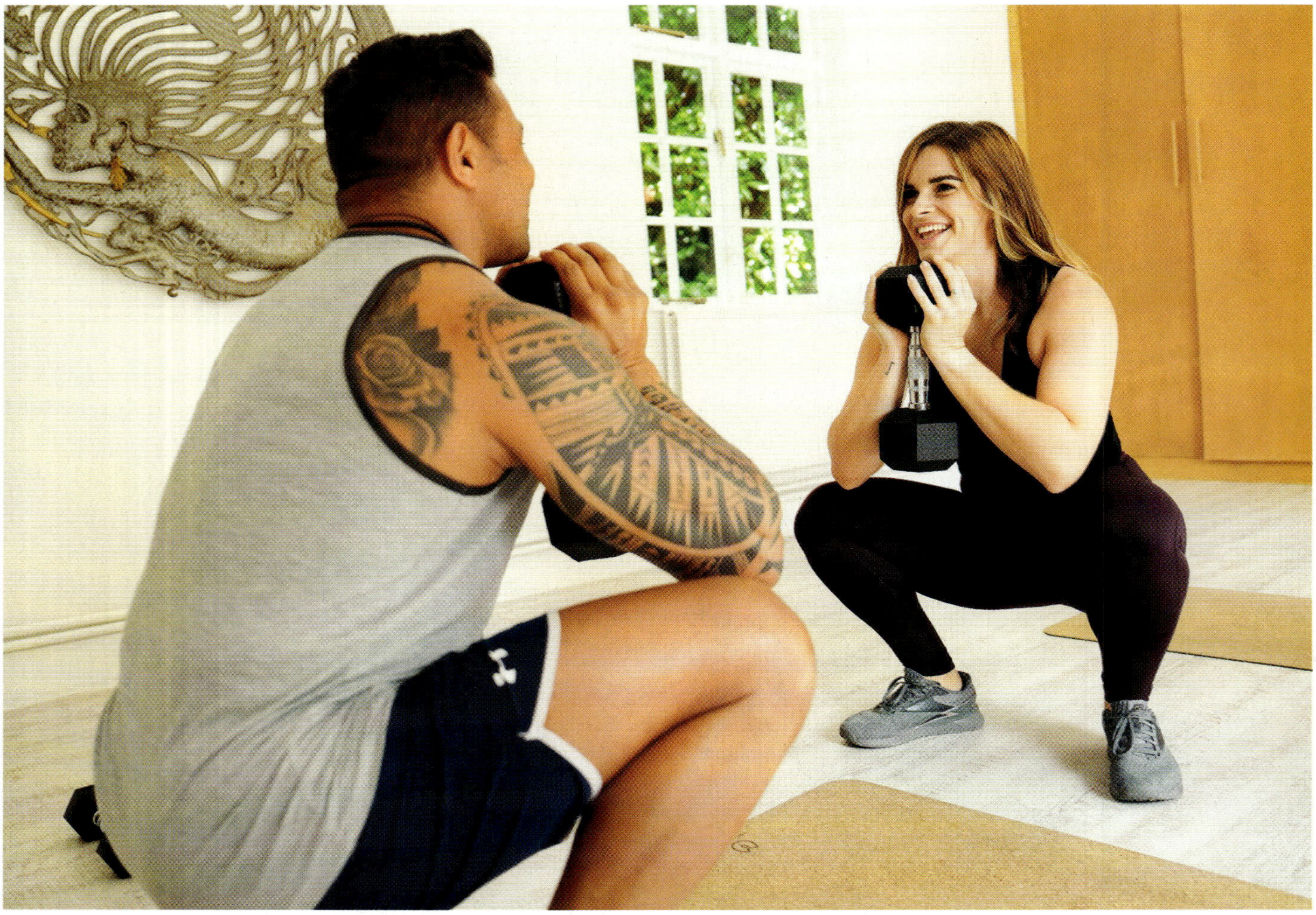

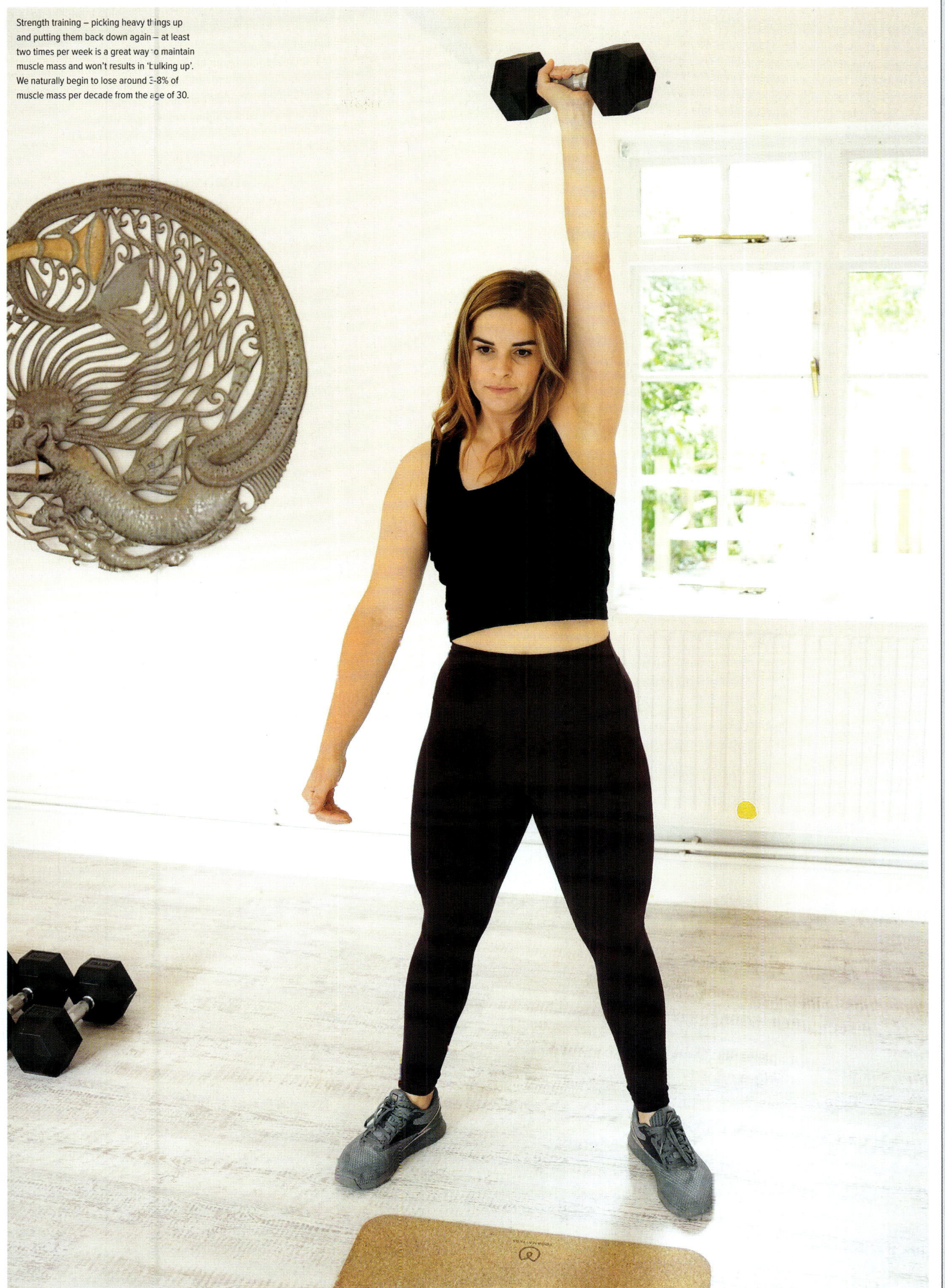
Strength training – picking heavy things up
and putting them back down again – at least
two times per week is a great way to maintain
muscle mass and won't results in 'bulking up'.
We naturally begin to lose around 3-8% of
muscle mass per decade from the age of 30.

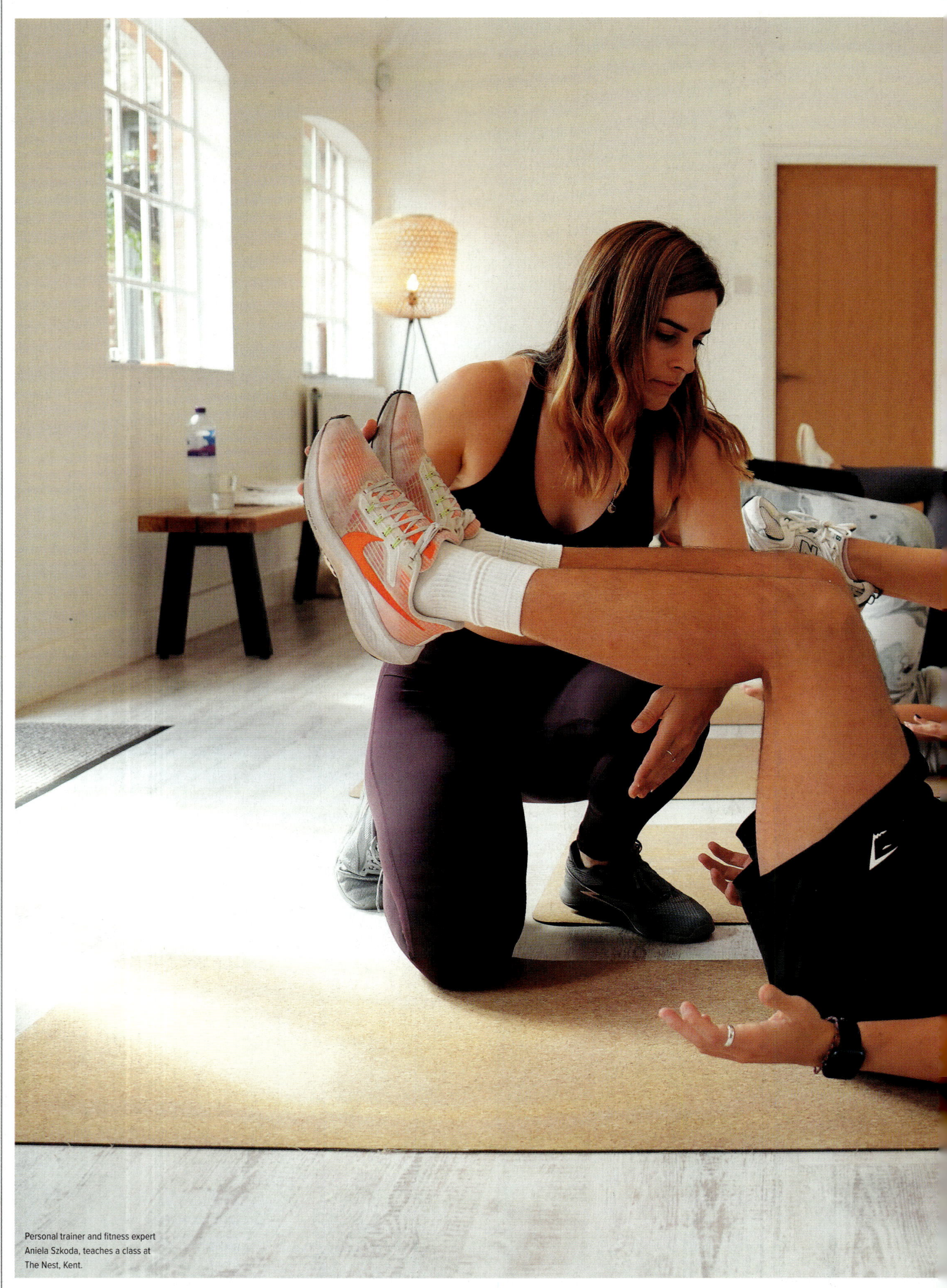

Personal trainer and fitness expert
Aniela Szkoda, teaches a class at
The Nest, Kent.

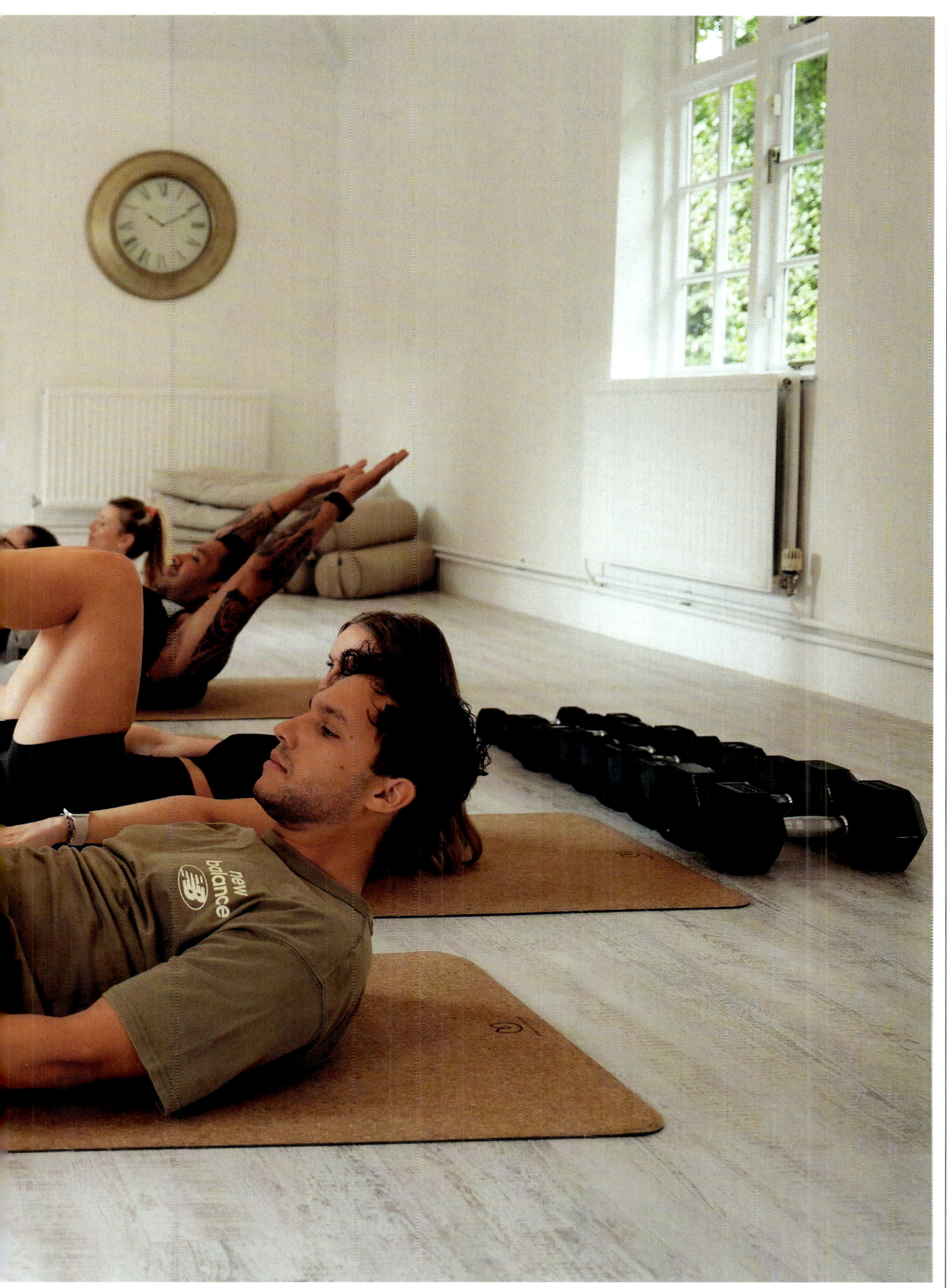

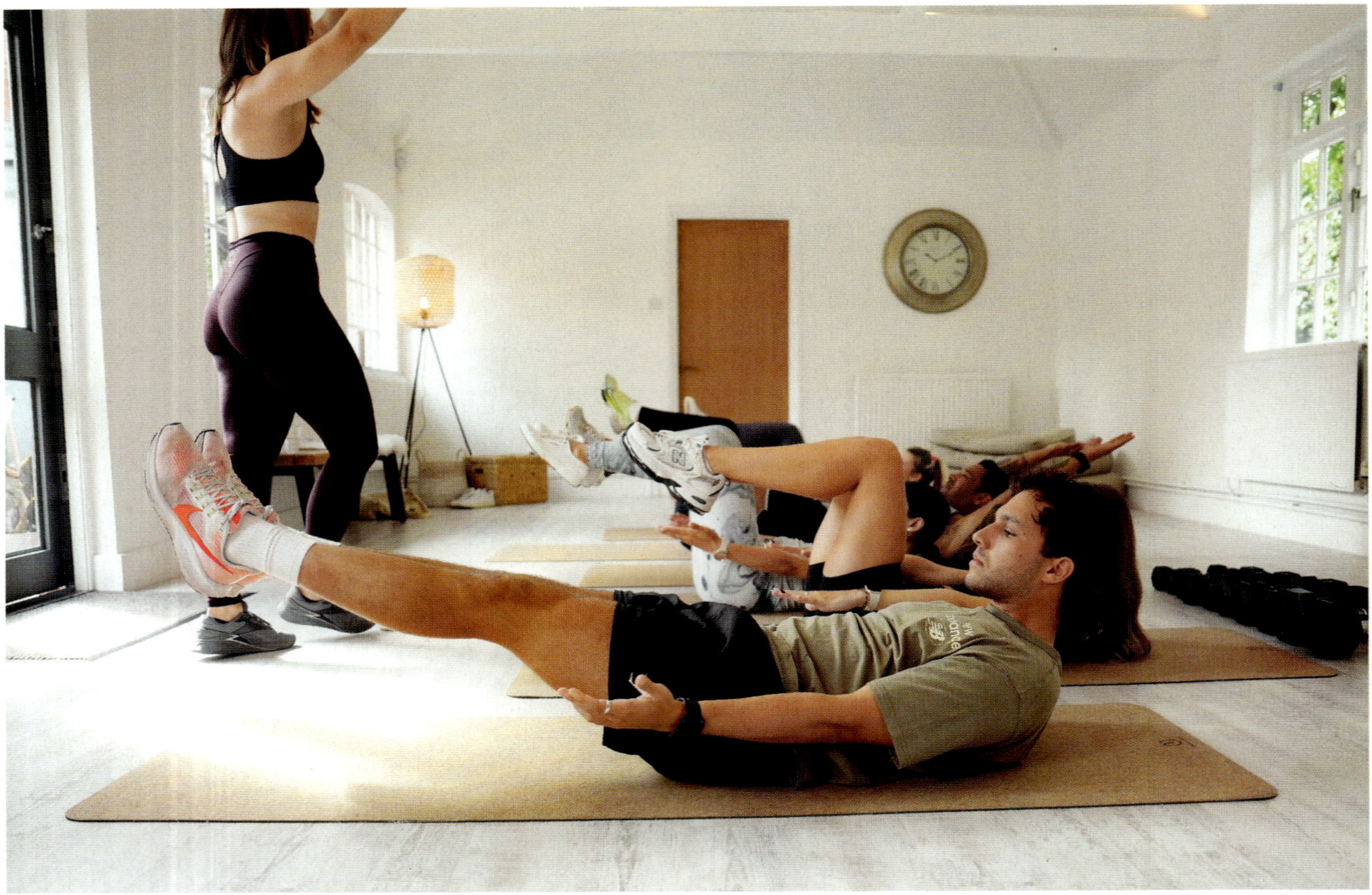

now been linked with lowering the risk of dementia. As a reactive response, regular exercise can also help manage the effects and severity of such diseases.

These physiological impacts of exercise are widely documented but what I don't think is spoken about enough is the benefits of resistance and strength training on both muscular and skeletal health. We naturally begin to lose around 3-8% of muscle mass per decade from the age of 30. This is especially prevalent in women due to hormonal changes increasing this rate of muscle loss even further, which also causes a reduction in bone density and a heightened risk of osteoporosis.

This bone and muscle loss, and subsequently loss of strength, can begin to significantly impede an individual's ability to perform simple everyday tasks such as getting up from a chair and walking, therefore impacting their quality of life and independence.

Strength training – picking heavy things up and putting them back down again – at least two times per week for each of the main muscle groups, is a non-negotiable for all women. It is a common misconception women have

Even starting out with a daily 15-minute walk can instantly improve your mood.

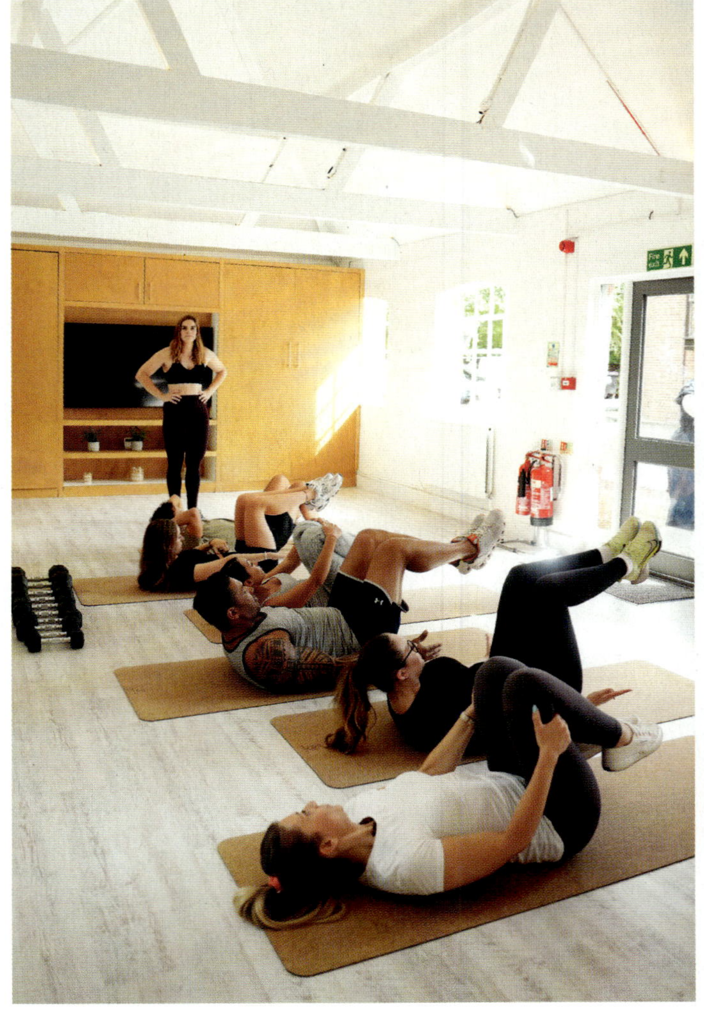

MIND & BODY

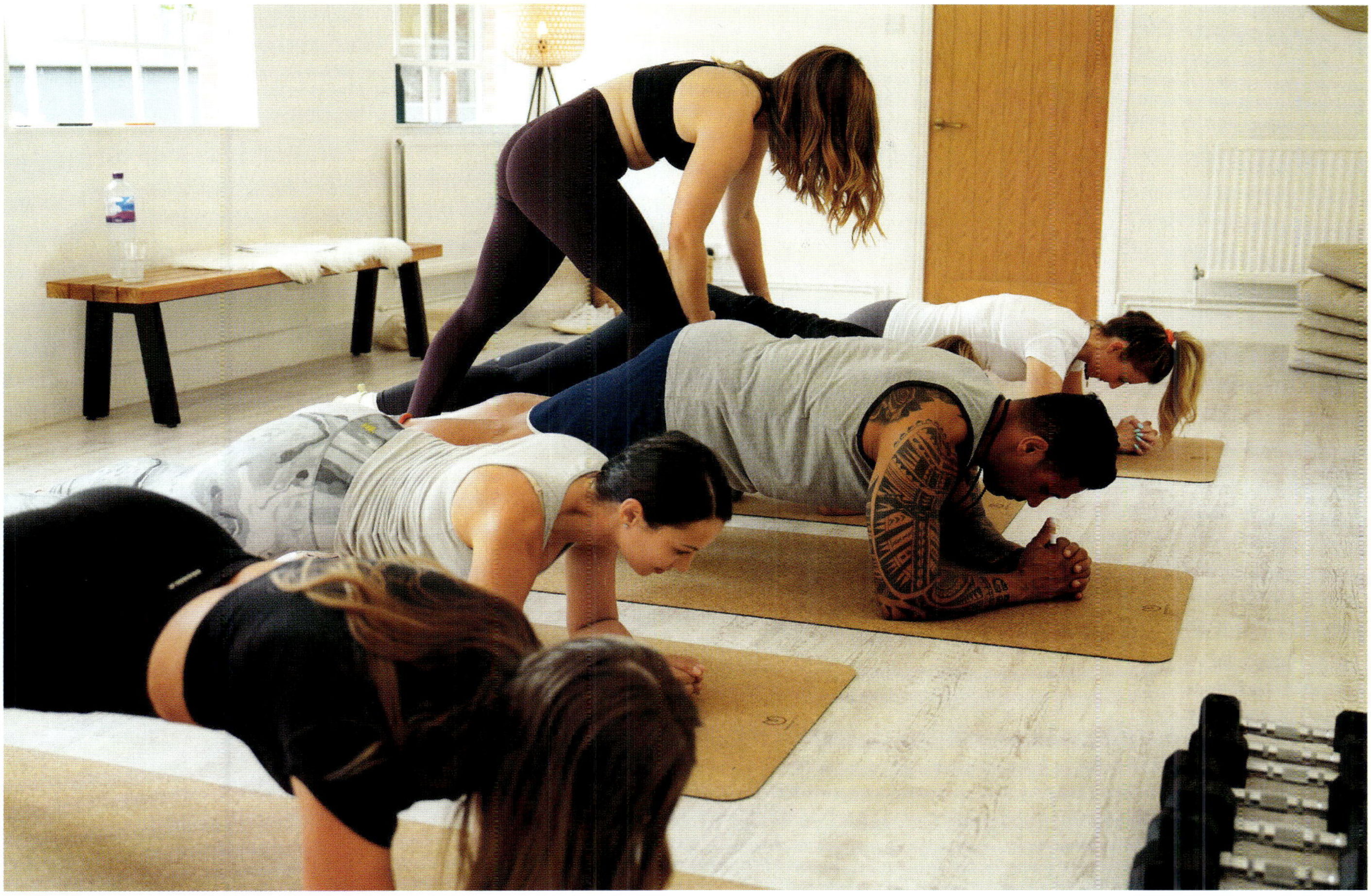

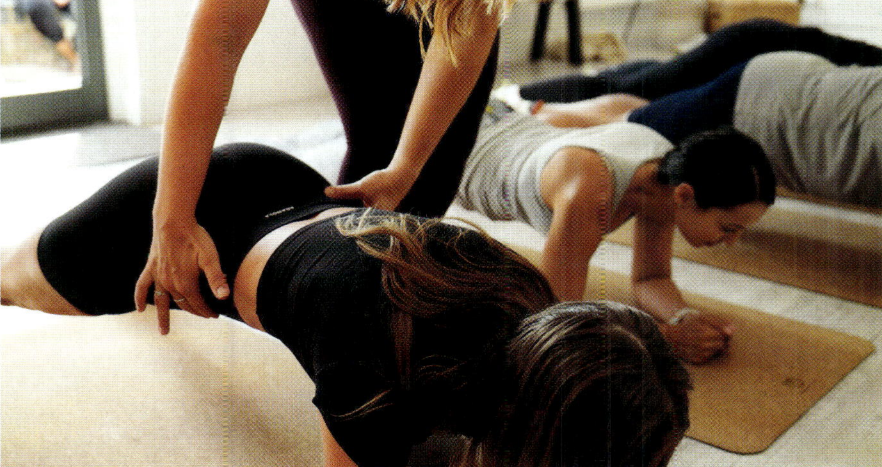

Above: Holding the pose "the plank" not only helps to build core strength, but also strengthens throughout the body: legs, back, shoulders, arms, hands and neck.

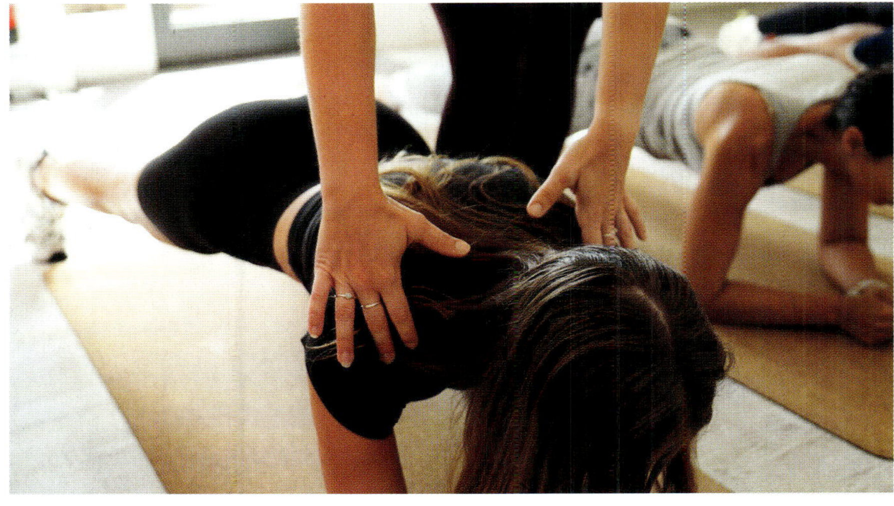

Left: Having a knowledgeable instructor guide you through different exercises can pay dividends, making sure you are holding your body correctly, to get the most out of the pose.

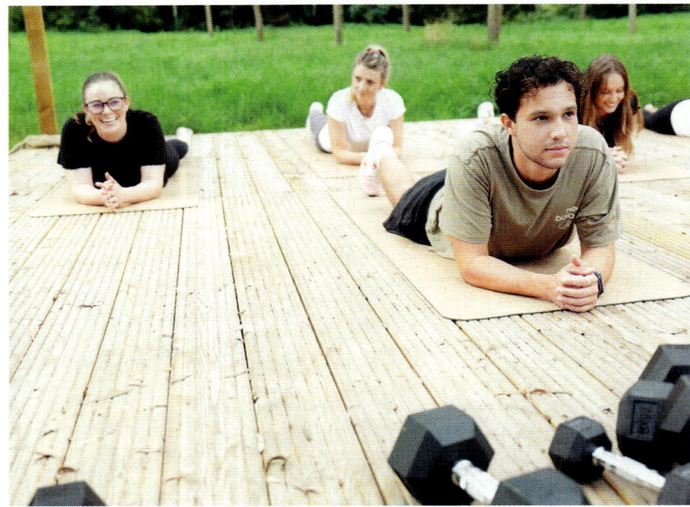

A staggering 74% of UK adults experience stress at least once a month, with one in five feeling stressed more days per month than not.

that lifting weights will make them bulky. Believe me, I wish it was that easy. It takes a lot more than a couple of sessions per week for us to build significant muscle mass. We need to stop worrying about this and instead remind ourselves that lifting weights is imperative to slow down the rate at which our muscles are depleting, improve our bone density and help us feel fitter and stronger even when we are in our 90s!

Mental health

Have you heard of the runners' "high"? That absolute euphoria you feel when your heart is racing and you've got a good sweat on. The good news is you do not have to run to get this high. The feeling people get from physical activity, notably more-so aerobic exercise, is your body experiencing a boost in production of endorphins; the feel-good hormone which acts as a natural pain-killer and improves your mood almost instantly.

Even starting out with a daily 15-minute walk can instantly improve your mood and help with symptoms associated with depression and anxiety such as fatigue, brain fog, headaches and tension throughout the body. Studies have shown adults who participate in daily physical activity have a 20-30% lower risk of depression than those who are sedentary.

Endorphins are just one of the chemicals in our brains that have an impact on how we feel, think and subsequently act in certain situations. Some of the other hormones released during exercise also improve our mood and overall mental health, such as serotonin, dopamine and adrenaline.

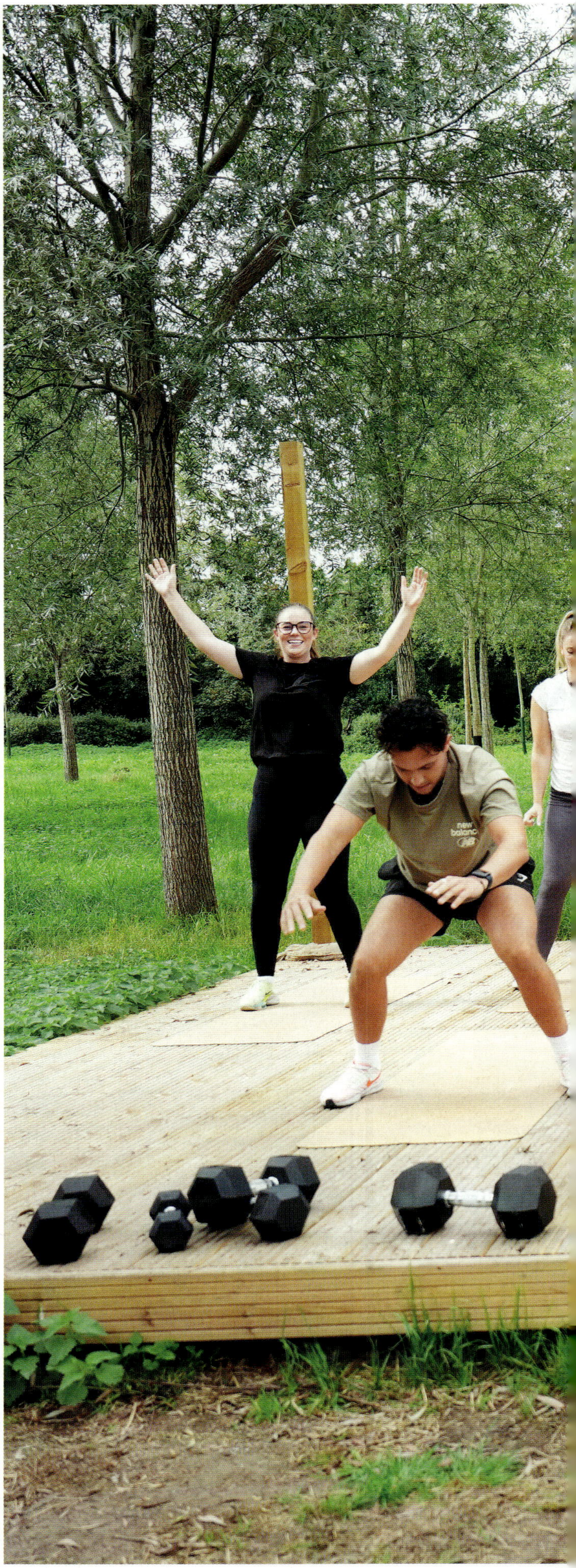

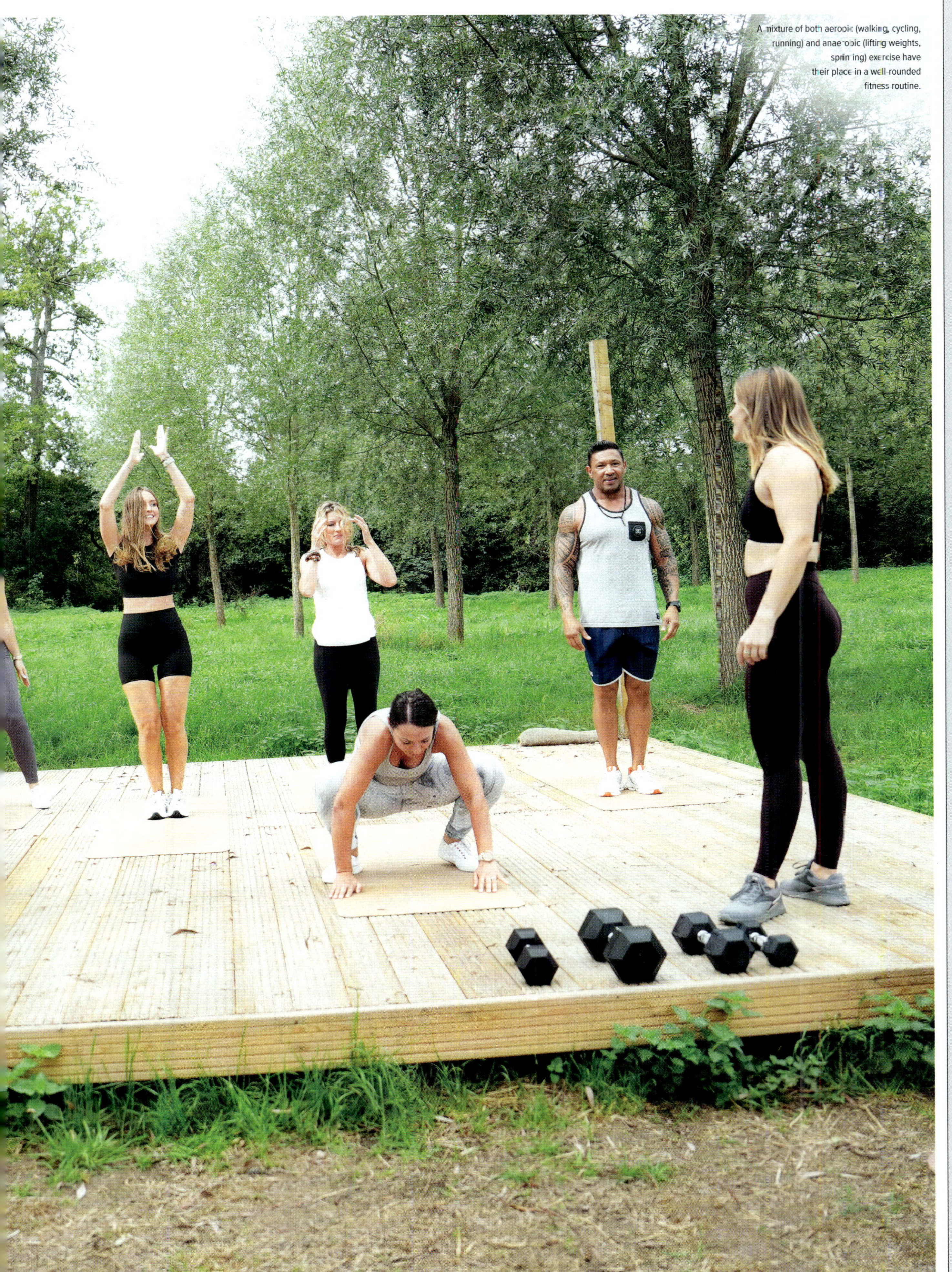

A mixture of both aerobic (walking, cycling, running) and anaerobic (lifting weights, sprinting) exercise have their place in a well-rounded fitness routine.

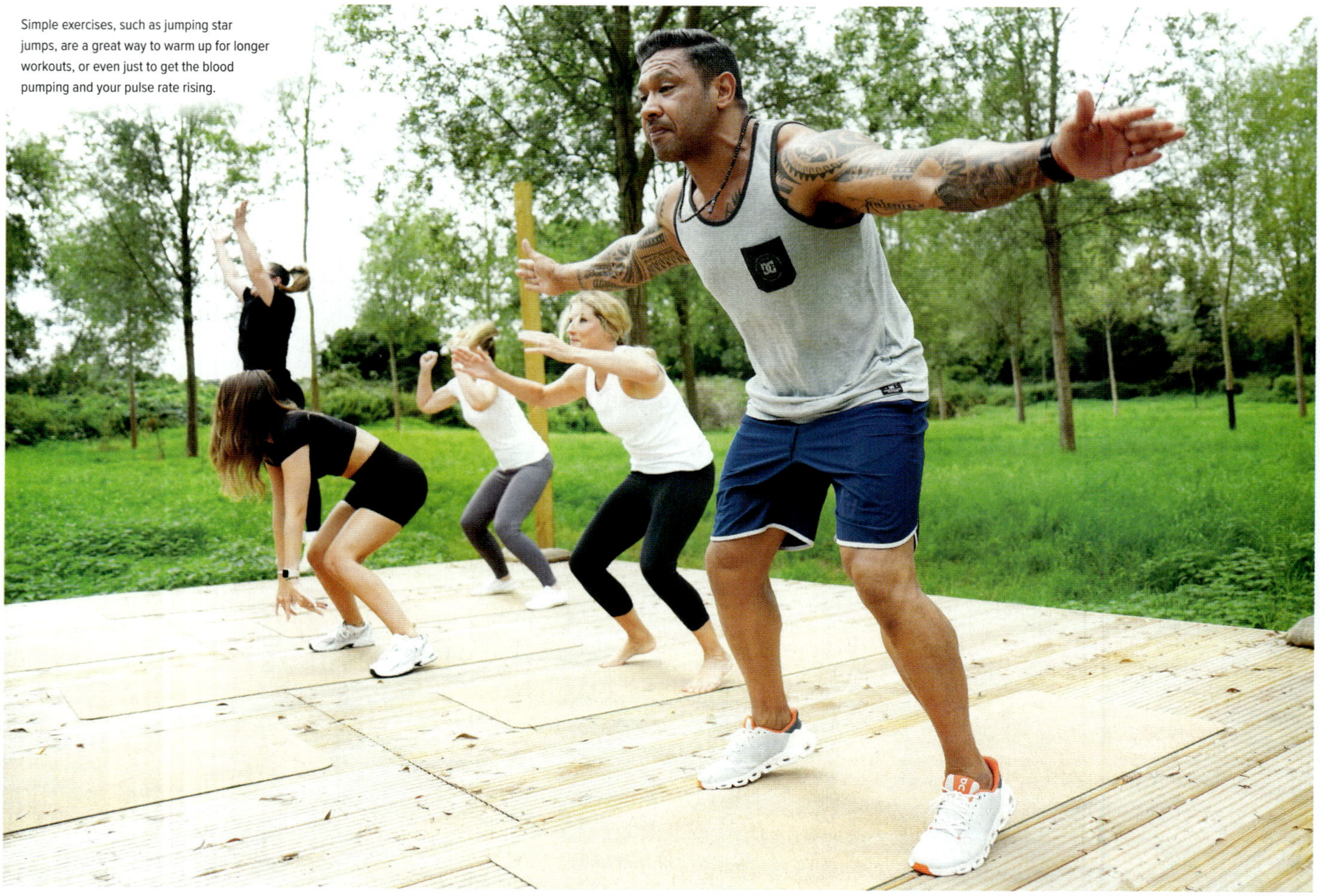

Simple exercises, such as jumping star jumps, are a great way to warm up for longer workouts, or even just to get the blood pumping and your pulse rate rising.

Both dopamine and serotonin are natural mood boosters. Dopamine gives us a temporary sense of pleasure, while serotonin creates longer lasting feelings of happiness and well-being. Boosting the production of serotonin is what most antidepressant drugs target, as low levels of serotonin are associated with depression and anxiety. With physical activity being proven to significantly increase levels of serotonin in the brain, it is no wonder why health specialists are beginning to incorporate exercise into treatment plans for an array of mental health conditions.

Stress

A staggering 74% of UK adults experience stress at least once a month, with one in five feeling stressed more days per month than not. Stress is a completely normal chemical reaction within the body and is in fact crucial for survival. However, the problems arise when we are in a heightened state of stress for a prolonged period of time. With the amount of 'stressors' we are now exposed to on a daily basis - from traffic on your way to work, to household bills, to what to watch on Netflix - it's no wonder many people feel overwhelmed and burnt out.

Exercise and getting your body moving, is one of the best strategies for combating stress. Not only does it release those happy hormones it negates some of the negative impacts stress is already having on your body. Physical activity has been proved to reduce the levels of both adrenaline and cortisol in the body which in excess can lead to depression, anxiety, fatigue, IBS and low libido.

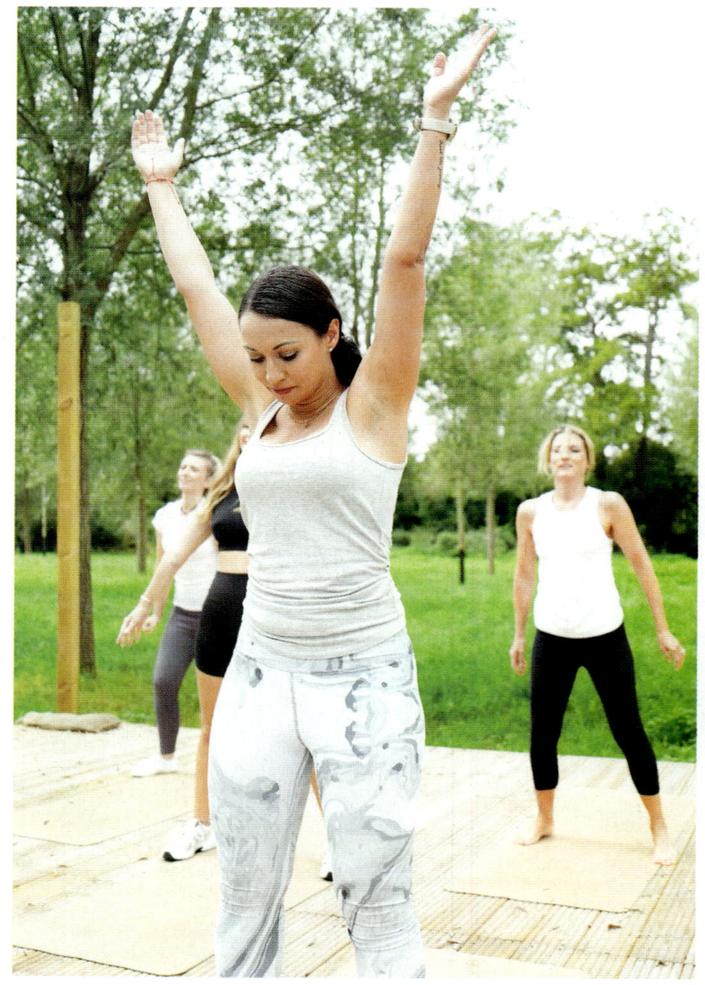

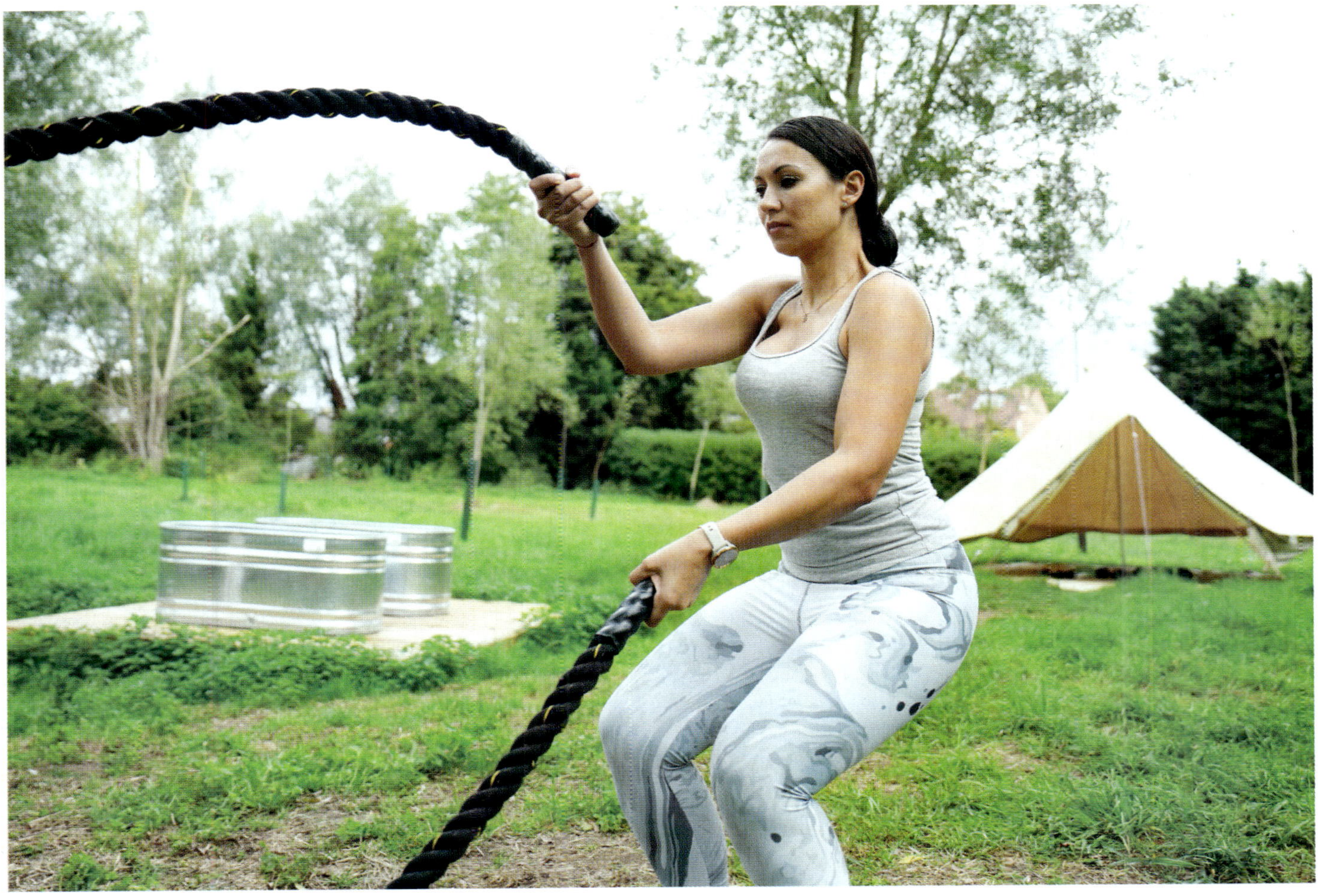

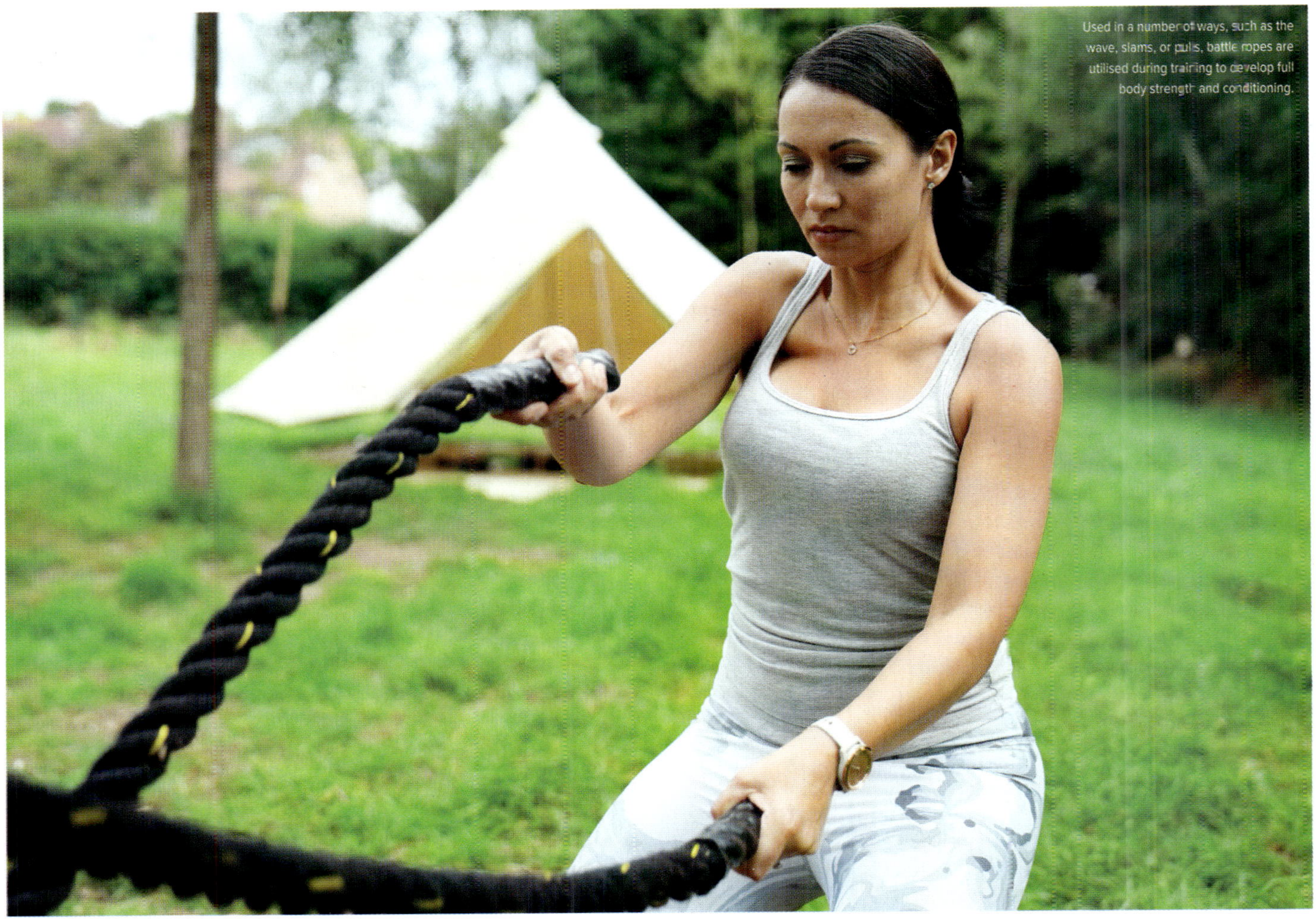

Used in a number of ways, such as the wave, slams, or pulls, battle ropes are utilised during training to develop full body strength and conditioning.

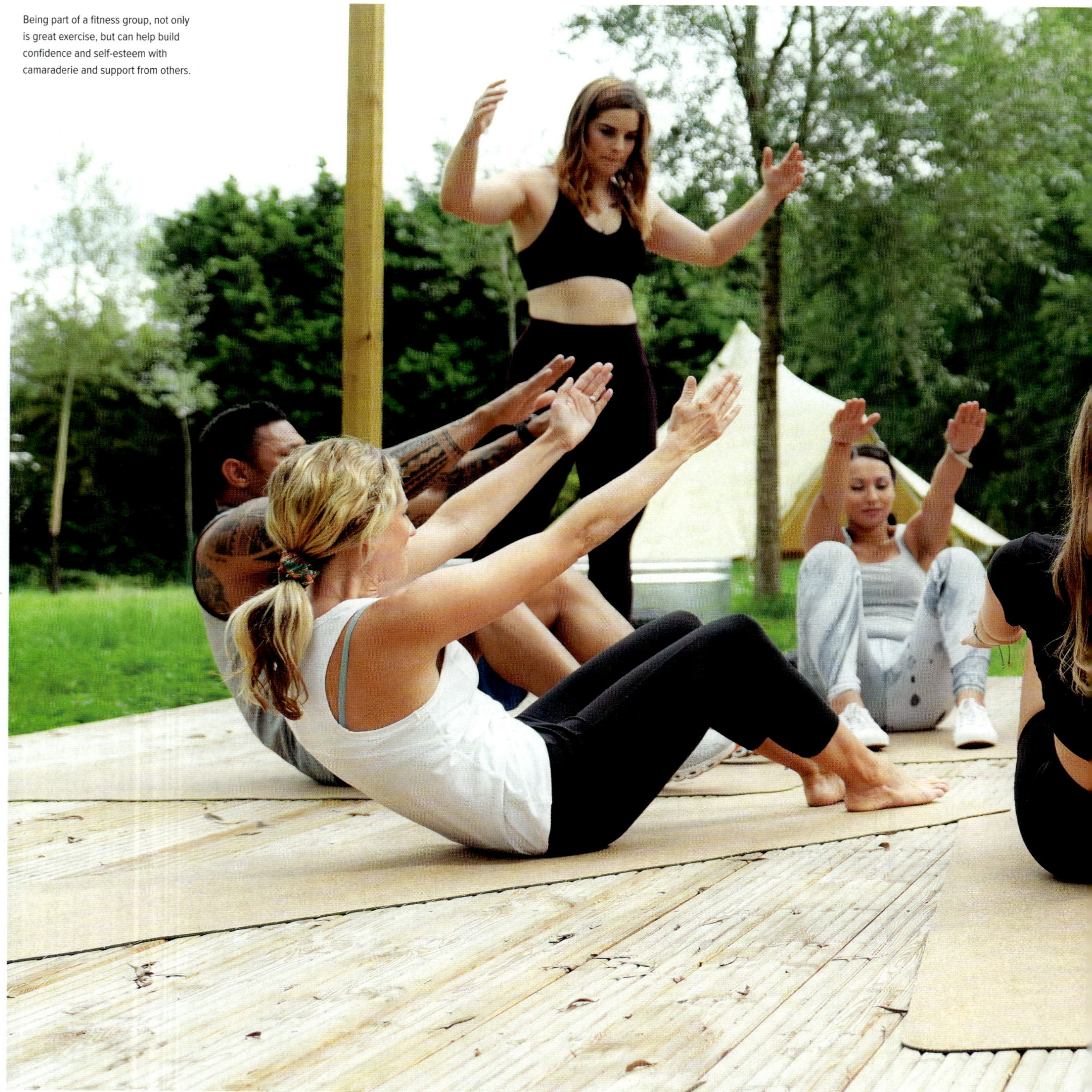

Being part of a fitness group, not only is great exercise, but can help build confidence and self-esteem with camaraderie and support from others.

Sleep

Getting less than seven hours of sleep on a regular basis puts your health at risk. Sleep deprivation has been linked to many chronic health issues including heart disease, diabetes, obesity, and depression. If you are someone who struggles to either get to sleep or stay asleep throughout the night, moderate to vigorous exercise has been proven to increase both sleep quality and quantity.

Sleep and stress go hand in hand with difficultly sleeping being a common symptom of stress and lack of sleep leading to higher levels of cortisol (the stress hormone) in the body. So, why not lace up those trainers and battle two problems with one solution: exercise!

Integrating fitness into your life gives you the belief in yourself that you can do harder things.

Self-esteem & confidence

Recently, a client I have been coaching gave me the best feedback I could possibly ask for as a fitness and mindset coach. When speaking about how her confidence in her ability to do exercises that before she was too scared to even attempt, she said: "I used to be 95% scared and now I am only 10% scared and I am learning to not listen to those voices." How incredible is that?!

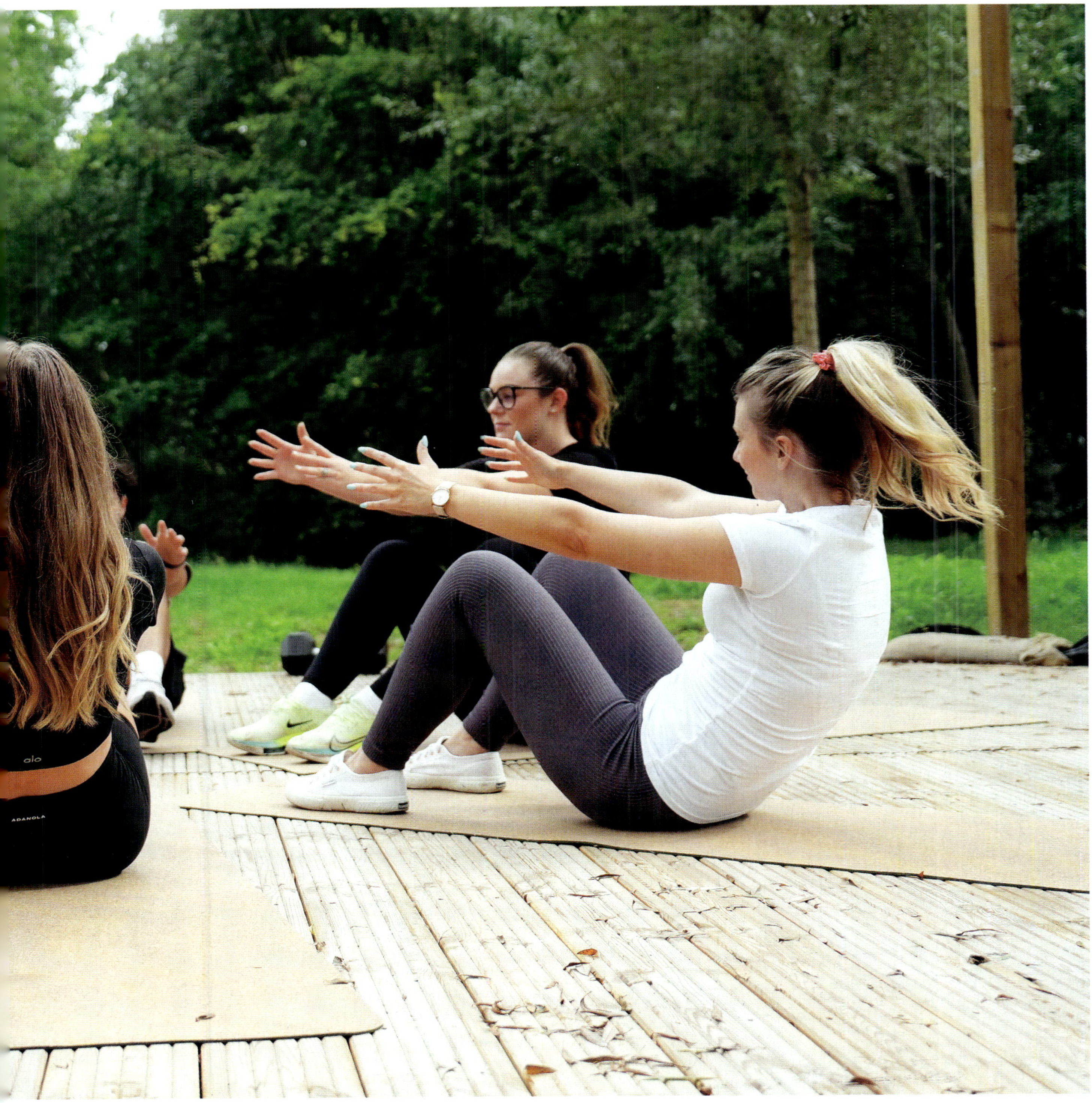

Integrating fitness into your life gives you the belief in yourself that you can do harder things, that you can push yourself out of your comfort zone and achieve things you didn't think were possible. You are able to show up for yourself daily, do what you set out to do and this in time builds on your self-confidence that begins to spill into to all areas of your life.

When I first started exercising I never in a million years thought I would be the girl who has completed three marathons, can squat over 100kg and has a successful career within the fitness industry. I, without a doubt, know that exercise has been a huge influence in forming the strong minded mentally resilient and confident woman I am today It was the catalyst for making

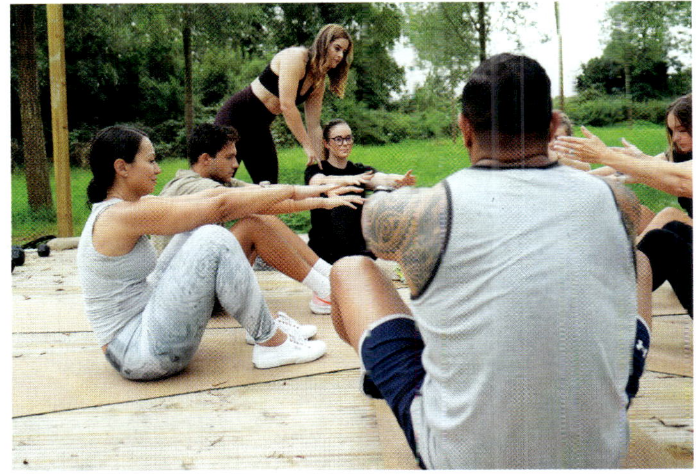

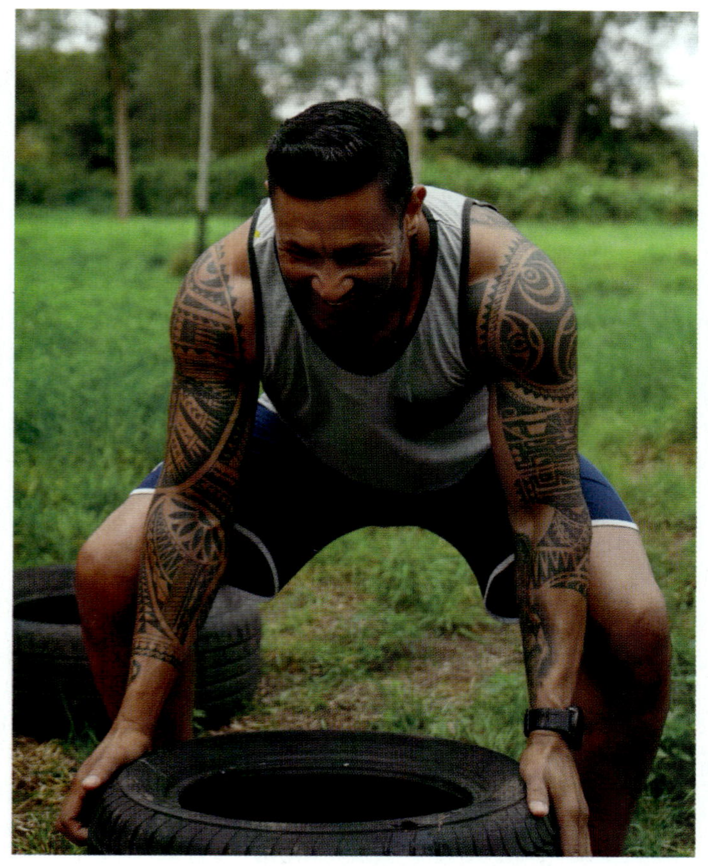

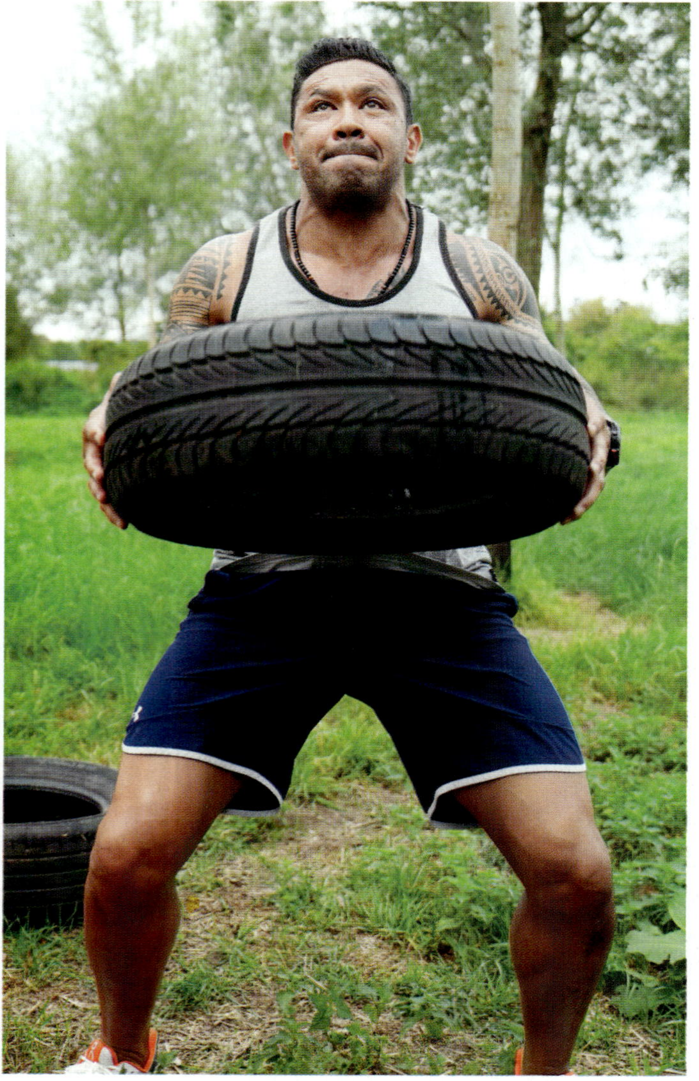

> ## What is important is to figure out what value fitness brings to your life.

positive and momentous changes in all areas of my life throughout the years, and I have seen it change the lives of the people I work with on a daily basis.

How to get going

If you're not sure where to start or how to integrate fitness into your life, here are a few of my top tips to help you get moving and to feel better in your body, your mind and your soul.

Find your WHY

There is absolutely nothing wrong with wanting to lose some weight. Yes, there is so much more to exercise than changing the way you look, but if that is your motivation to start right now then great. However, let's not stop there. How about you dig a little deeper and gain some clarity on the core desire that will be the driving force throughout your fitness journey.

I see it time and time again from clients. They come to me wanting to drop a few kilos, lose the belly fat or tone up and I always probe a little further to find that real reason for their desire to change.

Some of the most common whys are:
• I want to be more confident.
• I want to be able to run around with my kids.
• I want to be healthy in my old age.
• I want to be agile and mobile and keep
  my independence.
• I want to feel great about myself.
• I want to prove to myself I can.

Dig deeper, find your real why and let that be what drives you on the days when you really don't want to go out for that run or get to the gym.

Define what fitness means to you

For me, fitness isn't just about "doing exercise". For me it is about turning up and getting it done no matter how I am feeling. It is about proving to myself that I can do it no matter how far away or impossible "it" feels. It is about being able to actively participate in my life without any barriers to hold me back.

To some "fitness" is smashing through five-plus gym workouts a week, for others it is being present in mind and body during a yoga session. Both are valid. What is important is to figure out what value it brings to your life.

Do what you love

I am sure you have heard the phrase "Do what you love and you'll never work a day in your life". The same can be applied to exercise. We are lucky enough to live in a world where we have so much choice. Just because a

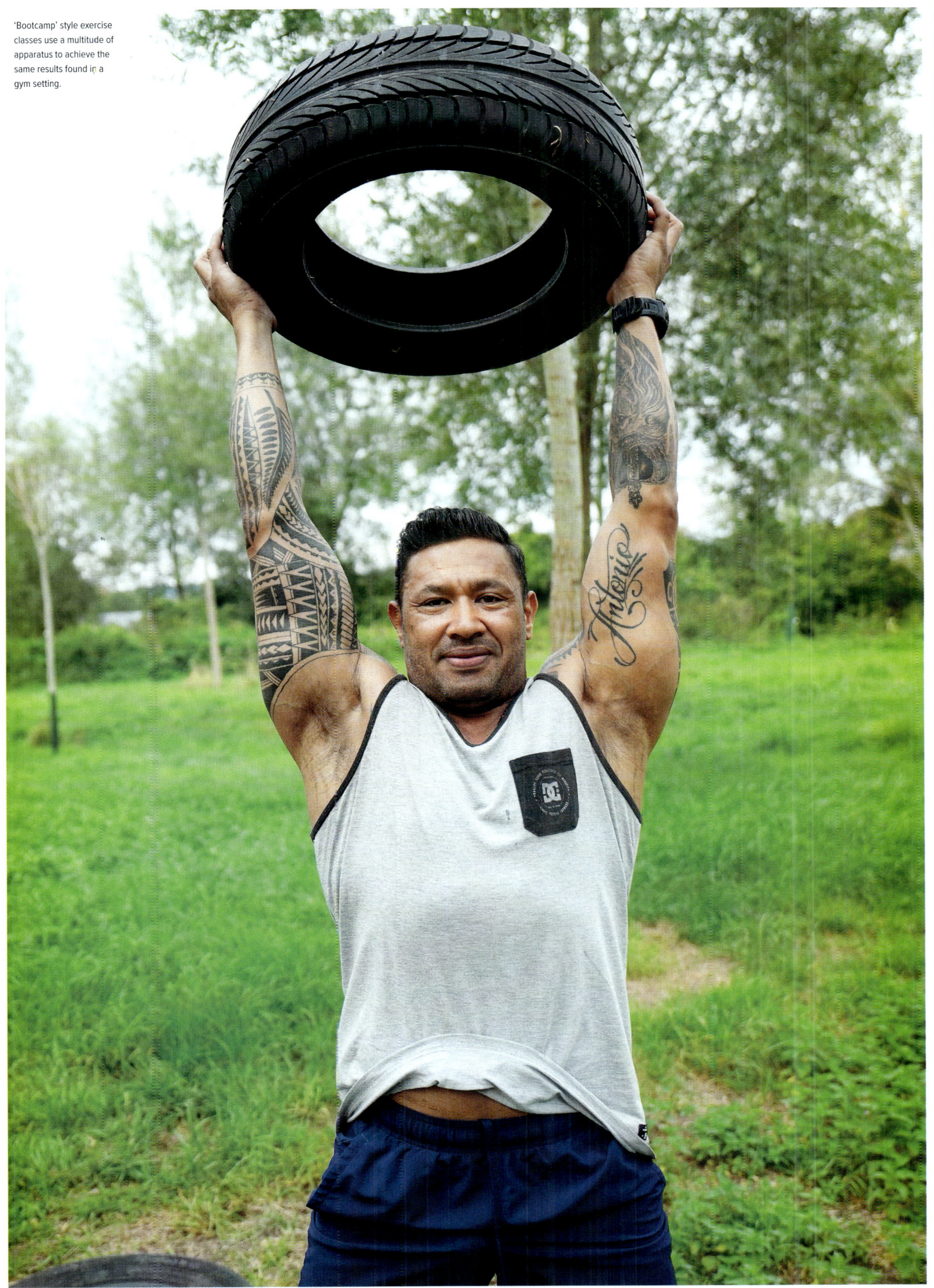

'Bootcamp' style exercise classes use a multitude of apparatus to achieve the same results found in a gym setting.

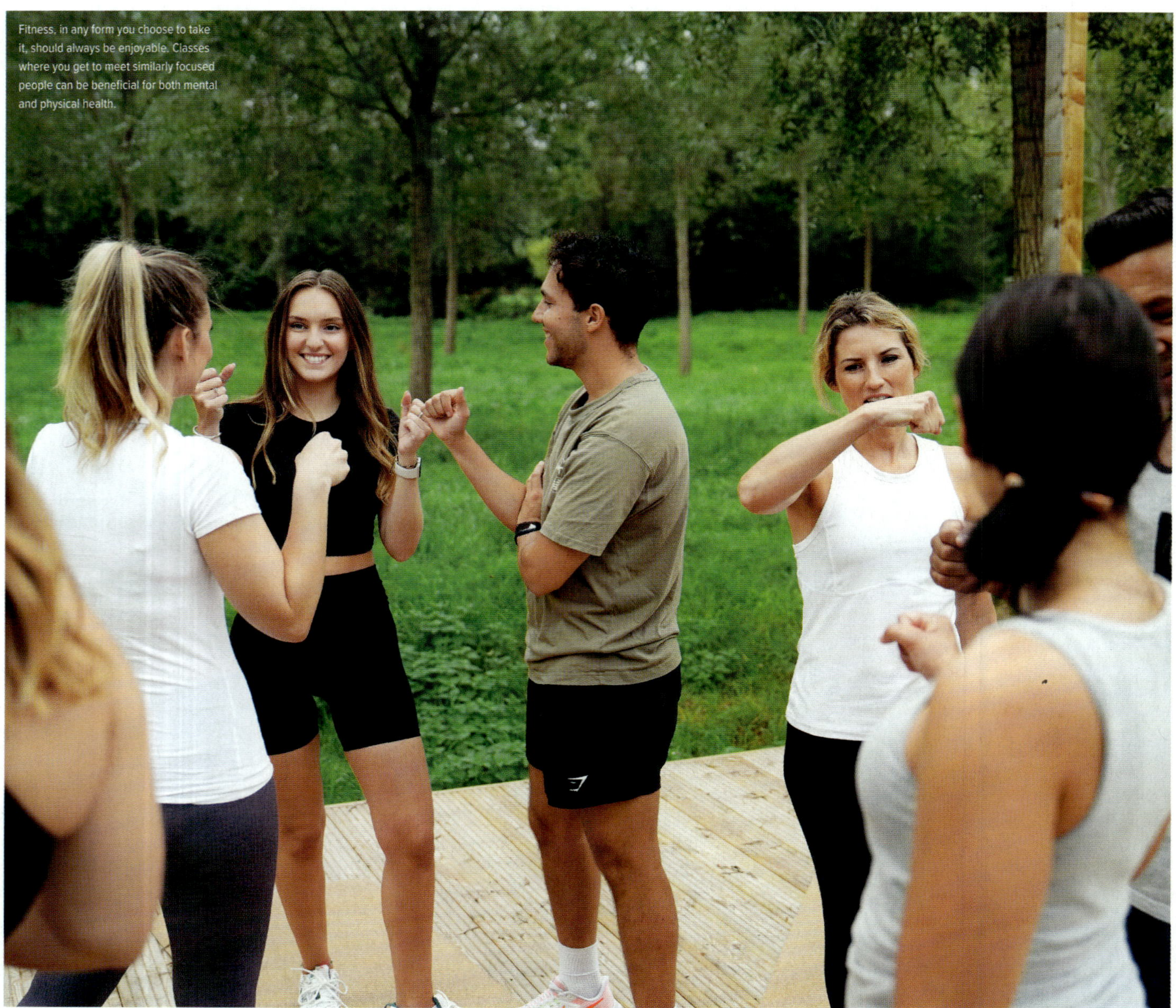

friend loves throwing weights around the gym, it doesn't mean you have to. By all means, give it a go but if it's not for you simply try something else. Find that thing that makes you feel good about yourself and has you wanting to go back for more. Join a club, try a class, go for a run or get yourself a personal trainer. Whatever it is, the most important thing is you are doing it, and you are doing it consistently.

Set goals

Once you know your why, what value fitness brings to you and what style you are going to try, the next thing to do is set yourself some goals. Goals are helpful to keep us motivated, hold us accountable, and give us something to work towards. They are a chance to see how far you've come and strive to go further; to prove to yourself what you are capable of.

Set yourself at least one performance goal, for example "to perform a pull up unassisted", and one process goal that will in some way help you get there, for example "I will pack my gym bag every night ready for the next day". The most important thing is to start small and build

up as you achieve each goal and to always revert back to you core "why".

Be patient

It won't be easy at first and it might take you some time before you notice the benefits of your exercise regime but I promise you it will 100% be worth it. Stick with it and you'll look back in a month, six months, a year's time and be so proud of what you've achieved, the person you've become and how incredible you feel both physically and mentally.

Finally...

I hope by now you have an understanding of how integrating fitness into your wellness routine can improve your life. Let's stop viewing exercise as a form of punishment for eating too much or not looking a certain way and instead let it be a chance to explore, discover and celebrate everything your body and mind is capable of. *M&B*

Aniela Szkoda
@coach.szko

SIGN UP TODAY

TO RECEIVE DETAILS ON **NEW** AND **UPCOMING** **PROMOTIONS** ON YOUR **FAVOURITE** PRODUCTS

shop.keypublishing.com

KEY Shop

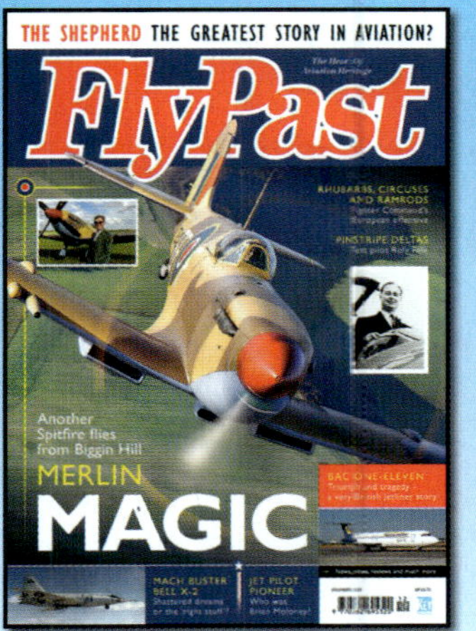

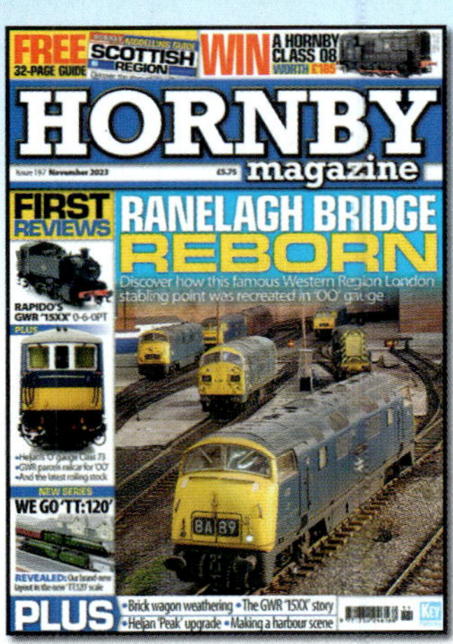

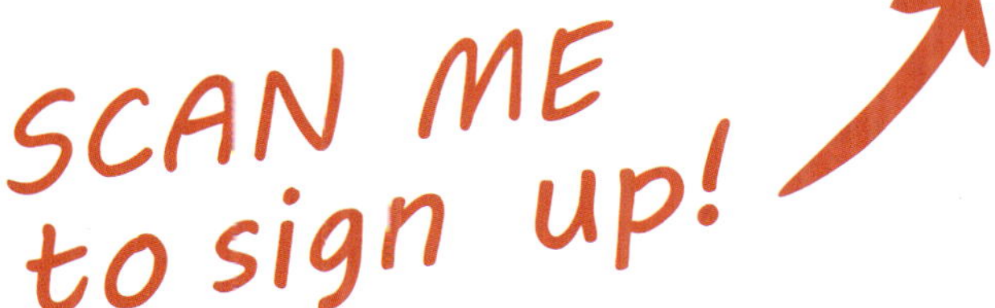

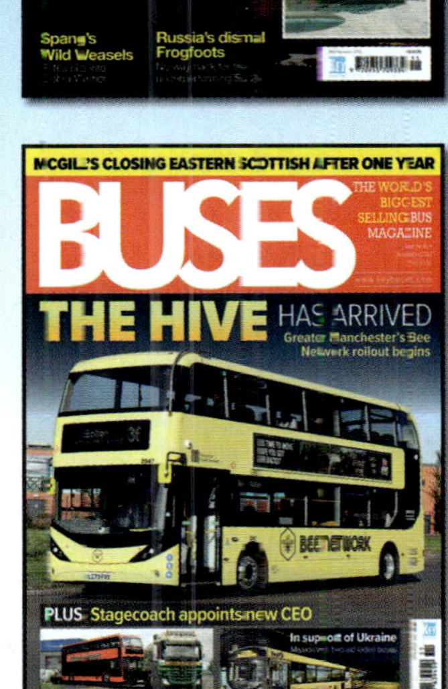

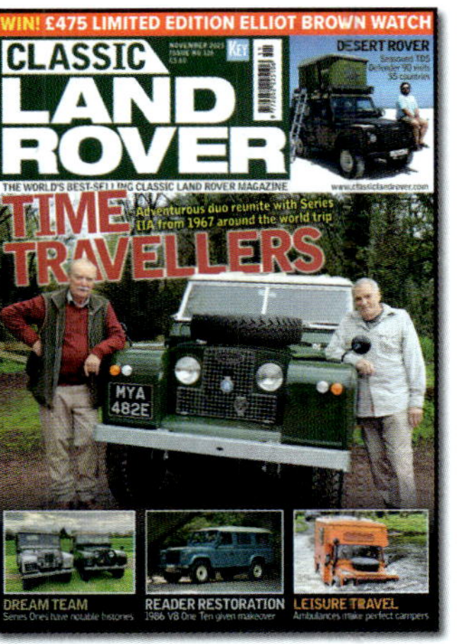

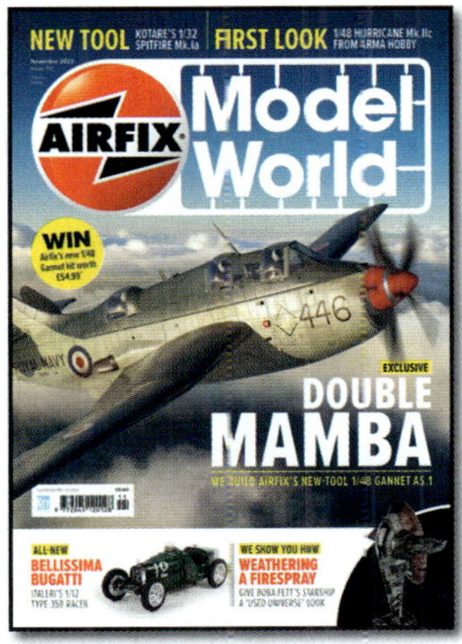

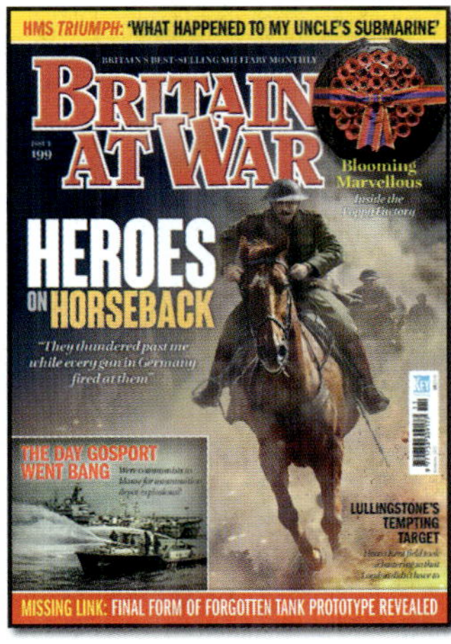

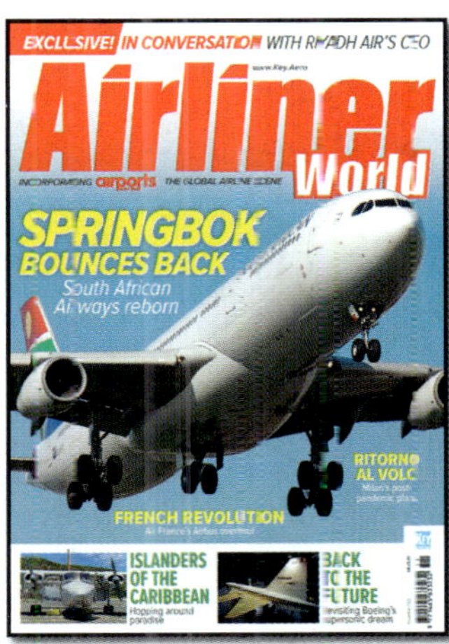

FREE P&P * when you order online at...

shop.keypublishing.com

Call +44 (0)1780 480404 (Mon to Fri 9am-5.30pm GMT)

*Free 2nd class P&P on all UK & BFPO orders. Overseas charges apply.

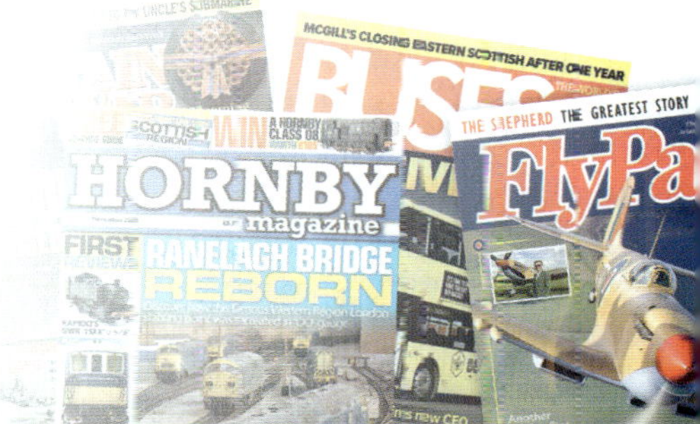

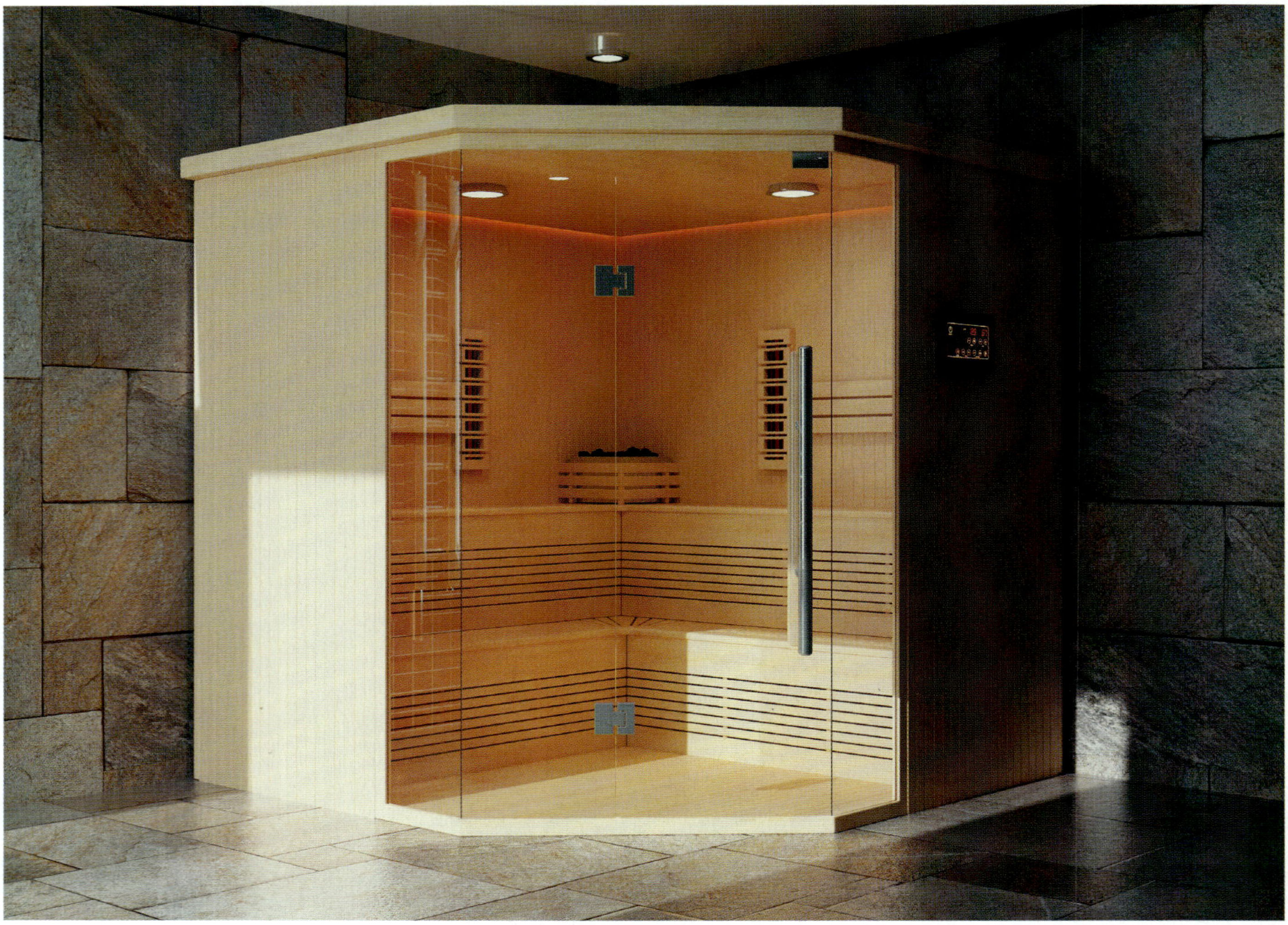

Heat Waves...

Hannah Sander explains what an infrared sauna actually is, and how experiencing regular use of one into your weekly routine can pay dividends for both your physical and mental health.

The buzz around infrared saunas appears to be very 'now', however, they have been around since 1965 when the first ceramic infrared saunas were used in Japan. Although mainly used by medical practitioners, they became more mainstream in the USA in the 1980s.

Moving to the present day, these saunas can be found all over the world in gyms, wellness centres, medical settings, hotels and even with people investing in versions for their own homes.

What is an infrared sauna?
Similar looking to a traditional sauna with their wooden cabin design, they do not have the initial moist yet dry heat, as you enter. Instead, they have infrared heat lamps around the walls and under your feet to raise your core body temperature directly. Giving you the sense you are lying on a beach with the sun heating you up softly and gently.

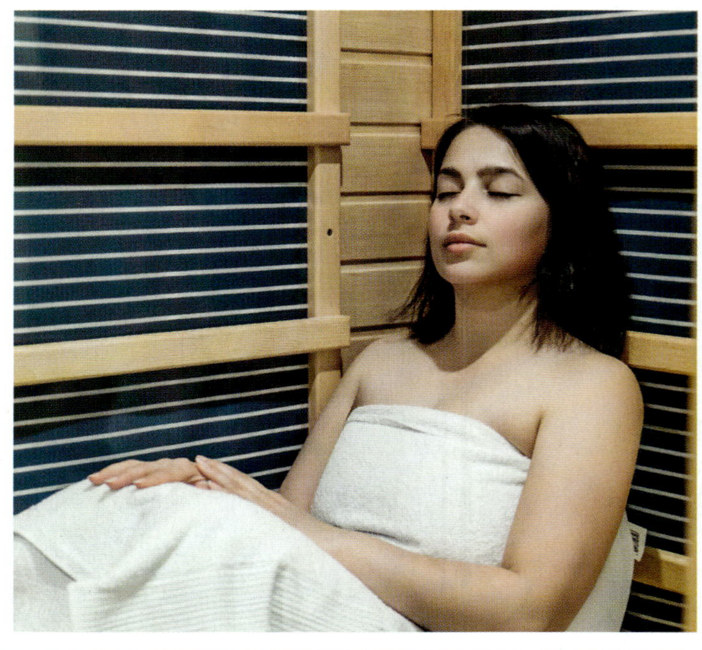

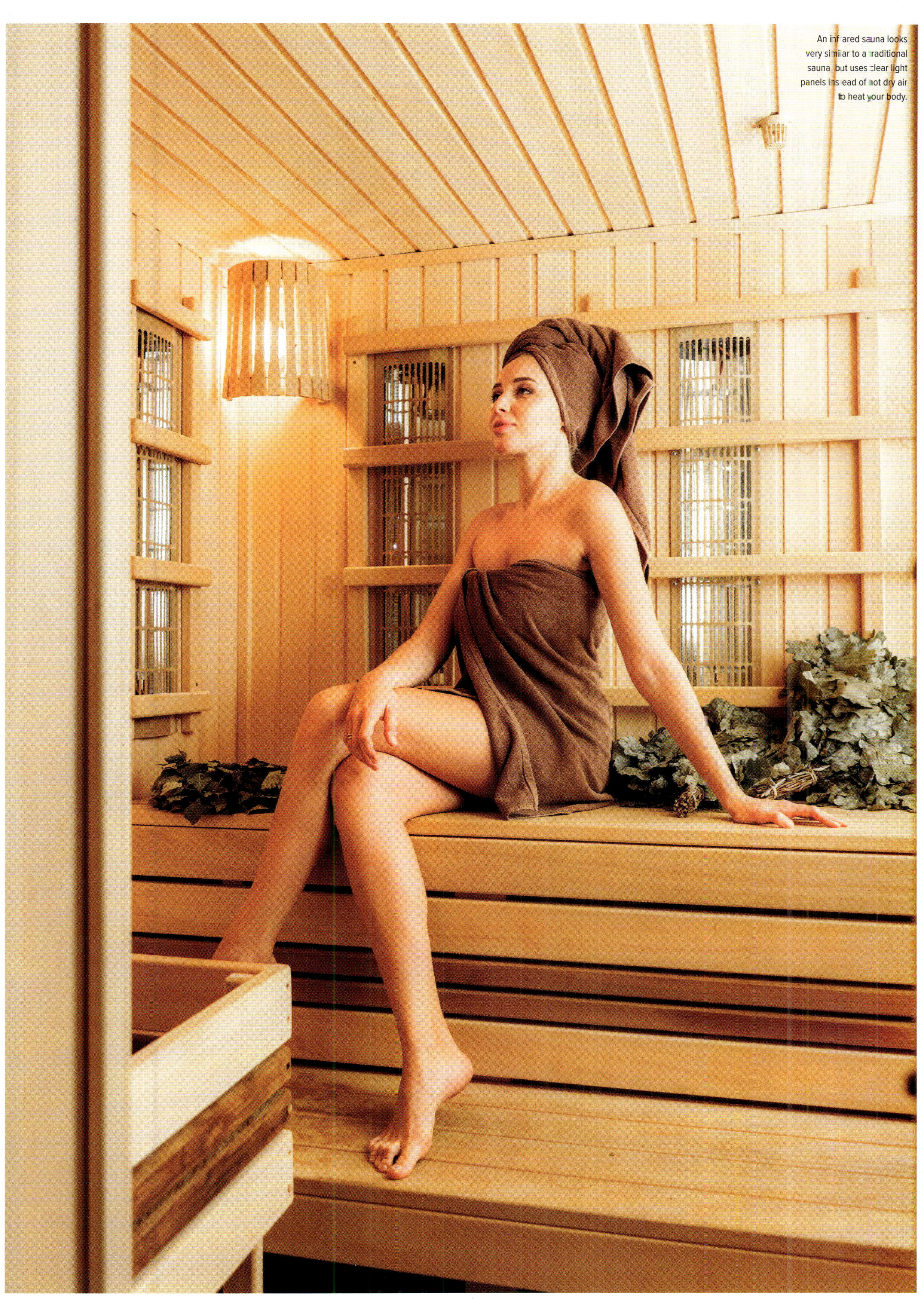
An infrared sauna looks
very similar to a traditional
sauna, but uses clear light
panels instead of hot dry air
to heat your body.

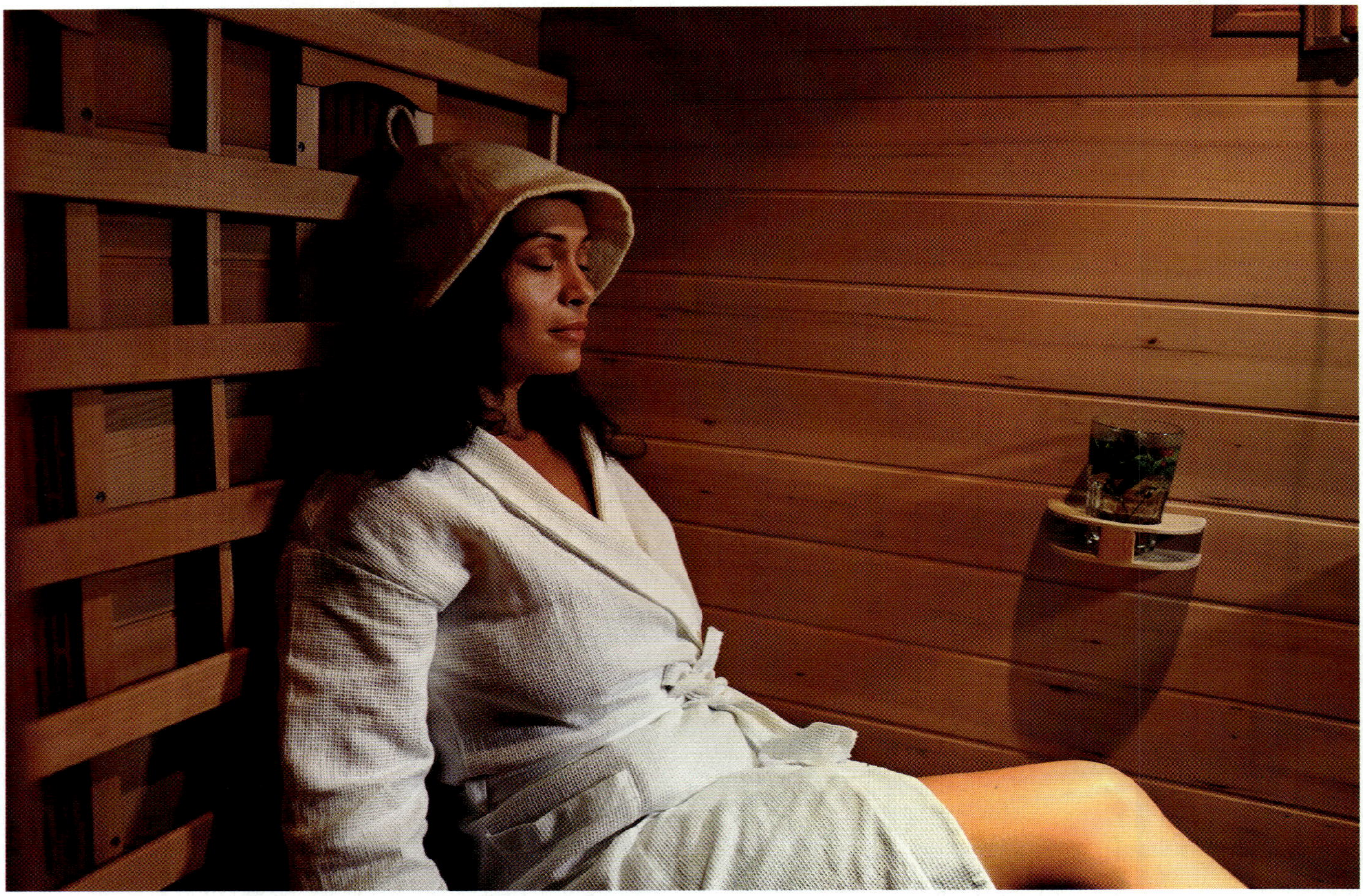

An average session tends to be around 30 minutes and can leave you feeling with a comfortable sweat/inner heat from start to finish, instead of the initial hot skin sensation that can leave you with shortness of breath and a desire to jump out and get in a cool shower. However, much the same as a traditional sauna you are guaranteed to still get that radiant post sauna glow. The other difference with these sauna sessions is that you continue to feel that internal warmth and heat for 30 minutes after, while it does its work.

A traditional sauna is one of instant heat and sweat. This is because the space is likely to be around 80-105°C. When you step into an infrared sauna the cabin temperature is much lower, more like 50-70°C, instantly giving you a feeling that you can stay in longer while the heat penetrates much deeper into your body tissue. Another great side factor is because the sauna is heating you and not the space the lack of moisture means you can take a book and a drink in with you for the ultimate chill factor.

The ideal is to experience these saunas at least three times per week for optimum benefits, however, with all of these things as often as busy lives allow us is still a better than not option.

A infrared sauna comes with a plenitude of benefits for both body and mind. The physical benefits of the relaxing heat waves are they are proven to release the toxins in fat cells and have been shown to speed up metabolism when used on a regular basis. It improves circulation, reduces inflammation, lowers blood pressure, heals wounds, detoxifies metal and toxins from the body, improves your immune system, relieves muscle and joint pain and it can also boost collagen production.

The mind also benefits with reducing stress and fatigue, improving sleep and giving you that mood boost, especially as the days draw in and it is dark and cold. As with all of these moments, it is time for us to switch off the phone, give yourself half an hour for you to feel relaxed, time to think while improving your overall 360° health. *M&B*

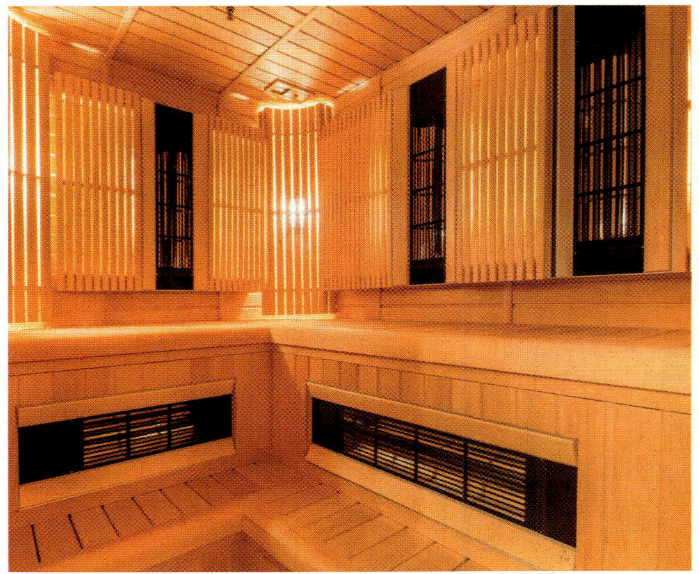

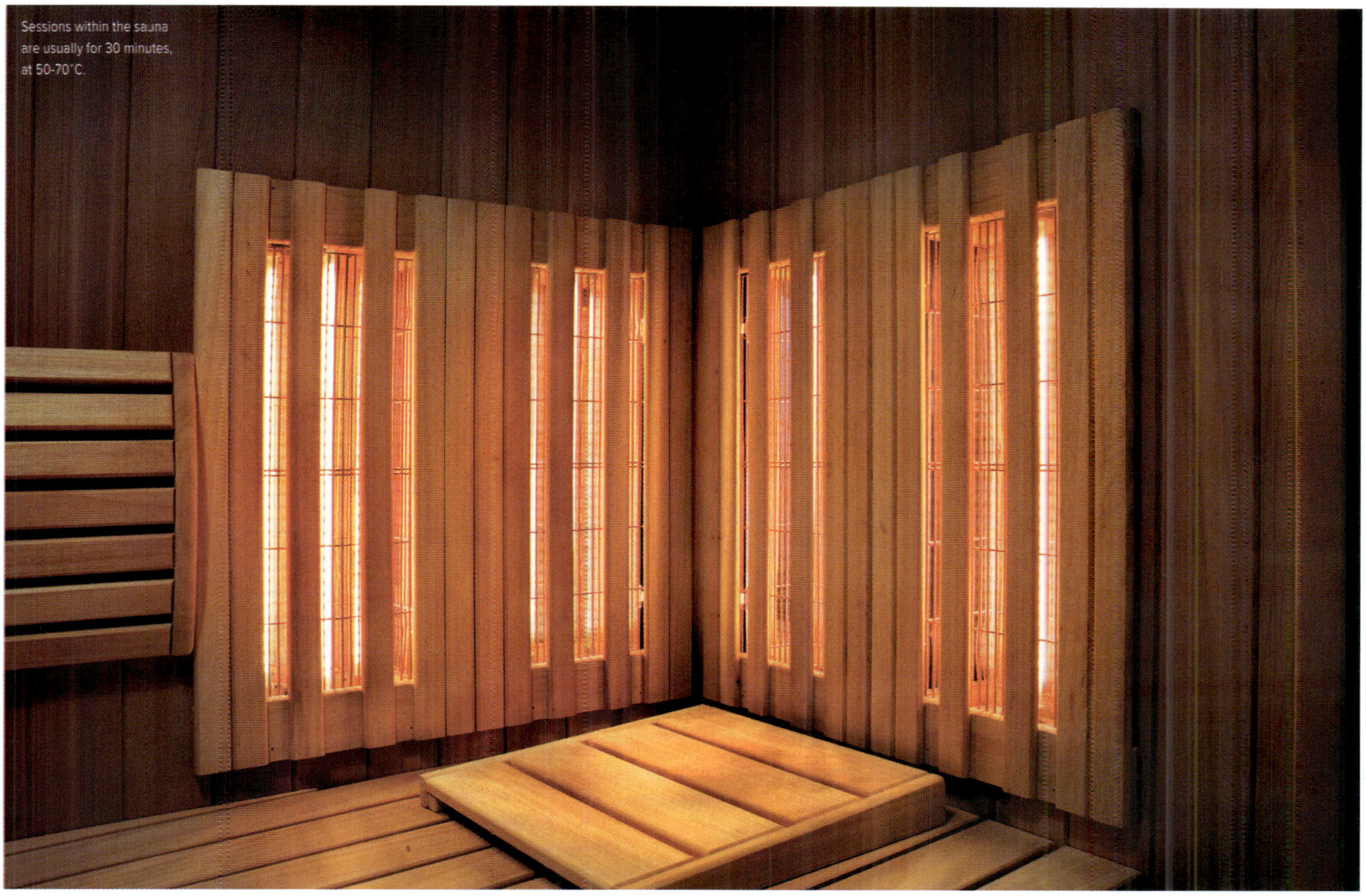

Sessions within the sauna
are usually for 30 minutes,
at 50-70°C.

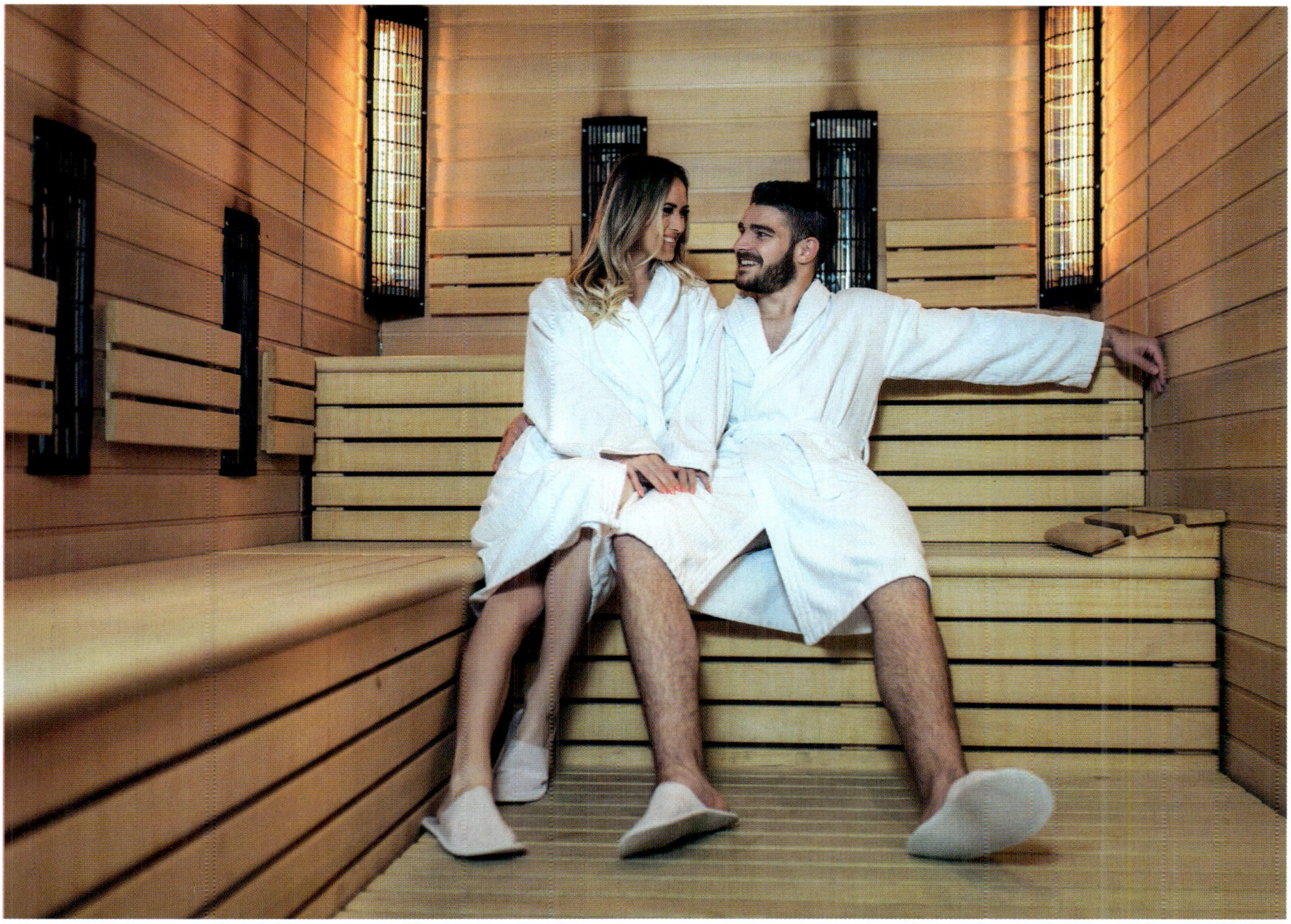

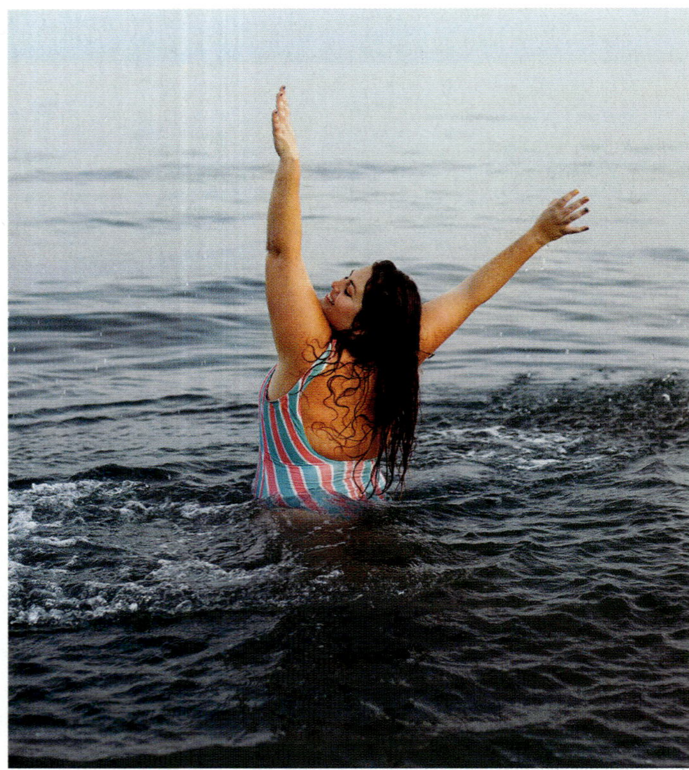

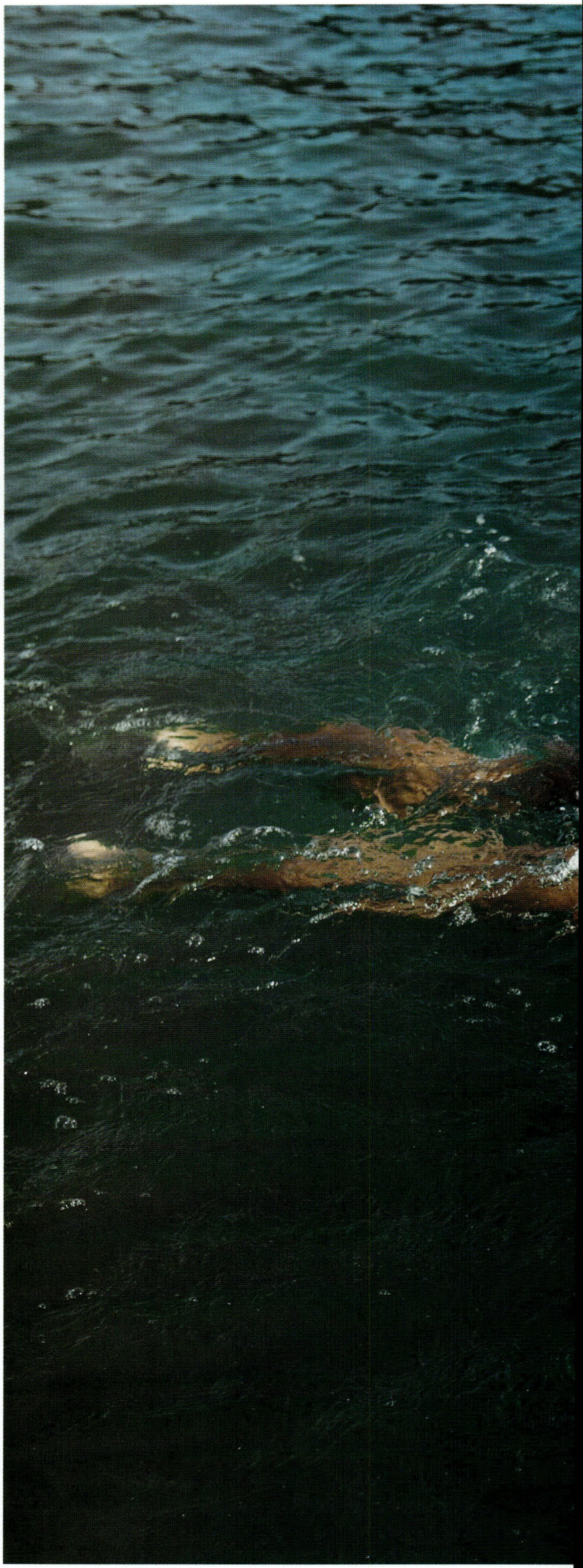

Plunging into a wild adventure

In this chapter open water swimming expert **Laura Ansell** gives you a beginner's guide to this health-boosting activity.

Open water swimming offers a world of adventure for beginners looking to explore beyond the confines of a swimming pool. When I was growing up, all I ever wanted to be was a mystical mermaid with an exquisitely shimmering tail and a beautiful singing voice. It was a sad day of shattered childhood dreams when my parents explained that no matter how much fish oil I consumed, I was never going to grow a mermaid tail and live 'under the sea' with my merman prince. But all was not lost since my parents embraced my love for all things water and promptly set about sending me to swimming lessons and encouraged my desire to be in the open water.

Roll forwards 30 years and I am now an elite pool and open water coach, an open water swimming venue owner, as well as an ultra-endurance open water swimmer. I've

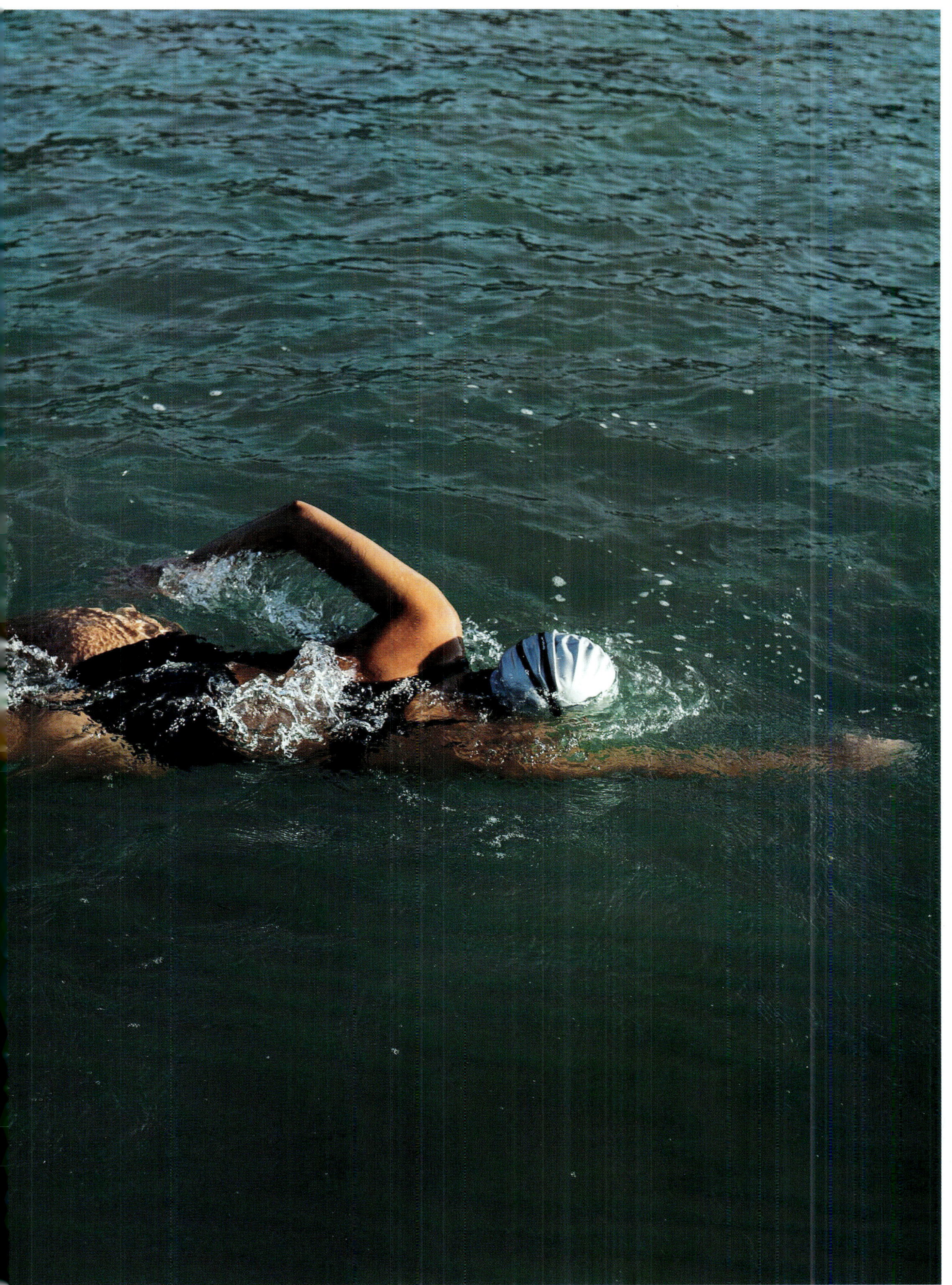

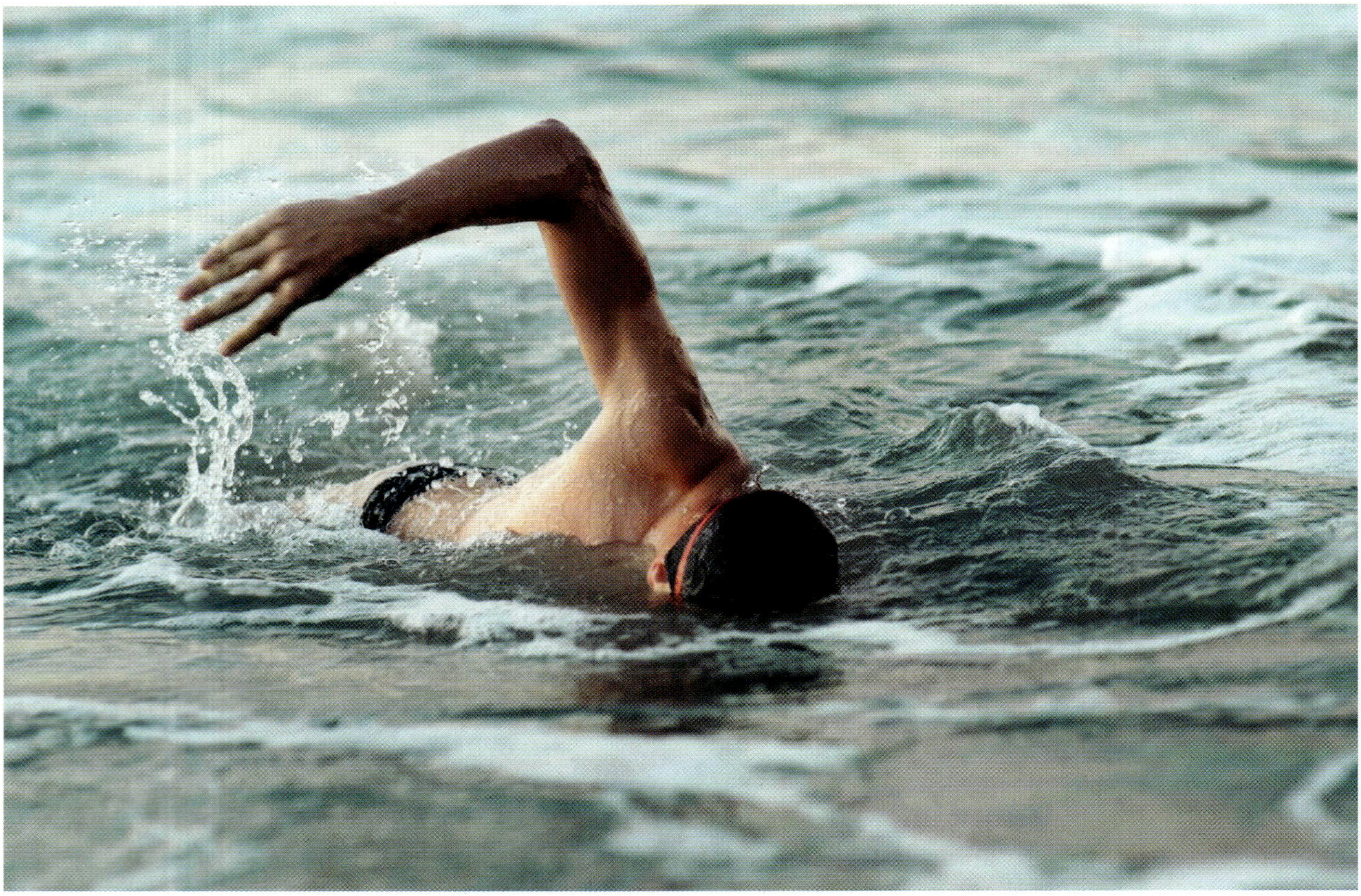

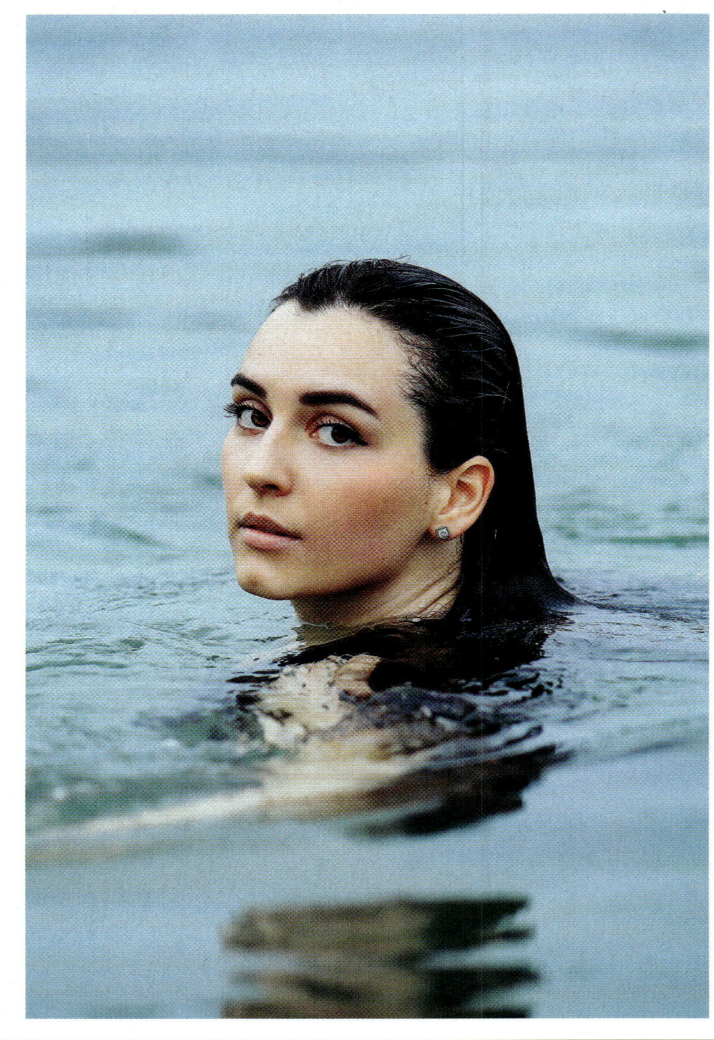

dedicated my life to helping others embrace the magic of swimming and to conquer their fears.

In this beginner's guide, I will take you through everything you need to know to start your open water swimming journey, from finding the right venues to understanding the necessary precautions for a safe and enjoyable experience. Together, we'll dive into a world of adventure, push our boundaries, and discover the joy and freedom of open water swimming.

The growth of open water swimming in England
Open water swimming has a rich history and a strong tradition in England. It has gained even more popularity since the COVID-19 pandemic, as people seek outdoor activities that offer both physical exercise and mental rejuvenation. Exploring the roots of open water swimming in England, from its origins in the 19th century to the thriving community it has become today, highlights the enduring appeal of this unique sport. By understanding its history, you'll gain a deeper appreciation for the passion and dedication that underpin open water swimming culture in England.

During the Victorian era, when seaside holidays were the height of fashion, open water swimming became increasingly popular. Bathing machines, resembling small huts on wheels, were used to shield the modesty of bathers as they ventured into the sea. As time passed, swimming as a leisure activity continued to grow, and the races and competitions in open water became a celebrated part of British sporting culture.

Today, open water swimming in England is far from just a fleeting trend; it is a deeply ingrained part of the country's recreational activities. People flock to stunning lakes, tranquil rivers, and breathtaking coastal stretches to experience the freedom and joy of swimming in nature. From seasoned swimmers seeking the thrill of racing against others to those simply looking to immerse themselves in the beauty of the environment, open water swimming caters to a wide range of interests and aspirations.

The growth of open water swimming in England is testament to the physical and emotional benefits that can be gained from this immersive sport. Whether it's the sheer exhilaration of swimming in untamed waters, the adrenaline rush of pushing personal limits, or the sense of freedom and connection to nature, open water swimming offers a unique experience that continues to captivate and inspire.

The physical benefits of open water swimming

Engaging in open water swimming goes beyond just a fun activity – it offers a range of physical benefits that can greatly enhance your fitness and help with weight management. When you dive into the open water, you're not just immersing yourself in a refreshing environment; you're also challenging your body in ways that can lead to improved strength and endurance.

The resistance provided by water is a fantastic way to work your muscles. Every stroke you take requires effort, engaging your arms, legs, and core. It's like a full-body workout, but with the added bonus of being surrounded by nature's beauty. As you swim against the current and navigate through the waves, you'll find your muscles becoming stronger and more toned.

But it's not just the physical exertion that makes open water swimming beneficial. The natural variations in water temperature create an extra element of challenge. When you swim in colder water, your body must work harder to maintain its core temperature, which can boost your metabolism and calorie burn. So, not only are you getting a great workout, but you're also helping to manage your weight.

Whether you're aiming to improve your cardiovascular fitness or simply looking for a low-impact exercise, open water swimming is a fantastic option to consider. It provides a refreshing change from the monotony of the gym and allows you to connect with nature while reaping the physical rewards.

The positive impact on mental health

The therapeutic effects of open water swimming on mental health should not be underestimated. Immersing oneself in nature, away from the hustle and bustle of daily life, can have a profound impact on reducing symptoms of anxiety and depression. When you dive into the open water, it's like entering a whole new world. The calming rhythm of the strokes, the sensation of weightlessness, and the connection with

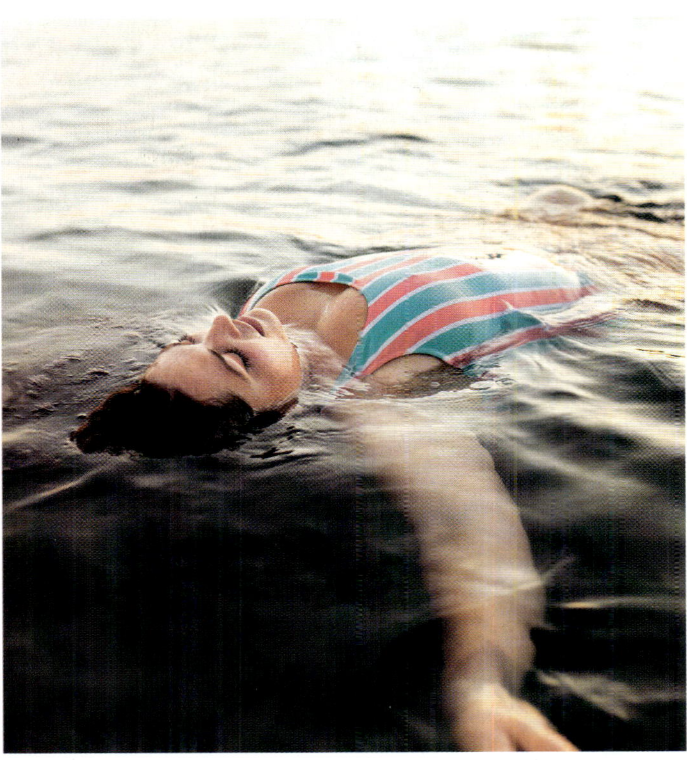

the natural environment all contribute to a sense of peace and tranquillity.

Picture this: you're floating on your back, gazing up at the sky as the sun's warm rays kiss your skin. The worries and stresses of the day seem to melt away, replaced by a feeling of serenity. The water cradles you, providing a haven where you can let go of your troubles and just be in the present moment.

Open water swimming offers an escape from the pressures of everyday life. It allows you to disconnect from technology, deadlines, and the constant noise of the modern world. Instead, you can focus on the gentle lapping of the water, the sound of your own breath, and the beauty of the natural surroundings.

Community and belonging

Joining an open water swimming community is not just about the physical act of swimming – it also provides a sense of belonging and social connection. Engaging with like-minded individuals who share a passion for the water can be incredibly rewarding. Whether it's participating in group swims, sharing tips and advice, or simply enjoying post-swim chats over a cup of tea, the community aspect of open water swimming adds an extra dimension to the experience. You'll find that the camaraderie and support within the community create

> Engaging in open water swimming goes beyond just a fun activity – it offers a range of physical benefits.

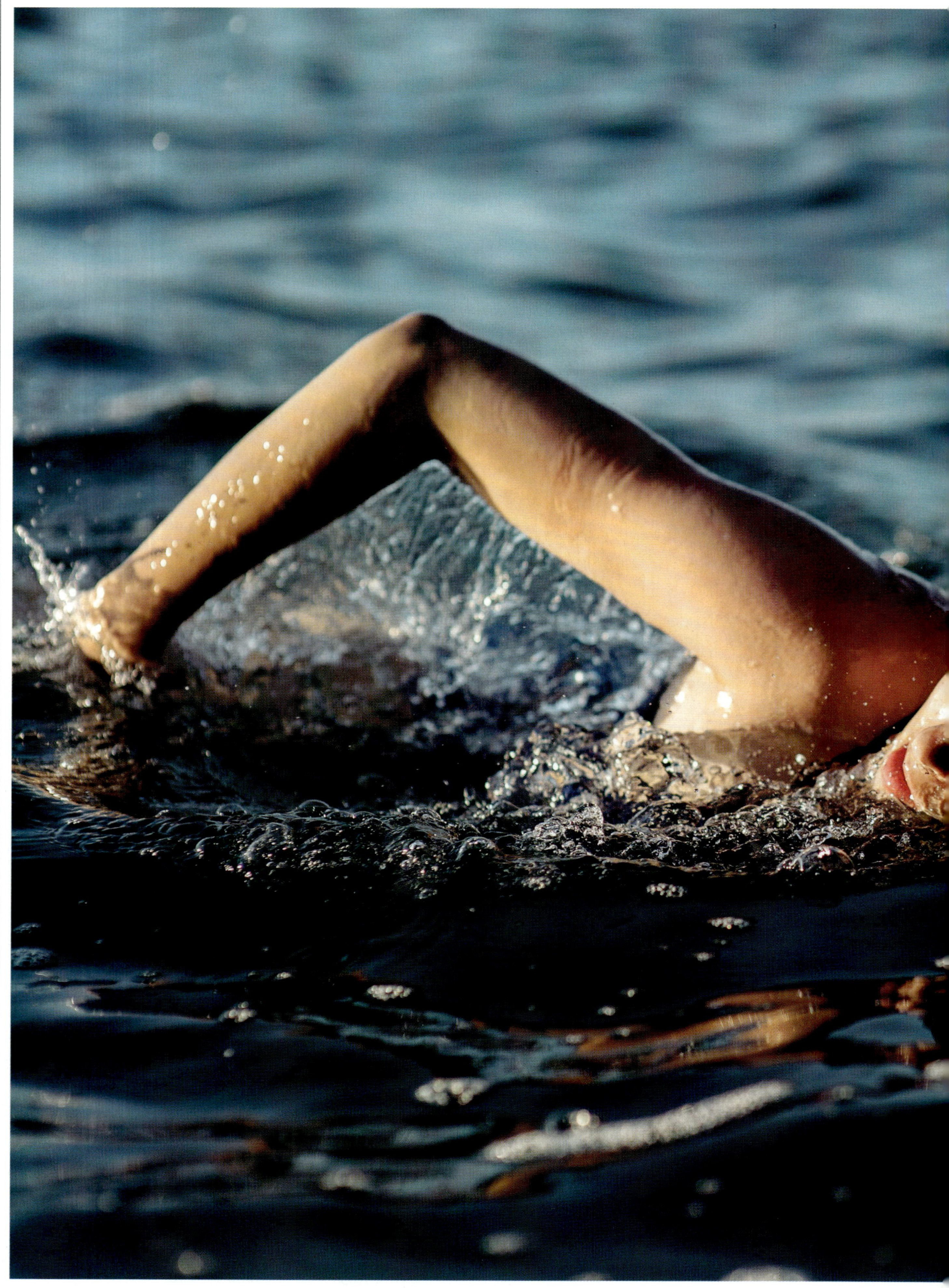

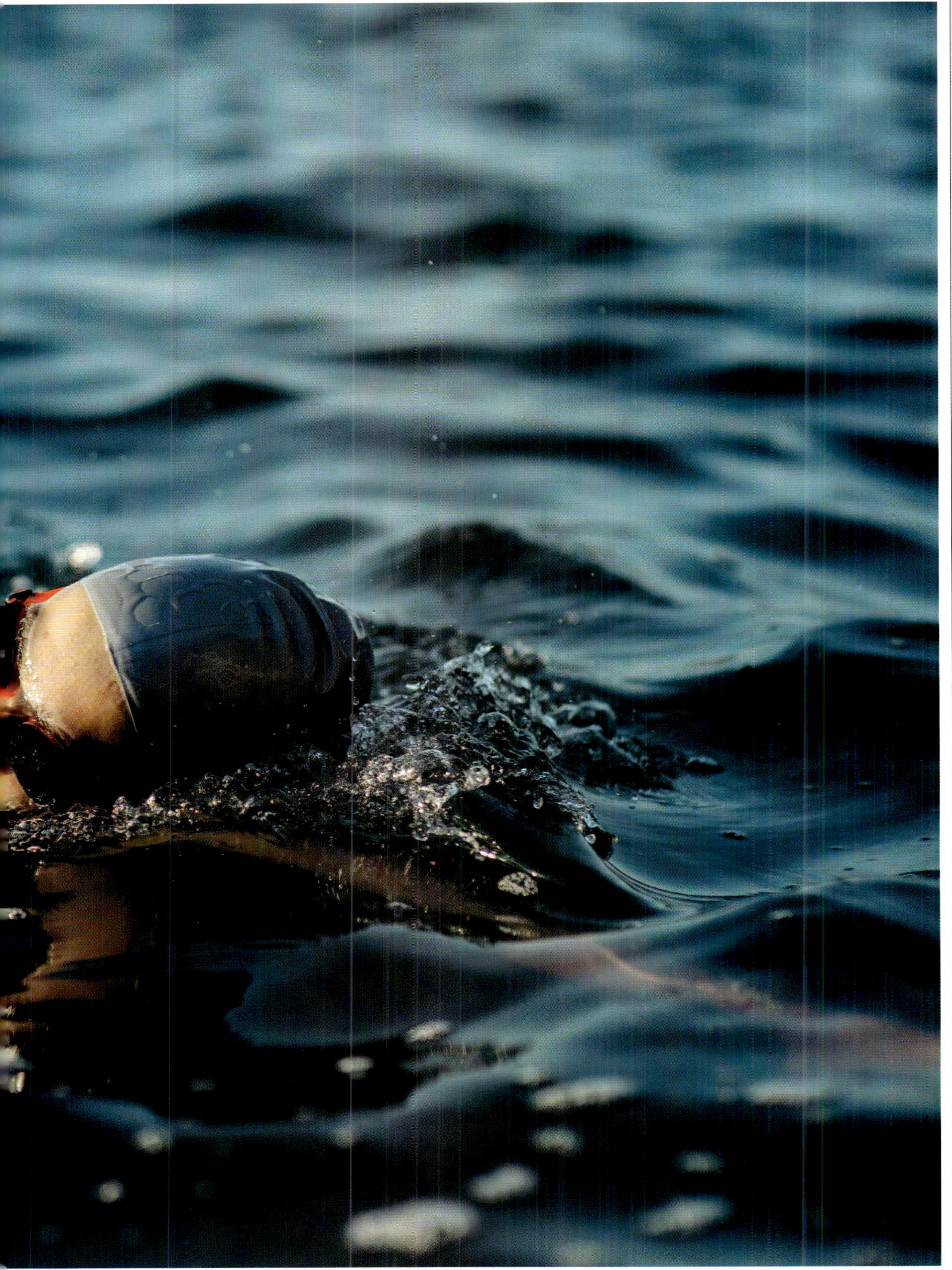

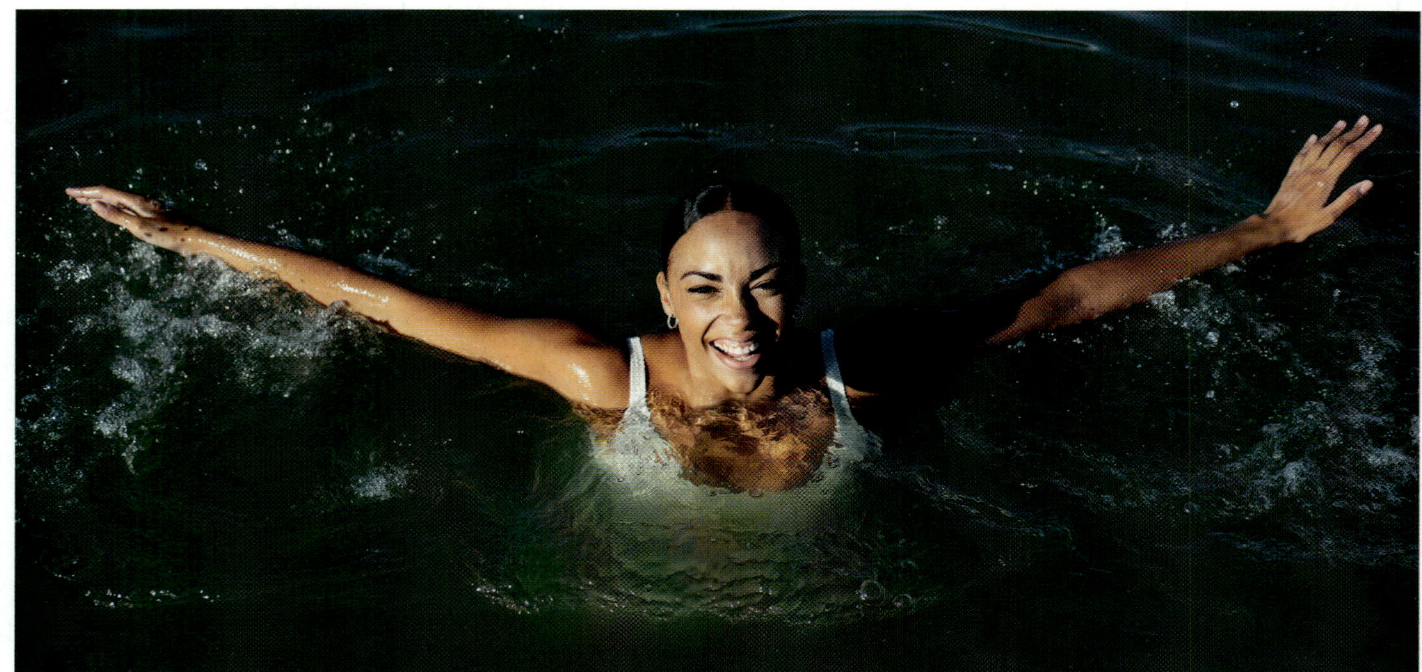

a welcoming environment for beginners and seasoned swimmers alike.

When you become part of an open water swimming community, you're not just joining a group of people who love the water; you're becoming part of a family. The shared experiences and challenges create a bond that goes beyond the surface of the water. It's a place where you can find encouragement, motivation, and a sense of belonging. Whether you're a beginner or an experienced swimmer, there's always someone to offer guidance, share stories, and celebrate your achievements.

Open water swimming communities are not just about the swimming itself; they're about the friendships and connections that form along the way. It's about finding your tribe, a group of people who understand your love for the water and share your passion for adventure. Together, you'll explore new swimming spots, challenge each other to push your limits, and create memories that will last a lifetime.

But where do I go?
Getting involved in open water swimming is easier than you might think. Lifeguarded venues provide a safe and supervised environment for beginners to take their first plunge. These venues often have designated swimming areas and safety measures in place to ensure a secure experience. It's like having your own personal lifeguard watching over you as you explore the open water.

Imagine the excitement of stepping into the water, knowing that there are trained professionals nearby, ready to assist if needed. Lifeguarded venues give beginners the confidence to discover the joy of open water swimming. Whether it's a lake, river, or coastal area, these venues offer a controlled environment where you can focus on improving your technique and building your stamina.

Open water swimming isn't just about swimming laps in a controlled setting. It's also about immersing yourself in the vibrant community. Participating in organized events, such as charity swims or races, allows beginners to connect with like-minded individuals and experience the electric atmosphere of these gatherings. It's a chance to celebrate your shared love of the sport and be part of something bigger than yourself.

Understanding the risks and taking precautions
While open water swimming can be exhilarating, it's crucial to be aware of the risks associated with swimming in natural bodies of water. The unpredictable nature of the open water means that understanding factors such as currents, tides, and water quality is essential for your safety.

Taking precautions is key to ensuring a safe and enjoyable open water swimming experience. One simple yet effective precaution is to wear a brightly coloured swim cap or using a tow float that attaches to your waist. This not only adds a touch of style to your swim attire but also increases your visibility to other water users. In the vastness of the open water, being easily seen can make all the difference.

But precautions go beyond just equipment. Familiarising yourself with common safety practices is vital. Always swim with a buddy, as having someone by your side can provide an extra layer of security. Additionally, knowing when to call it a day in challenging conditions is crucial. If the weather turns rough or you feel fatigued, it's better to err on the side of caution and head back to shore.

Choosing the right equipment
➤ First and foremost, a well-fitted swimming costume or swim shorts. Not only does it provide insulation, but it also enhances your overall comfort in the water. It's like a second skin, hugging your body and keeping you warm as you explore the depths. Trust me, a good swimming costume can make all the difference.

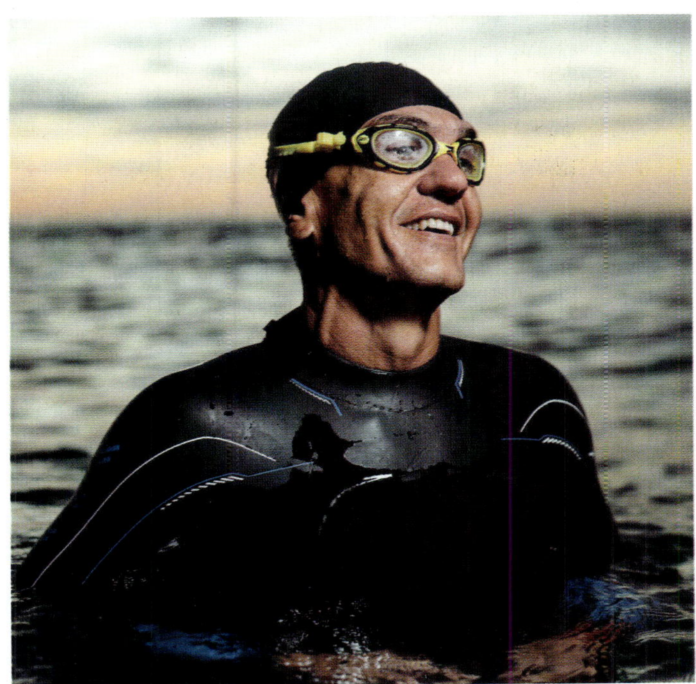

bathers, and this is totally fine! Equally, if you would like to explore using a wetsuit, this is also totally fine. Start at the lower end of the budget scale with an entry level suit of a minimum of 3mm thickness, and if you enjoy it and wish to progress, then you can invest in a more mid to high-range wetsuit of 5mm thickness.

➤ Next up, goggles. You'll want a pair with a wide field of vision and anti-fog properties. Clear sight underwater is crucial for navigating the open water with confidence. Imagine swimming through a school of fish or spotting a hidden treasure beneath the surface. With the right goggles, the underwater world becomes your playground.

Finally...

Open water swimming offers a thrilling and immersive experience for beginners, combining physical fitness, mental well-being, and a sense of community and belonging. Open water swimming is not just a sport – it's a lifestyle that connects you with nature, like-minded individuals, and your own potential. So, dive in, embrace the adventure, and let open water swimming ignite your passion for the water. **M&B**

Laura Ansell
Owner of www.triswim.org.uk

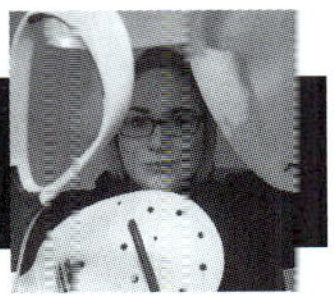

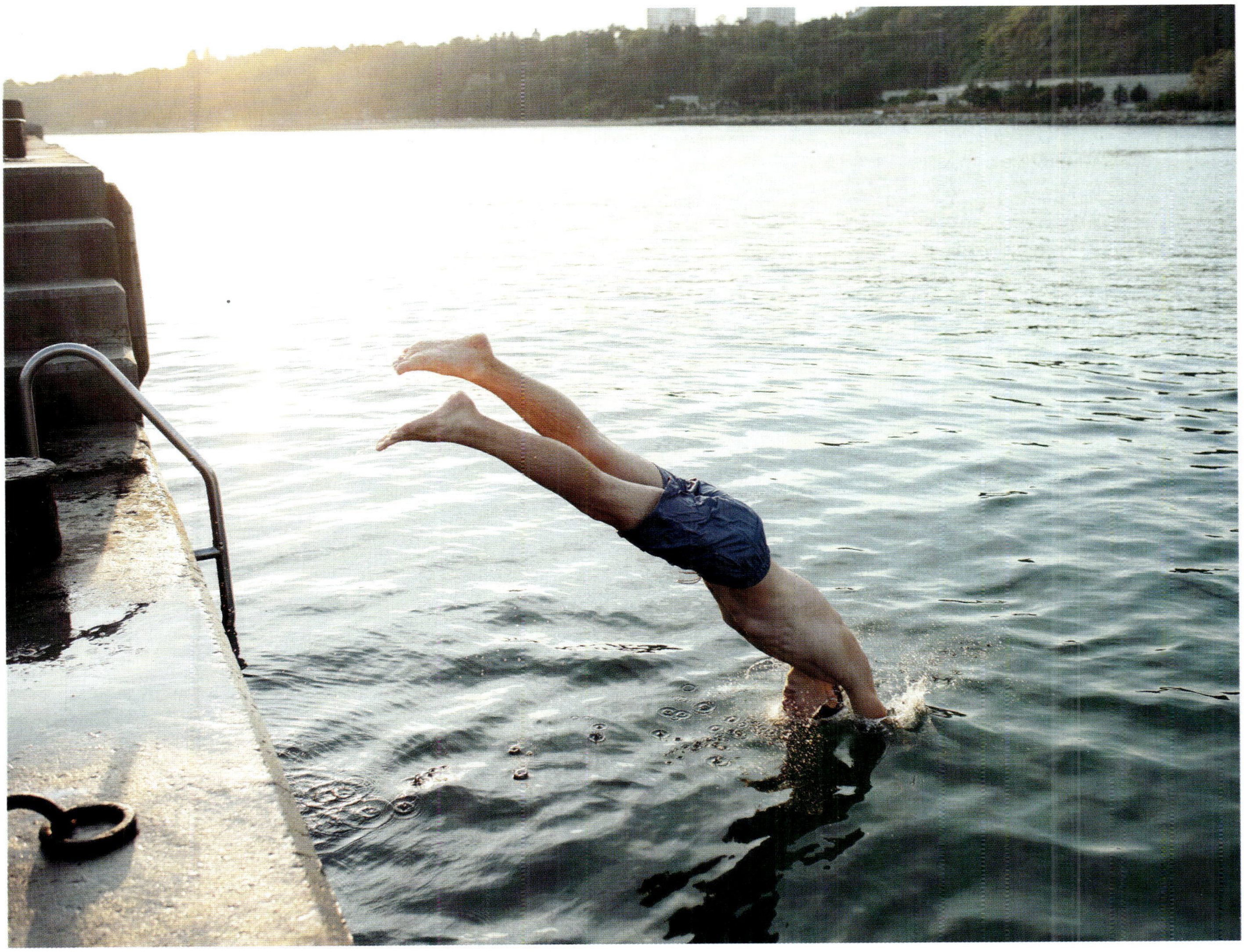

➤ Wetsuits are fantastic for those new and nervous about joining the open water swimming community, but they are not essential. There is no rule to say that to participate in this pastime you must have a wetsuit! Many people enjoy swimming year-round in just their

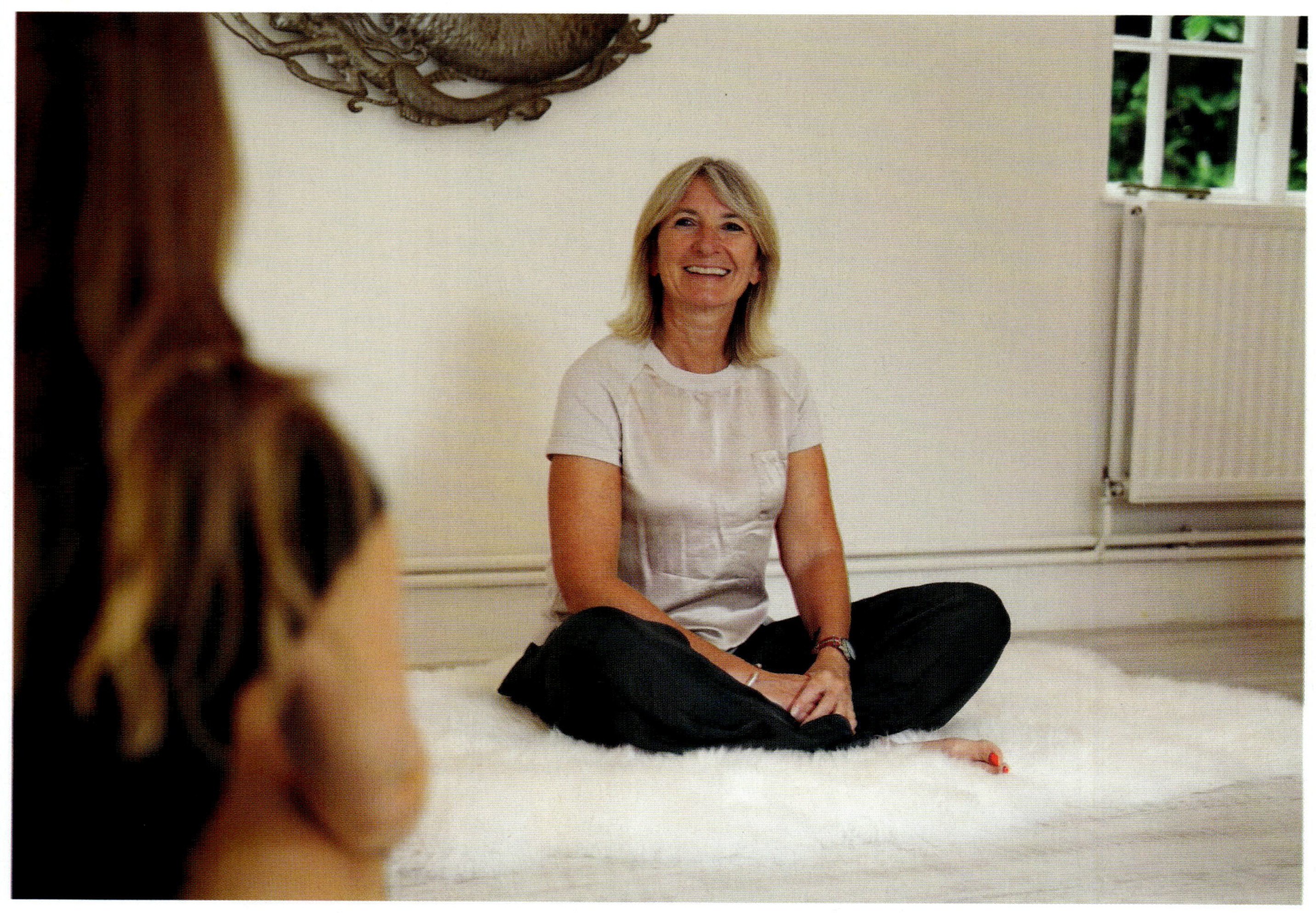

Breathe into Life

Justine Clemant has been teaching breathwork and its healing benefits for over a decade. Here she speaks about her experiences and practices you can bring into your own life.

've extolled the virtues of breathwork for many years now. I was first introduced to it when I took up Ashtanga yoga around 20 years ago, when it was all the rage with celebrities like Madonna and Sting. Getting up at some ridiculously early hour to bend it out on a mat in the dark in a small, sweaty London studio.

Fast forward to about 12 years ago, when I came across an article in the Sunday Times Style magazine about something called Transformational Breathwork. Reading about the possibilities for mental and physical wellbeing blew my mind and I immediately signed up to go on the retreat mentioned in the article, and over the years I went back several times. In-between, I had regular sessions and then, a few years ago, I had a lightbulb moment. I realised that I wanted to introduce others to this superpower. And so, I began training with that same breathwork coach in the article I'd read so many years before and became one myself. What's been the foundation of that journey is my belief that

breathwork is the most incredible life tool, available to us any time we need it.

Feeling disconnected

Along the way I noticed how I, and so many of us, have become disconnected from our bodies and have let our minds literally run every element of our lives. I've realised too, the importance of redressing this imbalance. breathwork helps enormously to get us back into a healthier relationship with our body. And for me personally, it's helped me move through all the different issues that invariably arise at stages in one's life. It's been a way of getting, and staying, clear of my emotional baggage. It's also been an invaluable tool in keeping my immunity boosted as much as possible throughout the year (and throughout Covid) and has proved to be life-changing in terms of my health and outlook on life, as well as my ability to respond to, and manage, stress in these challenging times (alongside cold-water swimming and Forest Bathing).

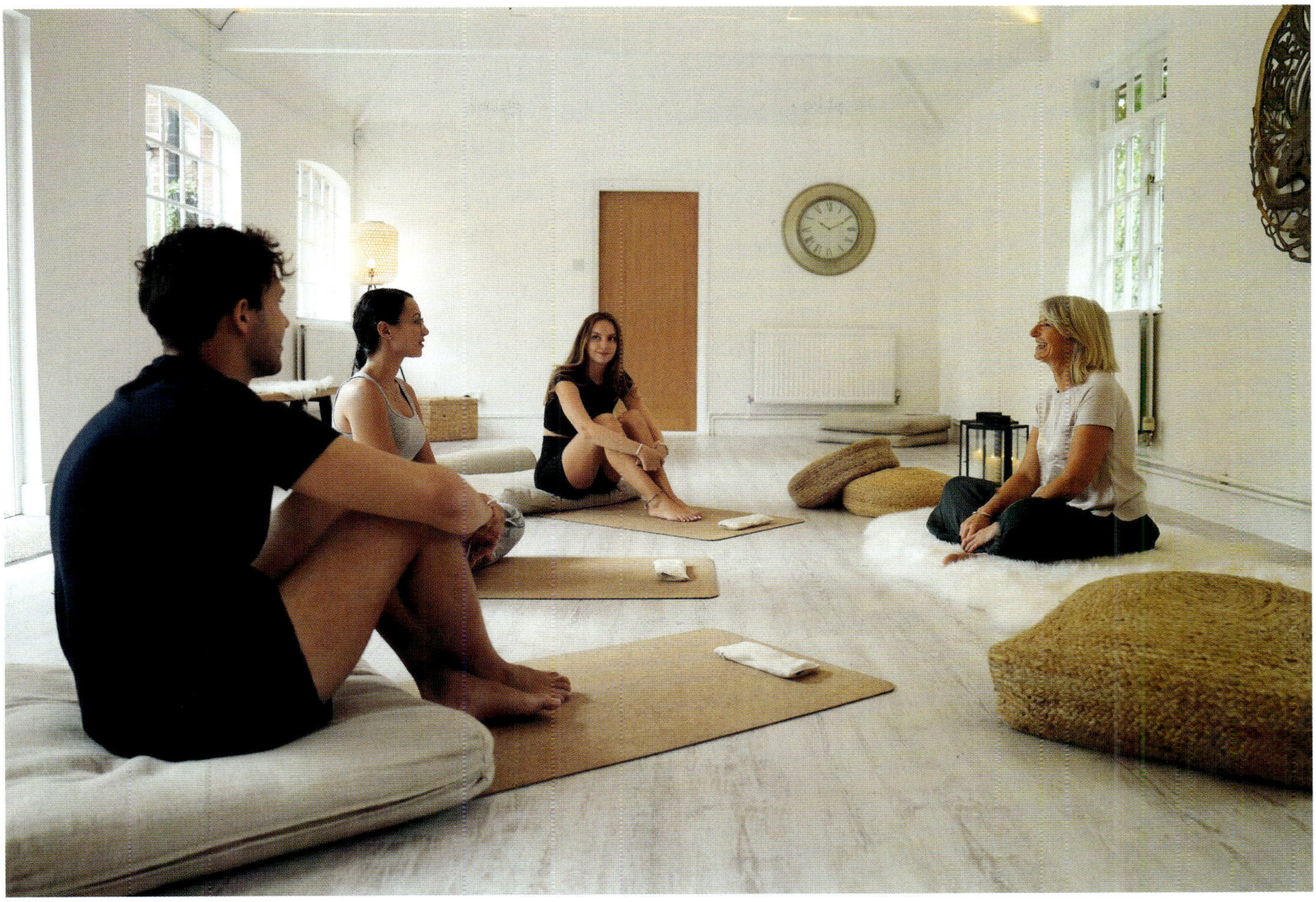

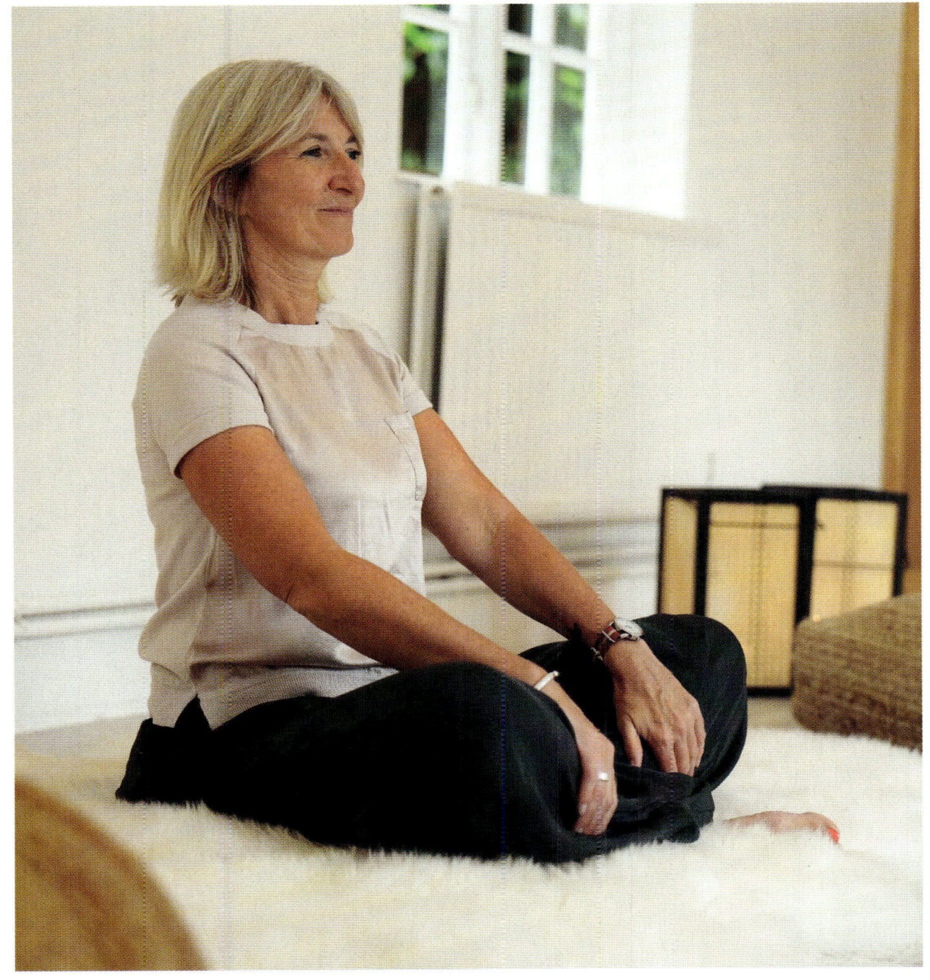

Above: Breathwork can be experienced at home of in a class setting, like here.

Left: With over a decade of teaching experience, Justine Clement specialises in a conscious connected breathwork.

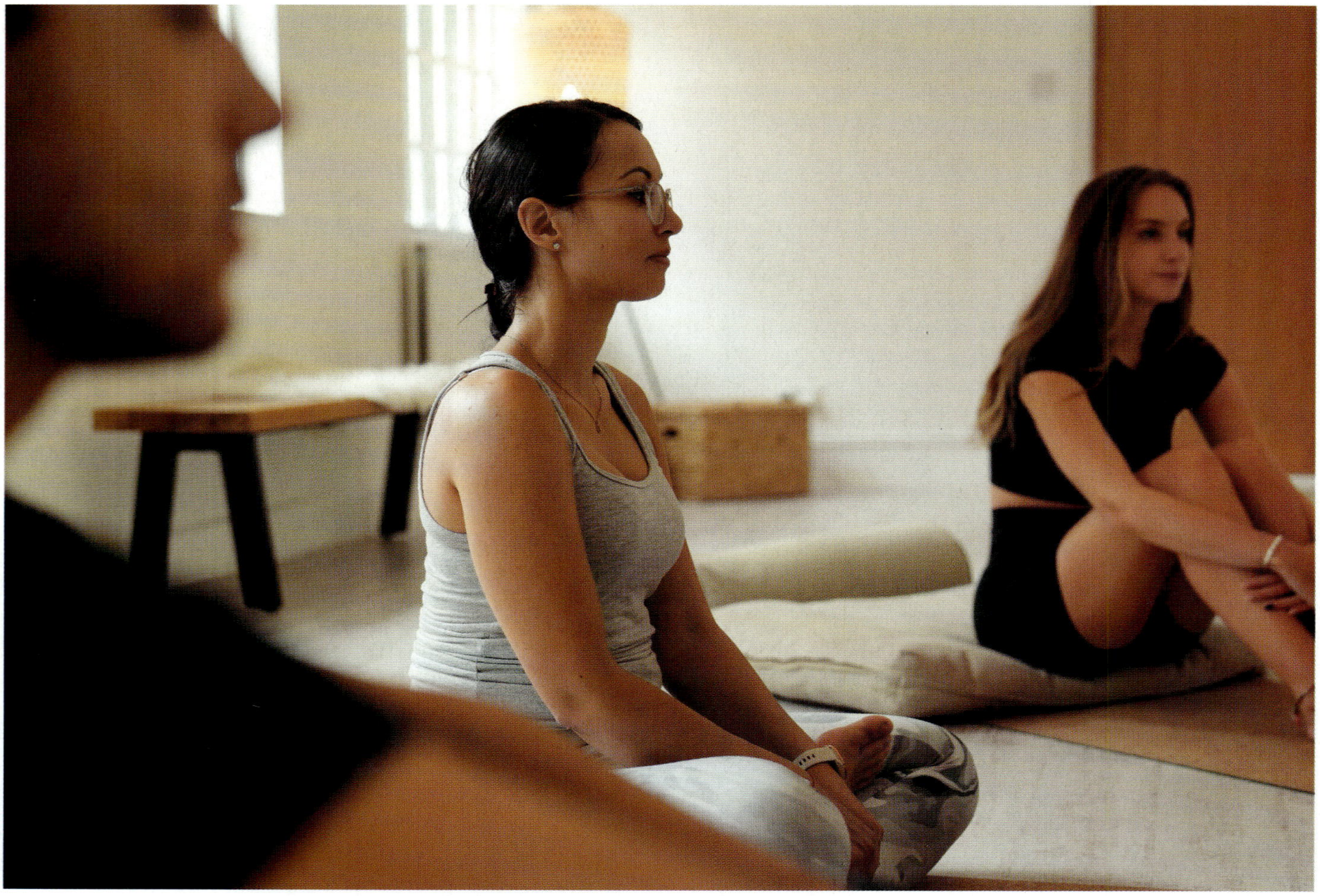

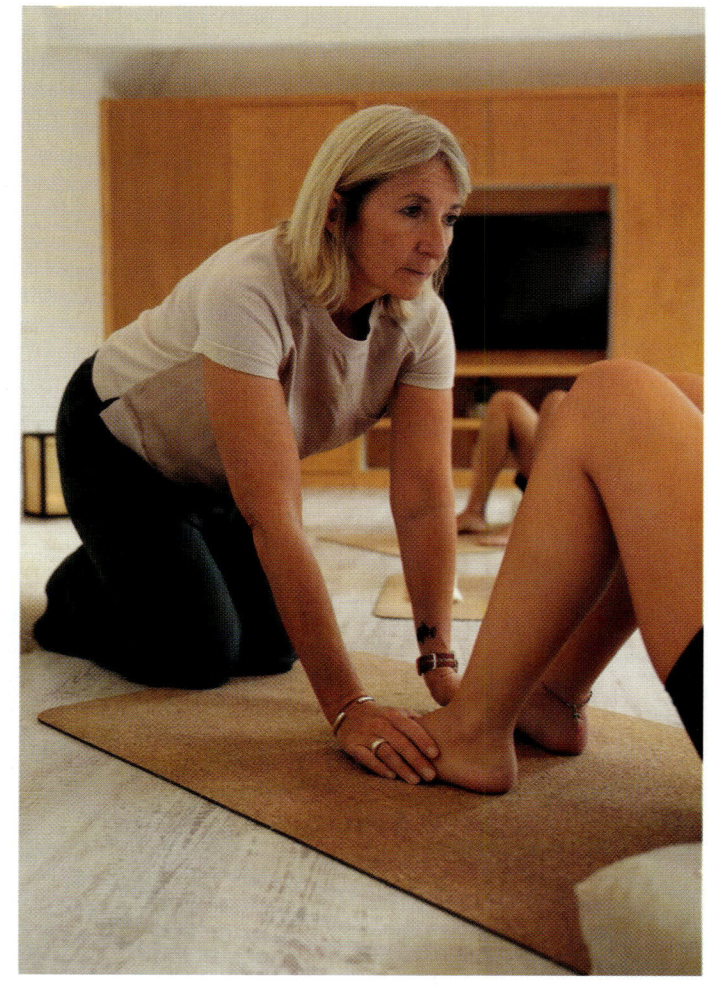

So, what is breath? In simple terms it's oxygen and energy – otherwise called Chi or Qi in China and Japan, or in Indian Sanskrit it's known as Prana - which literally means life, or life force. Think of it as bridge that connects us to all the parts of ourselves – so the physical, mental, emotional, and spiritual parts. Breathwork is a hugely effective tool for helping us self-explore who and what we are.

Most of us consider food as the place we got most of our energy from. But we actually get about 70% of our energy from our breath. So, if we're not aware, or conscious, of our breath, or aren't breathing properly or fully, or practicing some form of breathwork, then we're not getting all the energy we could be on a daily basis.

Variety is the spice of life
It wasn't always the case, but nowadays there's so much help available to us when it comes to physical and emotional health, isn't there? Whether you're looking out of curiosity or a real need, there's an amazing array of wonderful and talented folk enthusiastically sharing their wisdom and knowledge in order help us on our journey to better, if not optimal, health. I choose to stick with one type of breathwork mostly, but that's not to say I don't try other forms now and again, too.

My own experience has led me to believe that most of us need a little help to get going with a breathwork practice, as well as to maintain one. So, just as you work with a personal trainer in a new gym you join, who

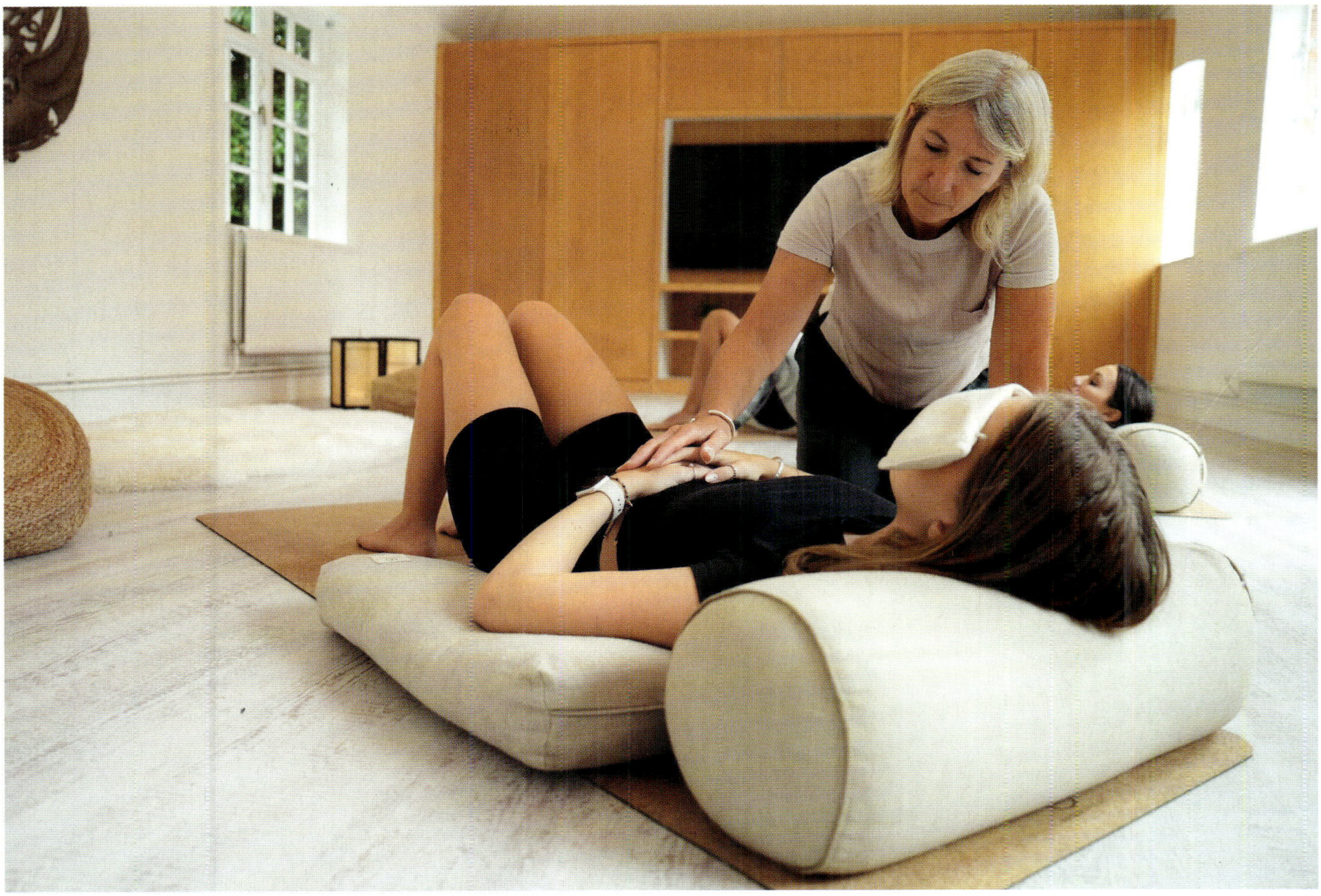

Above: Feeling the tummy rise and fall with each breath, indicates a 'full breath' is being experienced.

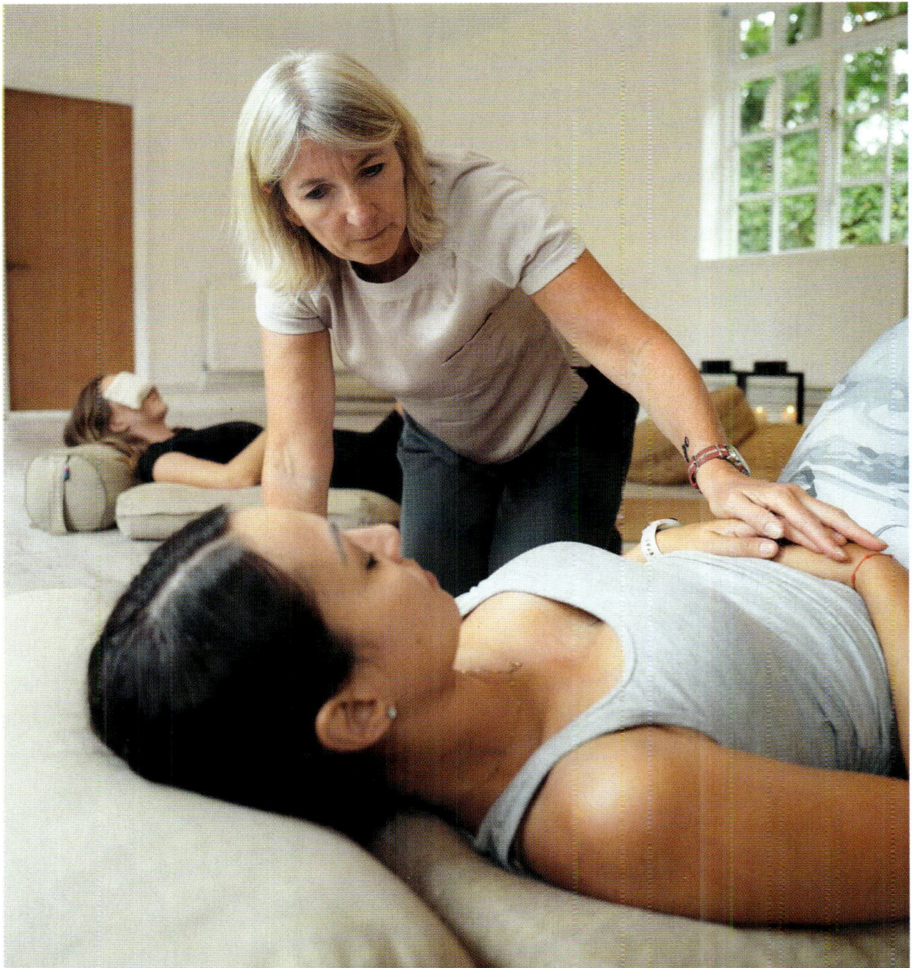

Left: Grounding the feet before an extended breathwork class allows our body and mind to reconnect, helping us relax into a steady rhythm.

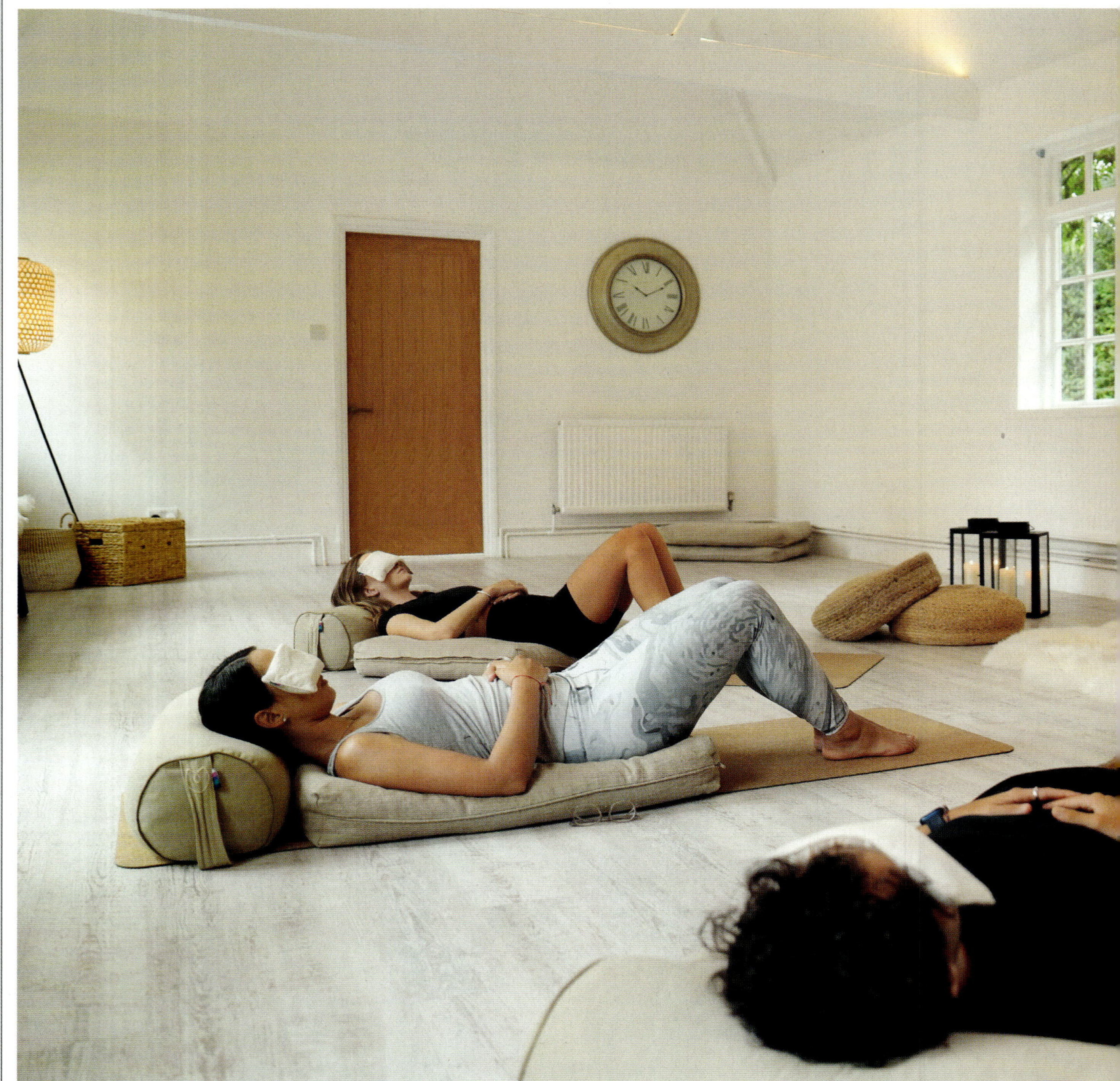

helps you kick start your training as well as deepen your skills and motivation, you may need a little help to get going with breathwork, as well as a top up 1-1, or group session every now and again. The key is to get into the habit of having a practice that you use to support you daily. Consider it as a key life-support tool. Having said that, you could just go to a breathwork practitioner for a specific issue, and with a goal in mind, and just have a few sessions and that's that.

In terms of types of breathwork that can help you with a particular issue, there are literally hundreds of breath practices to choose from. In the yogic tradition there are over 400 alone. Most yogic breathing (pranayama) is about breath retention (hold) and control. There's breathwork that enables your body to relax and others to heighten your state, or 'stress' your body, but in a good way. There are some to help you sleep, and others to bring greater clarity or answers to problems troubling you. Some forms of breathwork, such as the one I work with, a form of Conscious Connected Breathing, also helps with anxiety and depression, and I experience this benefit personally on a daily basis. It's not been called 'a detox for every level of your being' for nothing; boosting the autoimmune system, improving circulation. You name it, it helps in a wide variety of positive ways. I suffered from Raynaud's Disease (a type of vascular disorder which for me meant very poor circulation and often alarming blue fingers when I got cold) which did not go well with my love of sea swimming. Yet since having a regular breath practice each morning, in the same way people might build meditation into their early morning routine, it has disappeared.

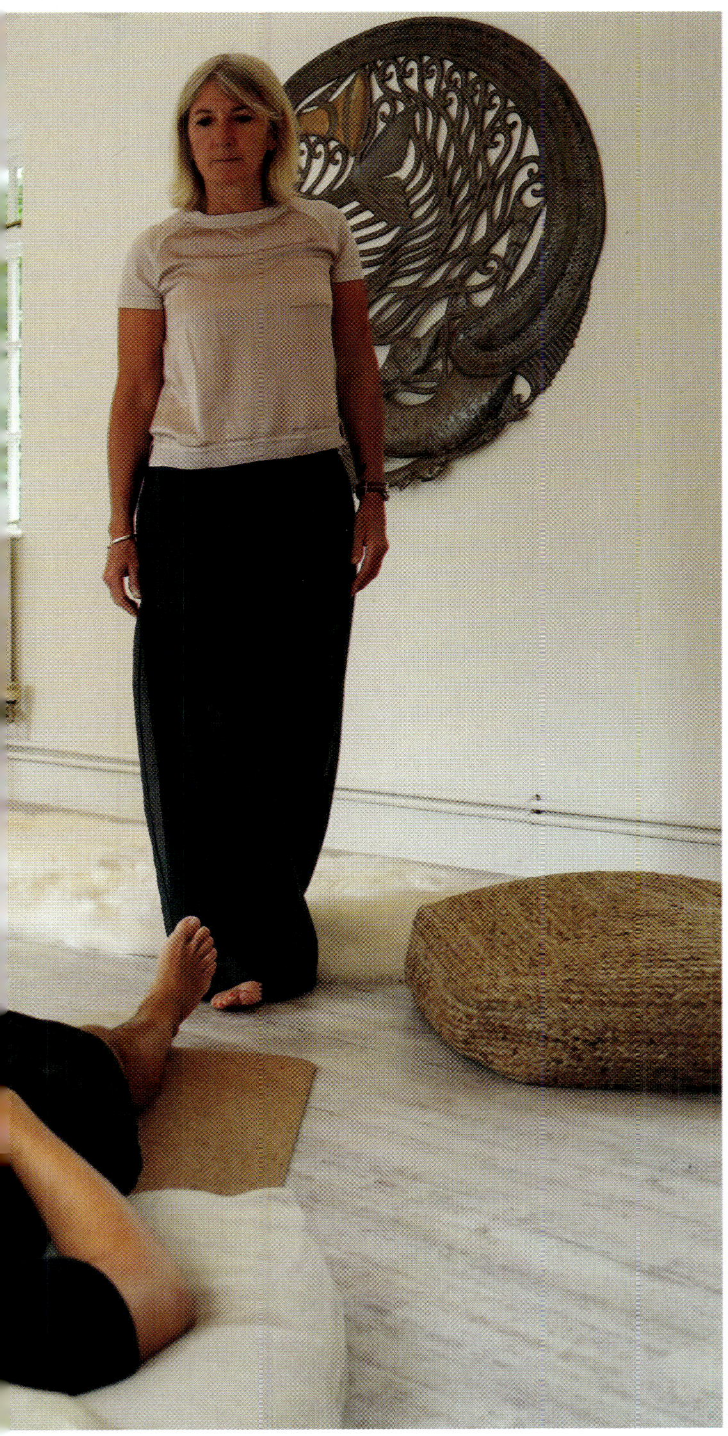

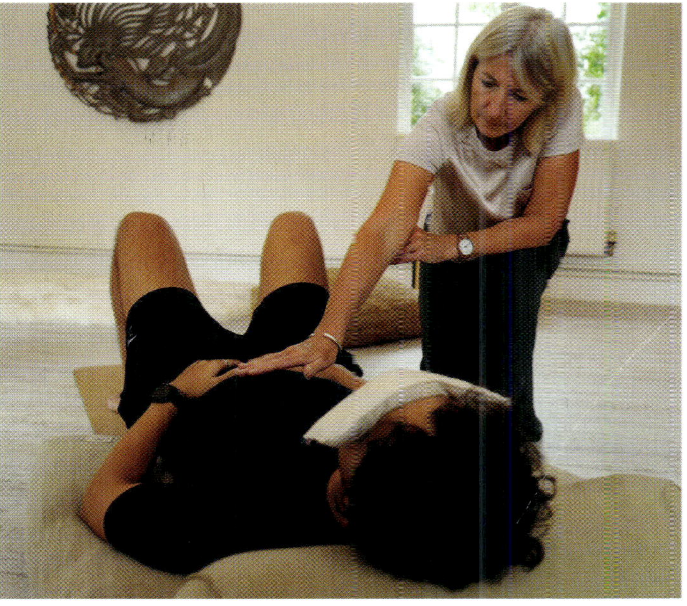

Google and you'll have an evening's worth of data to sift through and convince you.

Learning to breathe more effectively and taking up some form of breathwork practice will undoubtedly positively affect your mind, your emotions, your thoughts, and your physiology — even down to improving the density of your bones. It will increase your energy, your circulation, help you sleep better, help you feel calmer and reduce your anxiety. As I've said already, I know this because I've experienced it for myself. Noticing my breath patterns and practicing breathwork has improved my life in so many ways. It provides untapped potential that can transform every aspect of your life. And when you feel better, you live better.

Breath is wonder — a lost art and reclaiming your birth right

For me, breath is wonder. A wonderous, easy-to-access tool at our disposal. As babies, most of us are born as perfect breathers. Watch as a baby's chest and tummy slowly inflates and deflates; they are the

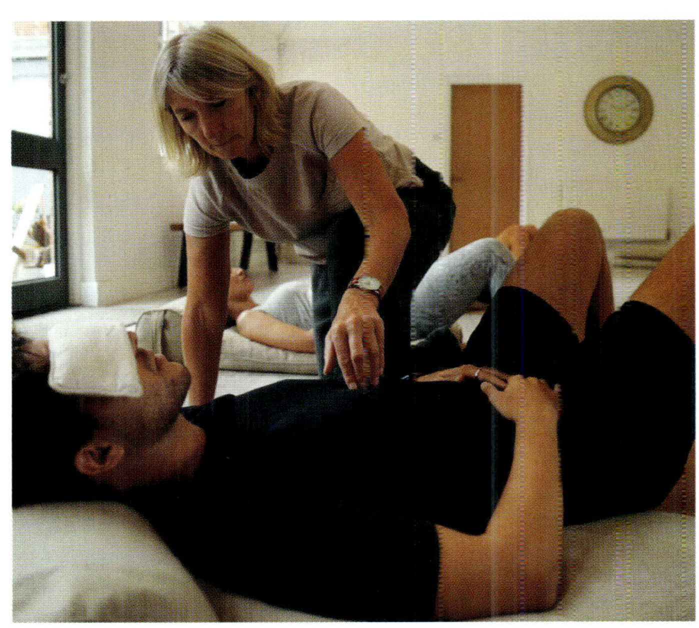

Why breathwork feels important right now

The stress of modern life, with its busyness, negative media bombardment and always-on culture, can lead to system overload for many people, to the point where you end up living with low (and often high) levels of anxiety as the norm. Believing that how we breathe doesn't matter to our wellbeing and to how well, or how poorly we live our lives, is no longer something we can afford to do. That's because how we breathe affects every single system in the body deeply, on a cellular level.

If you're new to this phenomenon and are not convinced by the well-publicised ancient wisdom from traditions in India and Tibet for example, then modern science is beginning to catch up and there's now much more scientific data to support this. Just pop it into

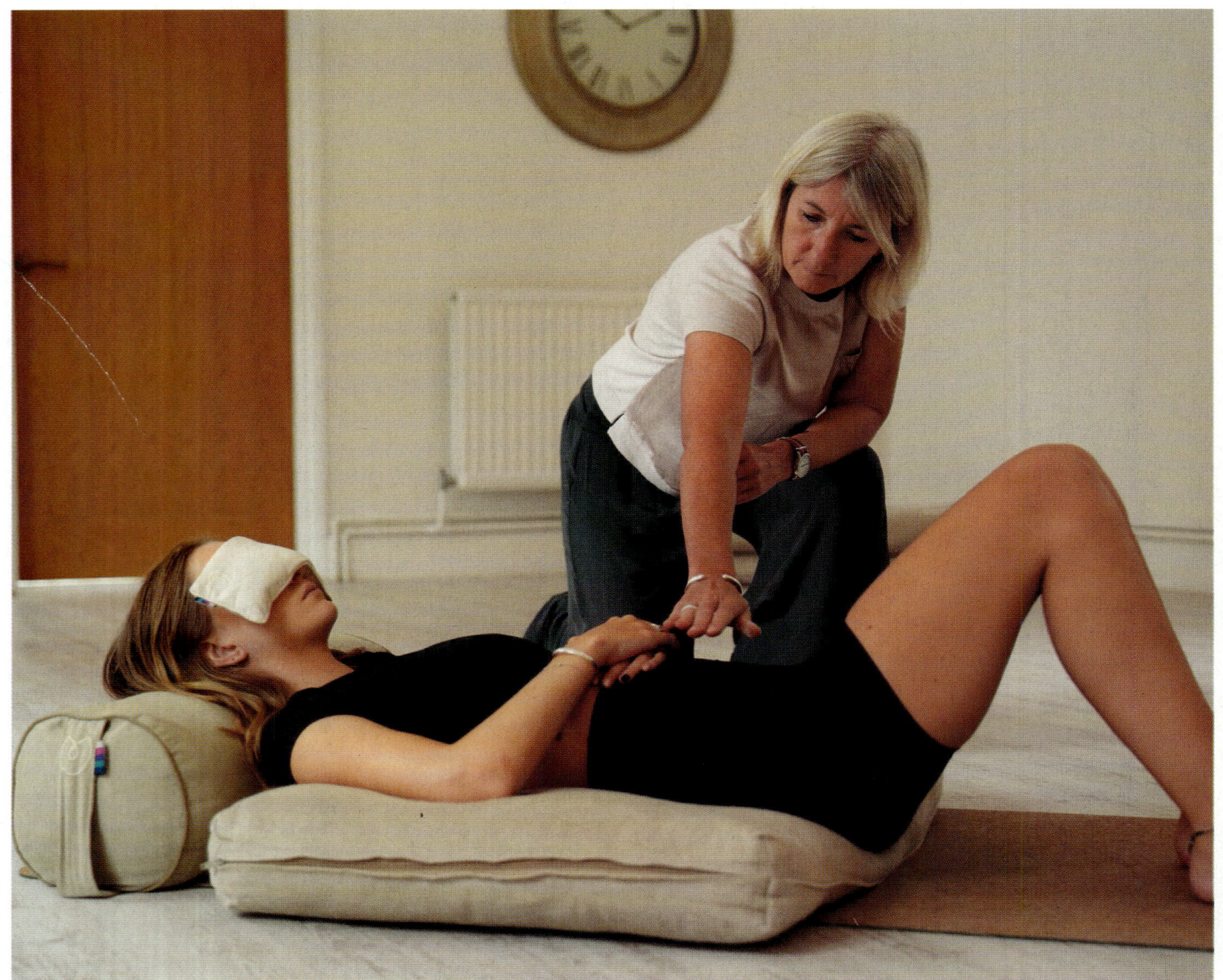

perfect breathing role models. But as the years go by, this perfect breathing changes, due to stress and, well, just generally not paying attention on it. And so, breathing properly becomes an art we lose; a lost art. Regaining this lost art has helped me to see the world in technicolour, when it might often have seemed a bit black and white.

Something has kept you reading this. You may just be curious. Or you may recognise some of the health challenges I mention above. Either way, most of us have something in us we want to change, or you could say, awaken. Whatever the reason, re-learning the forgotten art of breathing is suitable for pretty much everyone. You could call it reclaiming your birth right.

Simple 'morning' routine

A very simple breathing technique you can introduce into your daily life will encourage a sense of calm, increase focus and clarity, allowing you to have the confidence to tackle things head-on, where in times passed they may have been overwhelming. Maybe you can introduce this technique to your morning routine - set your alarm 10 minutes earlier - or maybe later when returning home, to help clear a busy day from your mind. *M&B*

For this technique we will use a nose in, mouth out method, in a seated postion:

- Place your hands on your stomach and take a full, deep breath. When inhaling, gently push the tummy out.
- Slowly release the breath through the mouth; imagine you are cooling something down with the exhale.
- Repeat the inhale and exhale another five times.
- On the sixth breath in, hold at the top of the inhale for 10 seconds.
- Repeat this technique four more times, giving you five rounds in total.
- Now feeling more relaxed, allow your breathing to come back into its own natural rhythm before standing up.

This technique is great to give ourselves a little reset. You may even find yourself doing it through the day, a few rhythmic breaths and hold to recharge that sense of calm.

Justine Clement
Wonderbreath

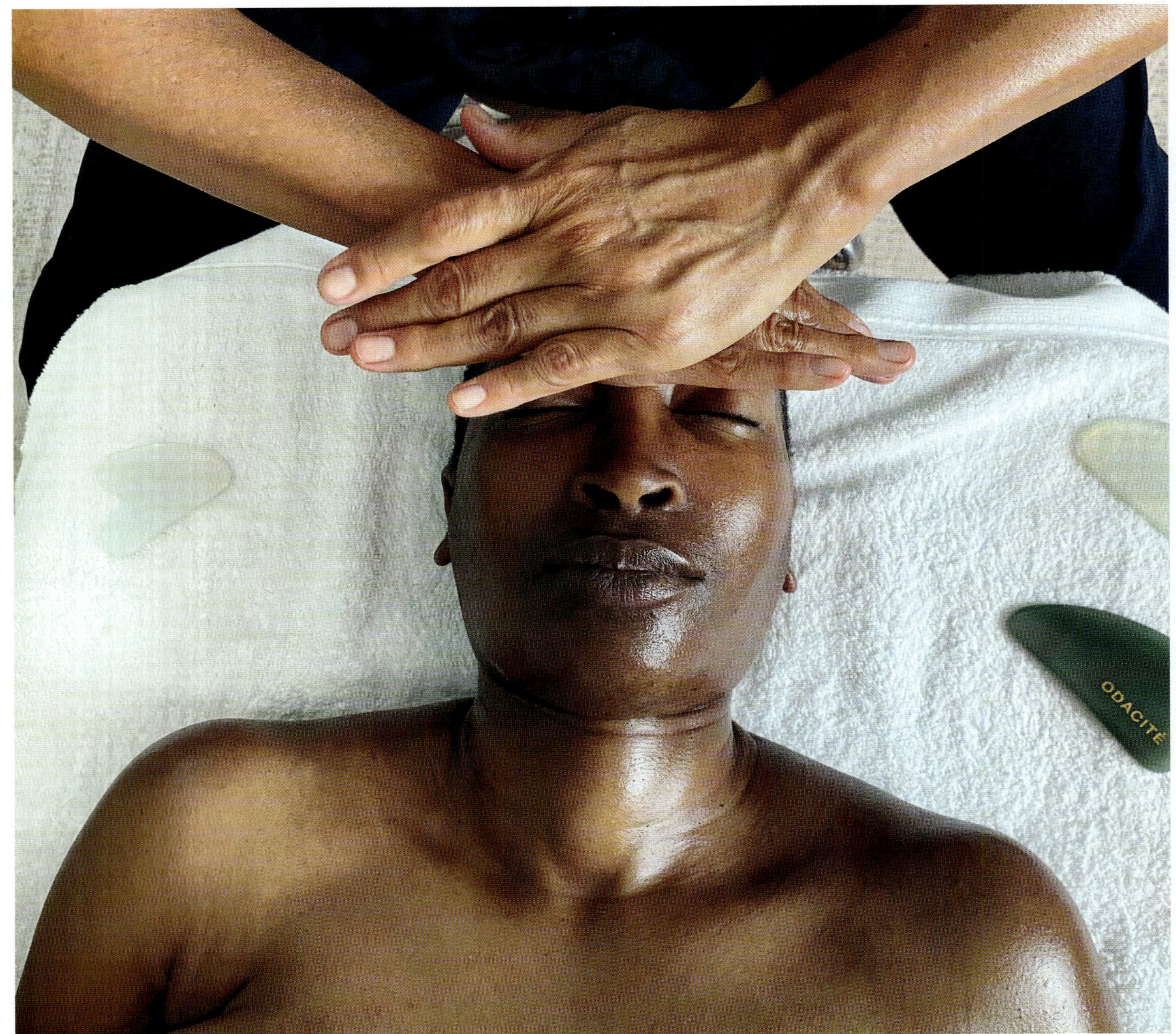

Energy Meridians

Bibi Bohorquez has been a manual therapist and postural alignment coach, providing deeply therapeutic treatments for over 15 years. With a strong belief in the mind/body connection, she is committed to helping people out of pain and achieving optimum health.

Traditional Chinese Medicine uses acupressure, an ancient form of massage, as one of its therapy modalities. Acupuncture and acupressure are closely related, and it's thought that these therapies date as far back as 2000 B.C. starting as Anma Massage. Over time, these practises spread across the world and different disciplines developed. Shiatsu, the Japanese version, developed to use palm pressure, stretching, kneading and other manipulation along with the finger and thumb pressure.

Acupressure's literal meaning is 'finger pressure, along with other forms of Chinese therapy, aims to promote the flow of qi (also known as "life energy") through the 14 channels (meridians), the electrical circuit board, in the body.' The acupoints and energy meridians that acupuncture targets are the same as in acupressure.

The entire body is mapped by these meridians, which extend from the soles of the feet to the top of the head. Each side of the body has twelve primary meridians that run parallel to one another.

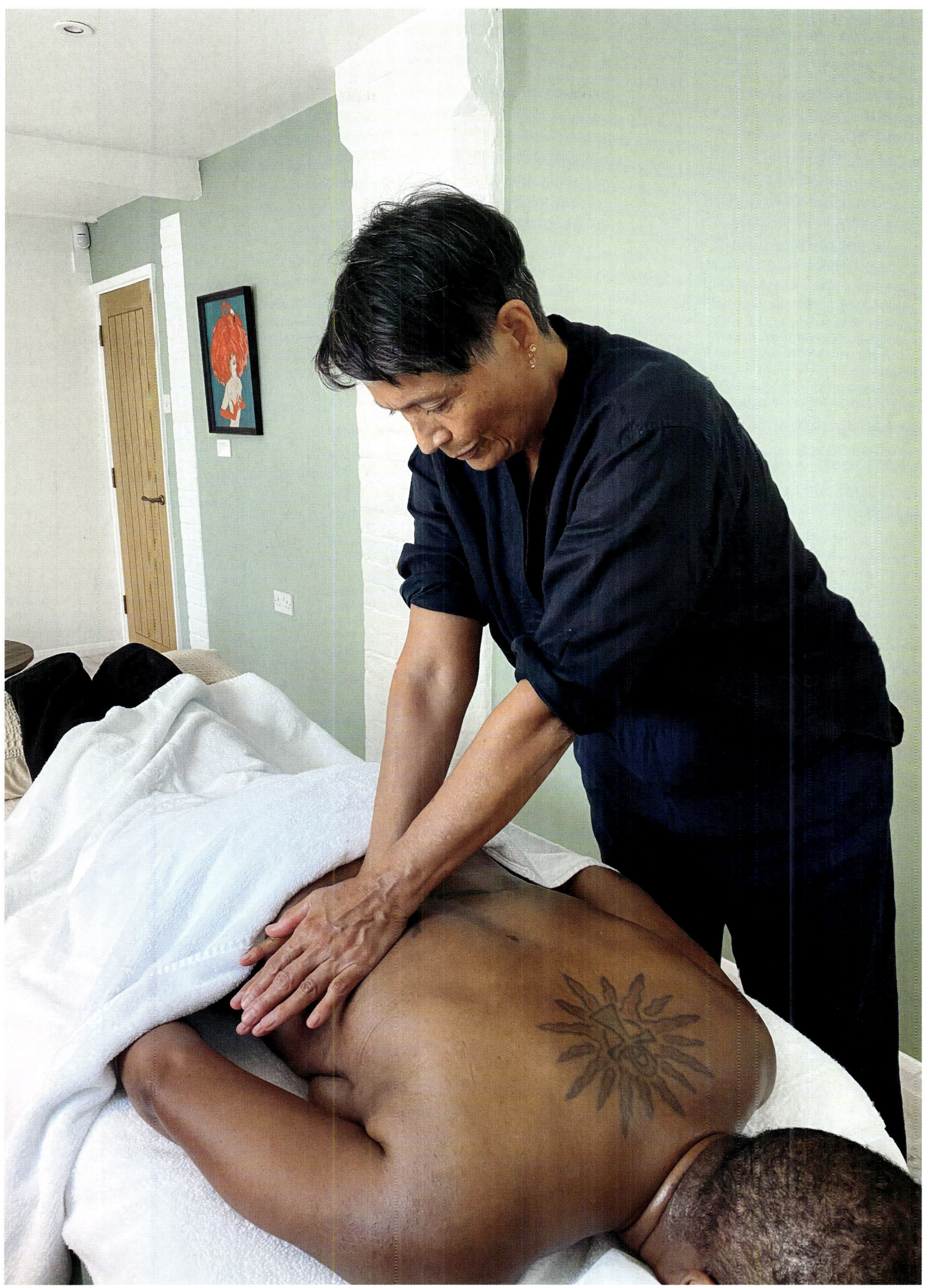

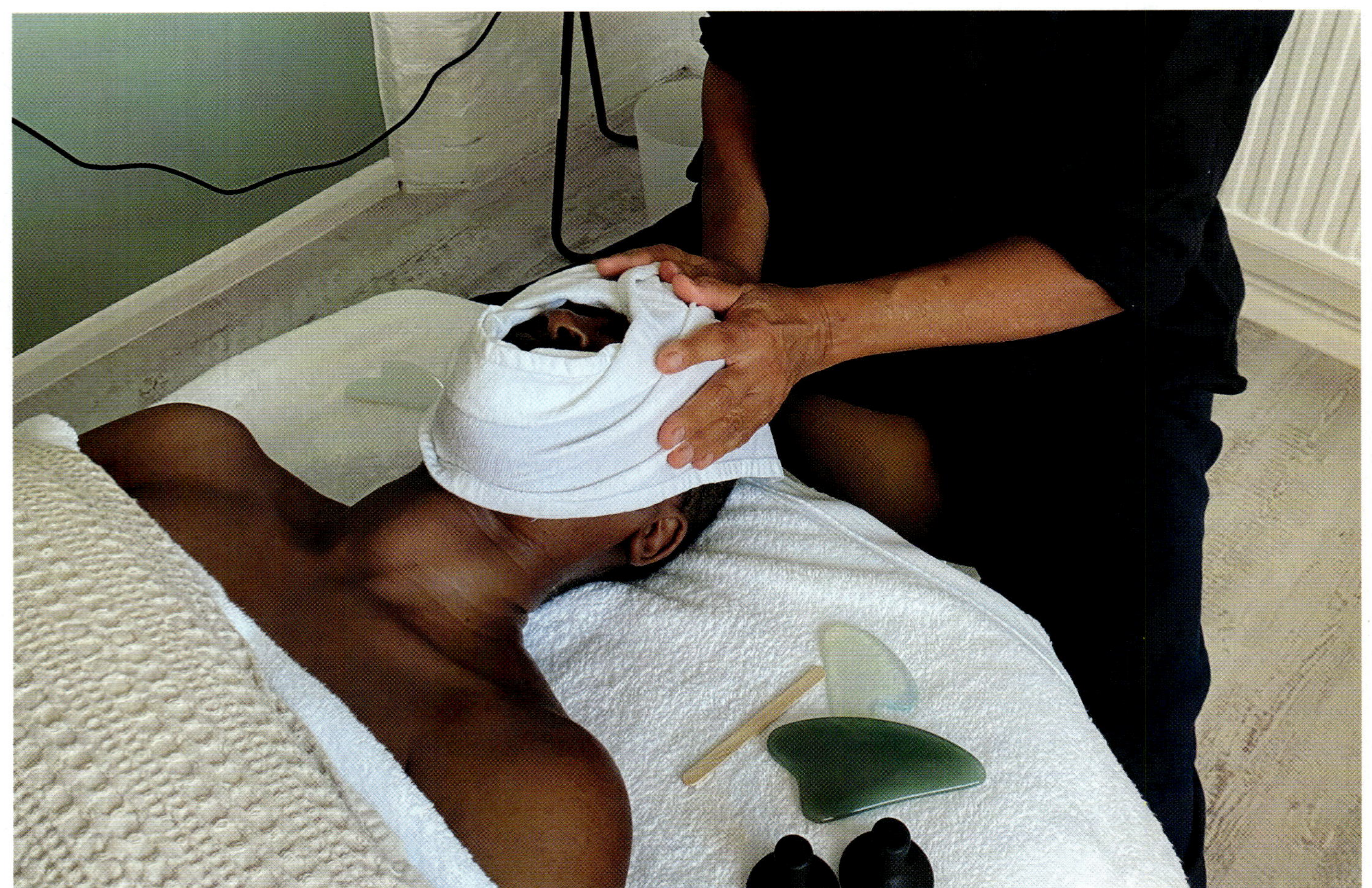

Each one corresponds to a certain internal organ, and each organ has its own physiological and intangible energy processes that link to the body's larger meridian network. Each meridian corresponds to a certain type of qi, a quality (cooling or warming), a set of sensory organs, a set of emotions, and a set of elements. Depending on whether they are located on the inner or outer parts of the limbs and torso, and the direction of energy flow. In the same way that energy-giving routes, the nervous system, connect and coordinate the body, these energy lines or meridians, maintain the body in balance and regulate all of its natural activities.

According to Chinese medicine theory, a person needs a steady flow of qi through these channels to maintain their health. Every system works on a balance of yin (calm) and yang (energetic). Traditional Chinese medicine (TCM) states that acupressure applies pressure to stimulate acupoints for therapeutic purposes. Stimulating these points can correct imbalances of yin and yang qi energy flow through channels, which will then treat specific ailments, musculoskeletal issues, and even emotional issues.

The body can no longer maintain the balance required to sustain high energy and manage health issues if this energy flow is stopped.

In a nutshell, acupressure involves applying pressure along the various meridians to encourage the free flow of energy for more vitality, focus and calm. In the

Acupressure involves applying pressure along the various meridians to encourage the free flow of energy.

days of the emperors of China, you did not go to a psychotherapist for emotional issues, you had a physical massage treatment.

Acupressure therapy can promote the body's hormonal, lymphatic, and circulatory systems. It lessens headaches and migraines, enhances sleep, helps to boost the immune system, relaxes your muscles, and joints, regulates digestive problems, and is especially helpful for back pain and menstrual cramps. Positive effects from acupressure massage accumulate over regular sessions, but the effects can be felt after just one treatment.

Acupressure treatments are very different from regular massages. Typically, it is carried out through clothing. A practitioner will use fingers and thumbs to press on specific acupoints on the body to trigger them. However, they may also utilise their feet, elbows, and knees.

We are piezoelectric beings; this means that pressure applied to one area of the body sets off electrical messages, via the nervous system and brain, that

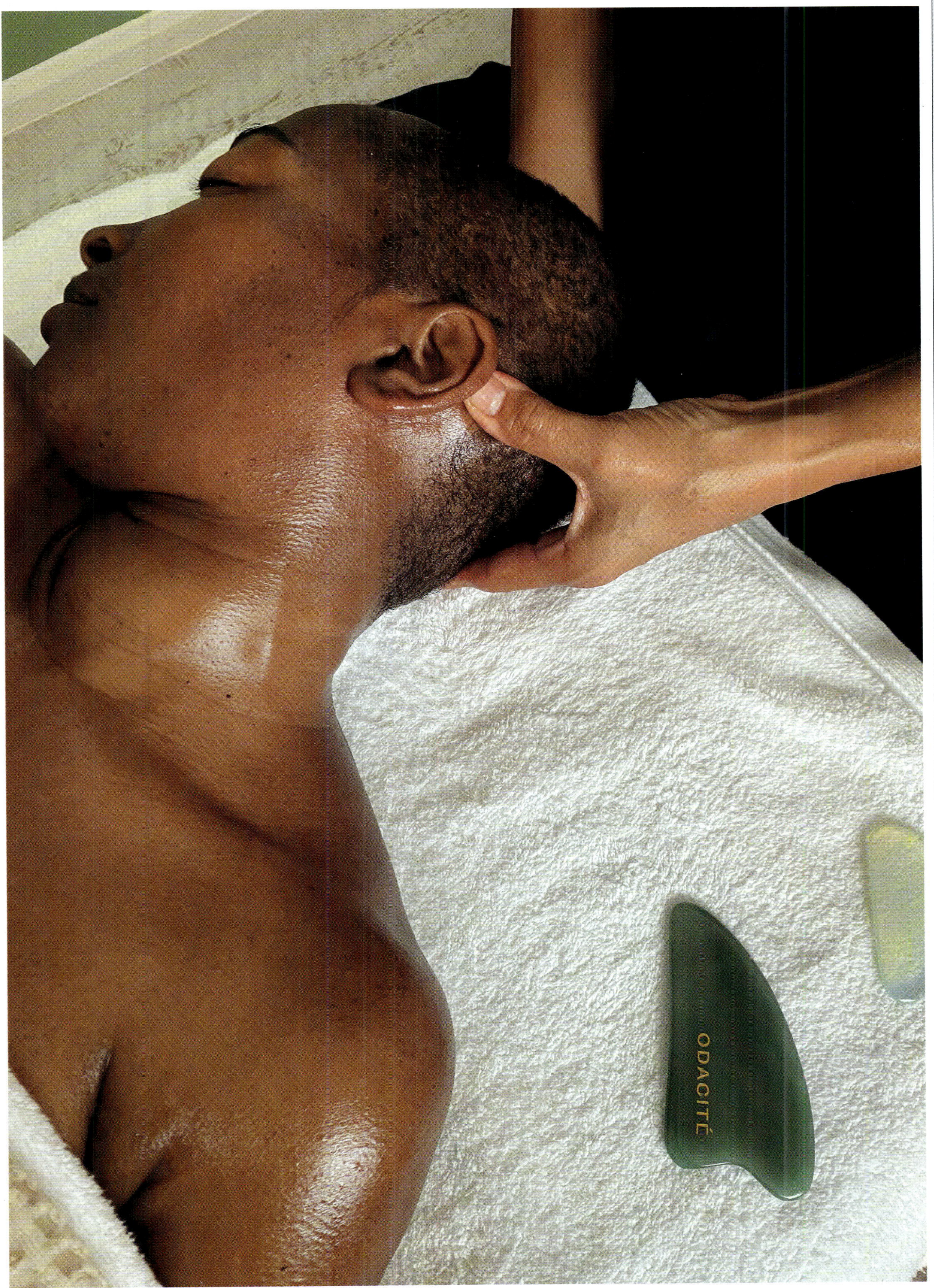
ODACITÉ

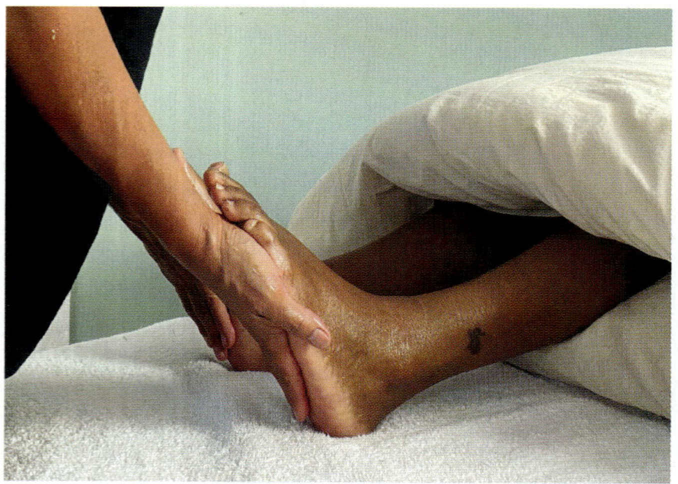

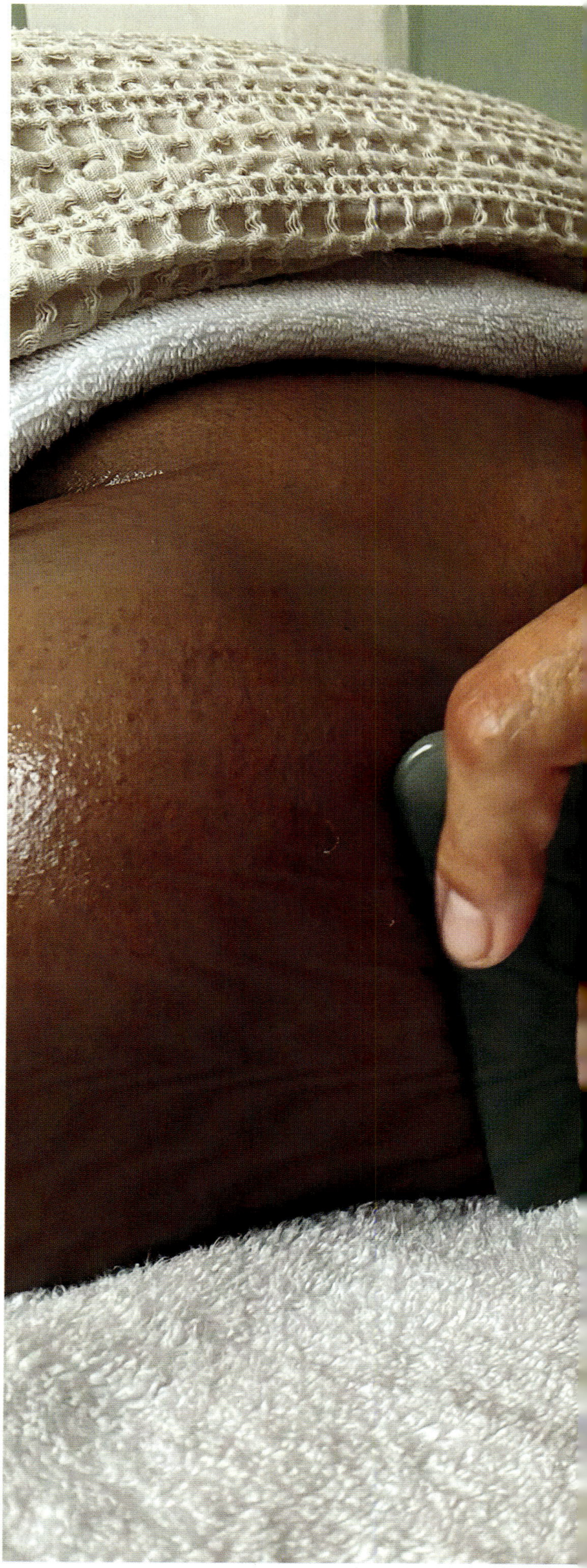

Even if you don't believe in meridians or qi, deep relaxation is a major benefit of acupressure.

transmit to various other parts. For instance, the energy of your kidney meridian is represented by a part of your foot.

It is also important during a session to stretch the muscles to allow for free flow of vital fluids, lymph, oxygenated blood etc. This essentially feels like you've done a session of yoga without making the effort yourself.

Overall, acupressure massage is a needle-free, non-invasive, cost-effective and non-pharmacological healing tool for the promotion of patients' overall well-being.

From the point of view of modern western medicine, which tends to focus more on identifying pathogens and treating diseases and their symptoms in isolation, it may all sound quite complex and mystical. But with 2500 years of history as well as numerous modern studies confirming the benefits of acupuncture and acupressure, it's not surprising that more and more people are turning to this amazing risk-free system of healing. Supporters of the method have claimed measurable pain reduction, decreased stress and anxiety, and an improved mood while following a protocol of monthly treatments.

Even if you don't believe in meridians or qi, deep relaxation is a major benefit of acupressure. And that's a win every time in this day and age.

Acupressure can help with the symptoms of IBS, any musculoskeletal pain, tight jaw, nervous tension, post-operative pain, sinus issues to name just a few. It can also be of great benefit throughout pregnancy to relieve stiffness in the joints, stress and fatigue. *M&B*

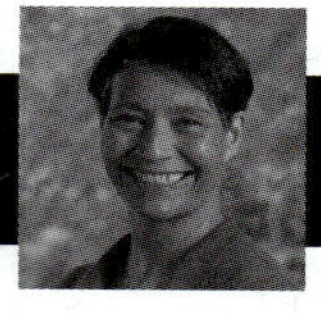

Bibi Bohorquez
www.thisisalign.com

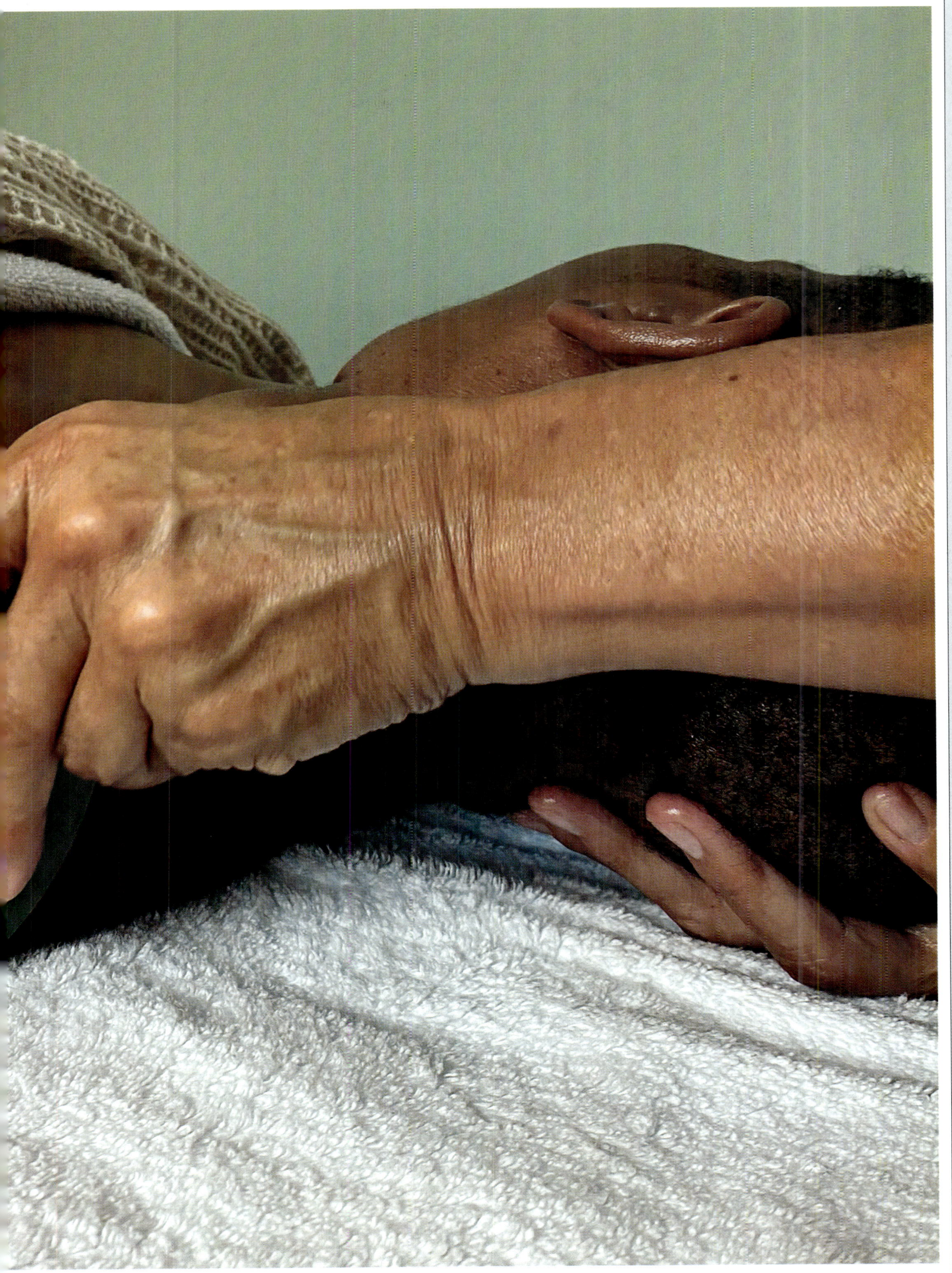

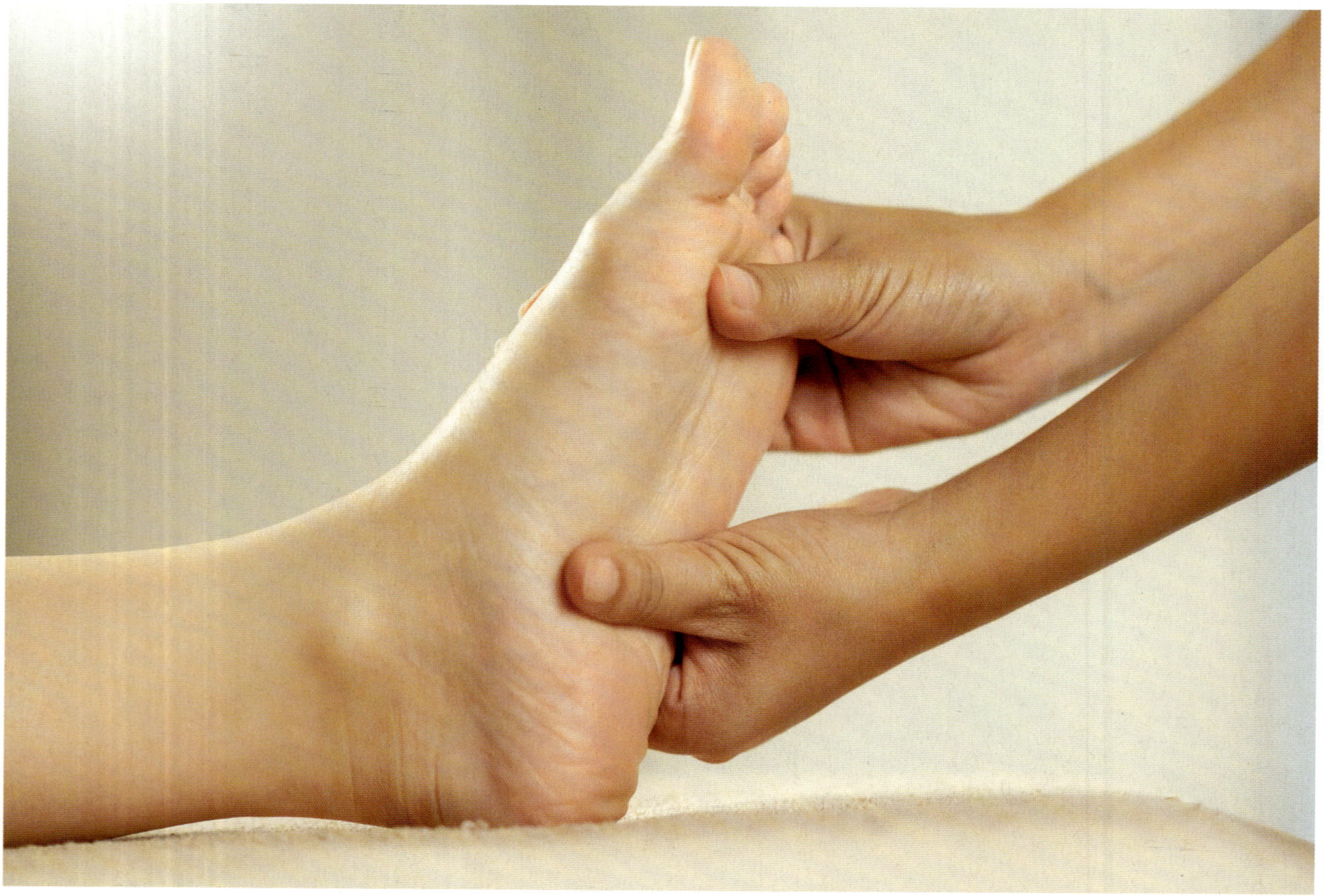

The Right Steps

Reflexology is a non-invasive, powerful healing therapy that can be useful in bringing about profound relaxation and balanced well-being. **Bibi Bohorquez** talks more about this ancient therapy.

The practice itself has been around for thousands of years, as evidenced by images found in the great tombs of Egypt. Even the Greeks were aware of the power and connection of the feet to the body (the body is a single unit and it is ALL connected after all). A saying often attributed to the Greek philosopher Socrates is as true today as it was then: "If the feet hurt, one hurts all over".

The development of reflexology as we know it began when American physician William Fitzgerald introduced the concept of "Zone Therapy" around the beginning of the 20th century. Fitzgerald proposed dividing the body into 10 equal vertical zones. The idea being that pressure and manipulation techniques applied to these zones in the hands and feet might have an impact on physio-pathological problems in corresponding locations within the same vertical zones.

It was believed that tenderness in certain zones of the foot was a sign of dysfunction in its interrelated area. After a brief period of massaging this area, the foot's

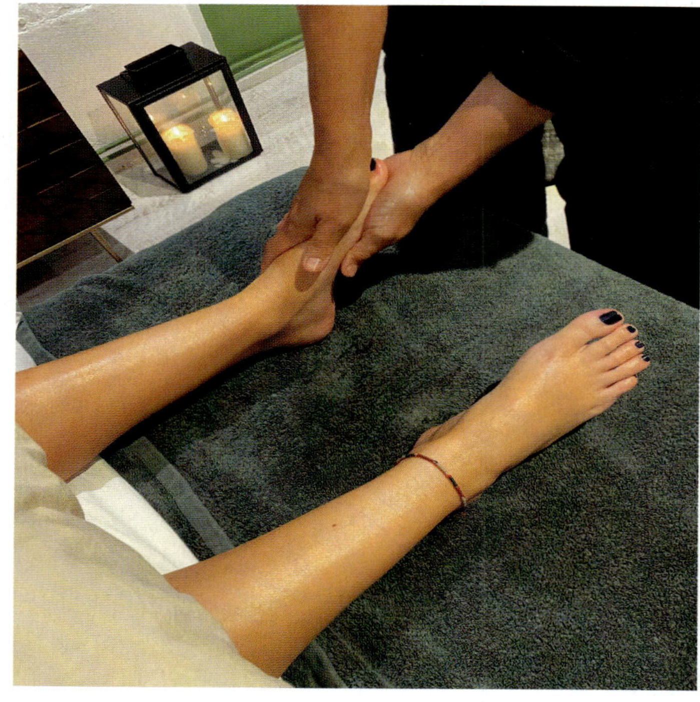

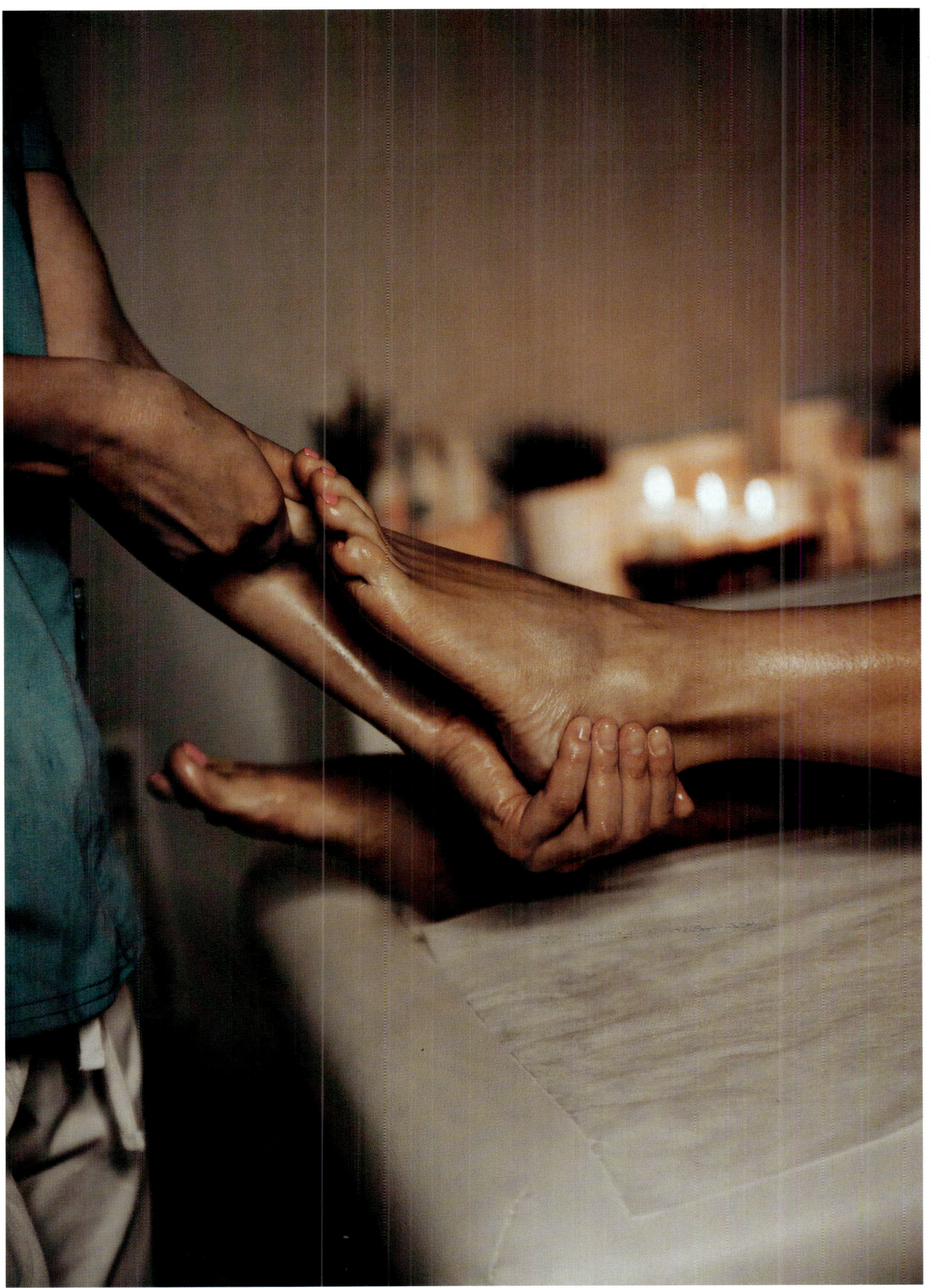

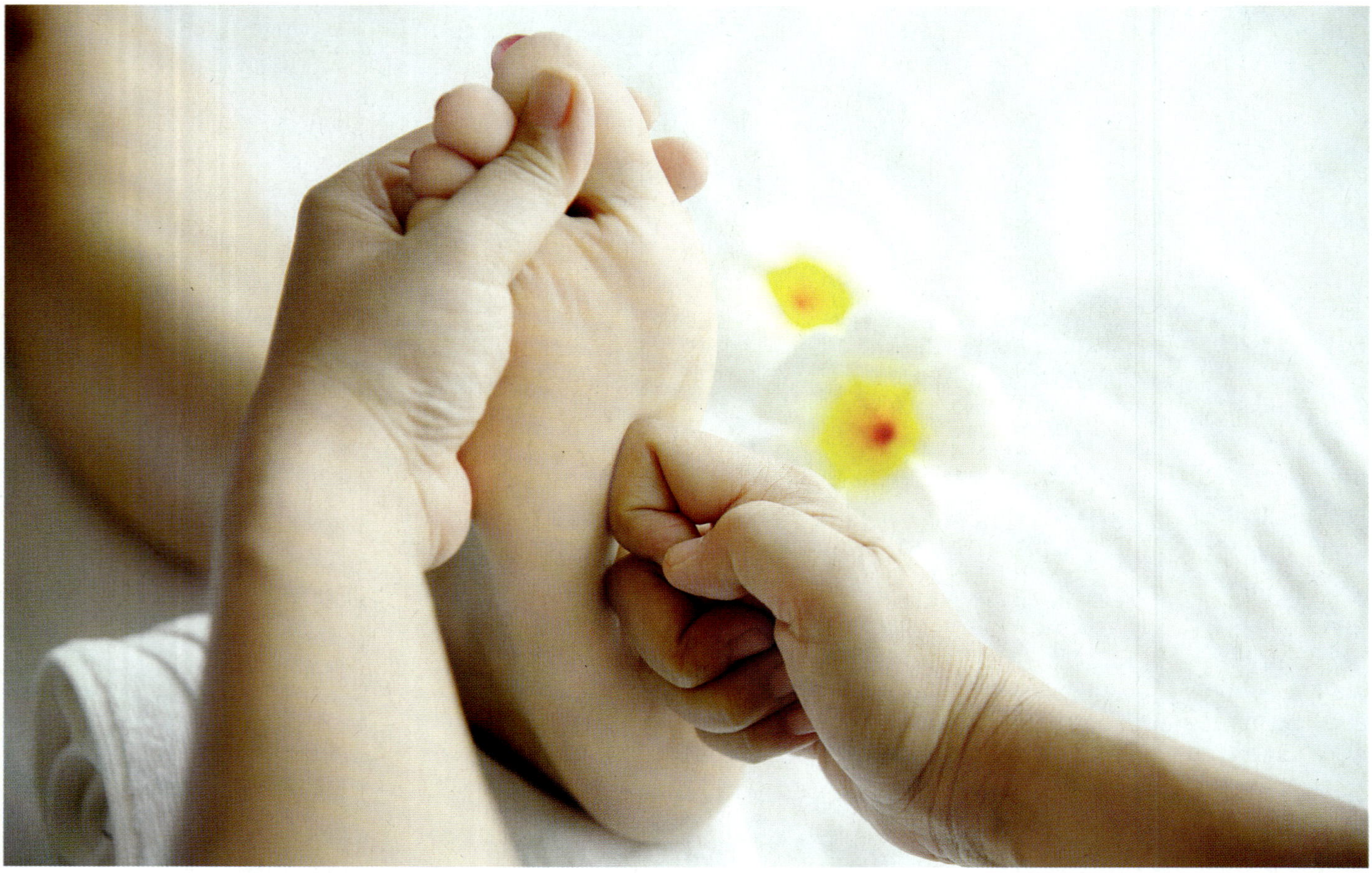

soreness subsided, and it was accompanied by an associated improvement in the corresponding zone's other impacted locations.

In the early 20th century, Eunice Ingham, a bodyworker and physical therapist working in the US, travelled throughout China and studied their massage techniques, further refining Zone Therapy into "Compression Massage" and later, "Reflexology."

Reflexologists work various points and areas using the principle of 'nerve theory' in that specific points on the feet, lower leg, hands, face, or ears correspond with specific body parts. By applying pressure to these points, an impulse is triggered, through sensory neurons, which reaches the brain via the spinal cord. The brain interprets these impulses and sends out messages, again via the spinal cord and clusters of nerve cells throughout the body, to the muscles groups it has received signals for.

In its basic form, reflexology follows maps of the body which are said to represent every physical area of the body via these 'reflexes' on the feet. As a basic rule, from

It was believed that tenderness in certain zones of the foot was a sign of dysfunction in its interrelated area.

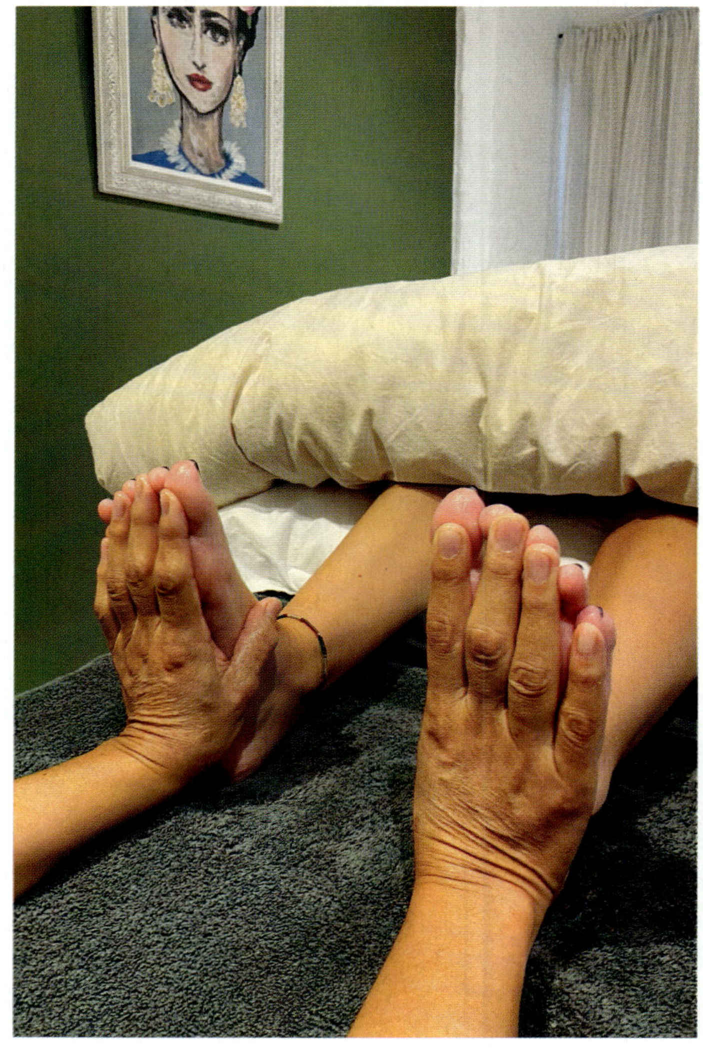

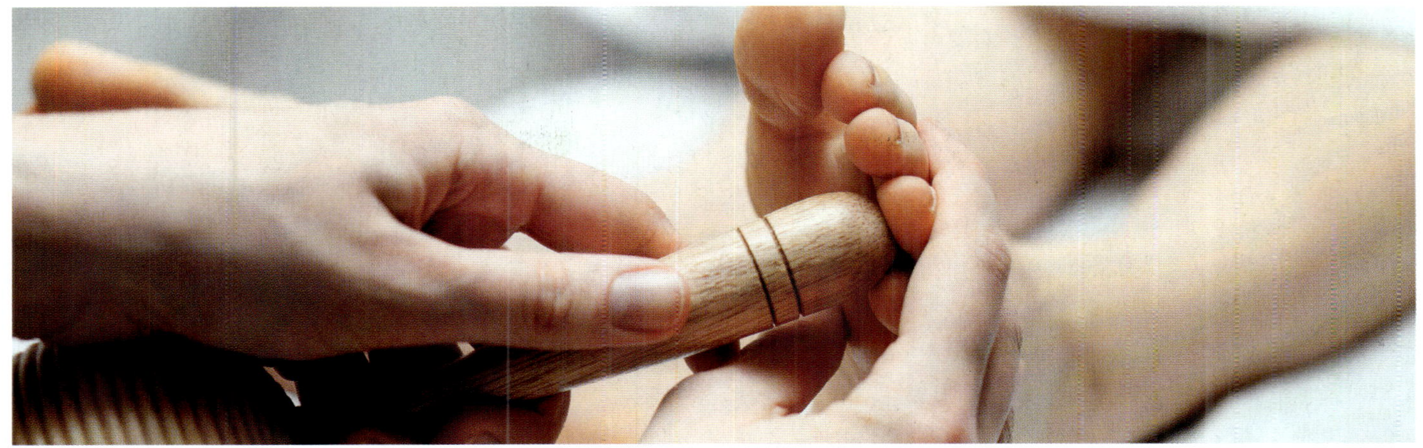

the practitioner's view of the feet, the sole of the right foot corresponds to the left side of the body and the left foot works the right side.

A series of precise pressure movements using thumbs and fingers (sometimes tools such as acupressure pens) are conducted until the whole surface area has been covered. Depth of pressure and patterns of movement allow for a variety of assorted styles of reflexology, all of which are descended from this common theory.

Reflexologists can specialise their treatments for specific client needs, such as lymphatic drainage, fertility assistance and pregnancy. All good therapists will have the ability to deal with these as well as assist with sleep issues, back pain, restless leg syndrome, headaches, endocrine imbalances along with much more, and most obviously, foot issues.

Reflexology was examined in a 2014 study with 80 participants for anxiety. Over four days, the reflexology group members reported much less anxiety. Reflexology was also found to lessen nausea, vomiting, and exhaustion in chemotherapy patients, according to

Reflexologists work various points and areas using the principle of 'nerve theory'.

a 2017 study. Recently, a 2020 study found Reflexology to significantly lower heart rate in people with stage 2 hypertension or high blood pressure.

Aside from all the above, a good reflexology treatment will provide a safe space to relax deeply. There is nothing more soothing than having our feet massaged. They carry us through all the journeys of our life and deserve your care and attention.

A reflexologist with a thorough understanding of this therapy will have had a year to one and a half years of training, completing approximately 100 thorough case studies, along with walk in client studies and research, and around 450 hours of theory and anatomy and physiology study. Always check if the person you are seeing has completed a full level three reflexology training. **M&B**

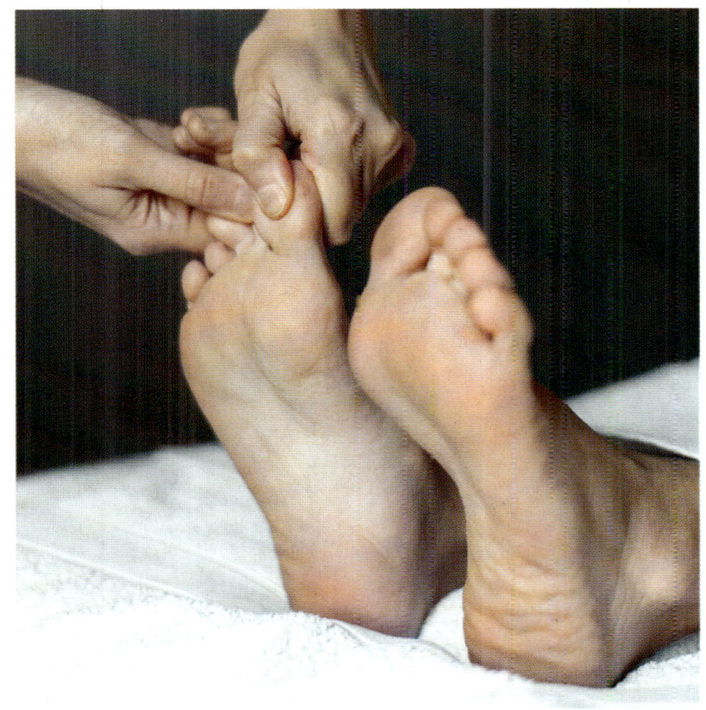

"Stand Up Straight"

Postural alignment therapy is a unique and effective modality designed to treat musculoskeletal pain without manipulation (massage techniques). It is a process which involves a series of stretches and specifically targeted exercises designed to restore full, natural function to muscles and joints. It does this by reminding all the muscles of the jobs they were specifically designed to do, thus decreasing chronic pain issues, and reducing energy wasted by the body in trying to compensate for the lack of designed function. **Bibi Bohorquez** continues...

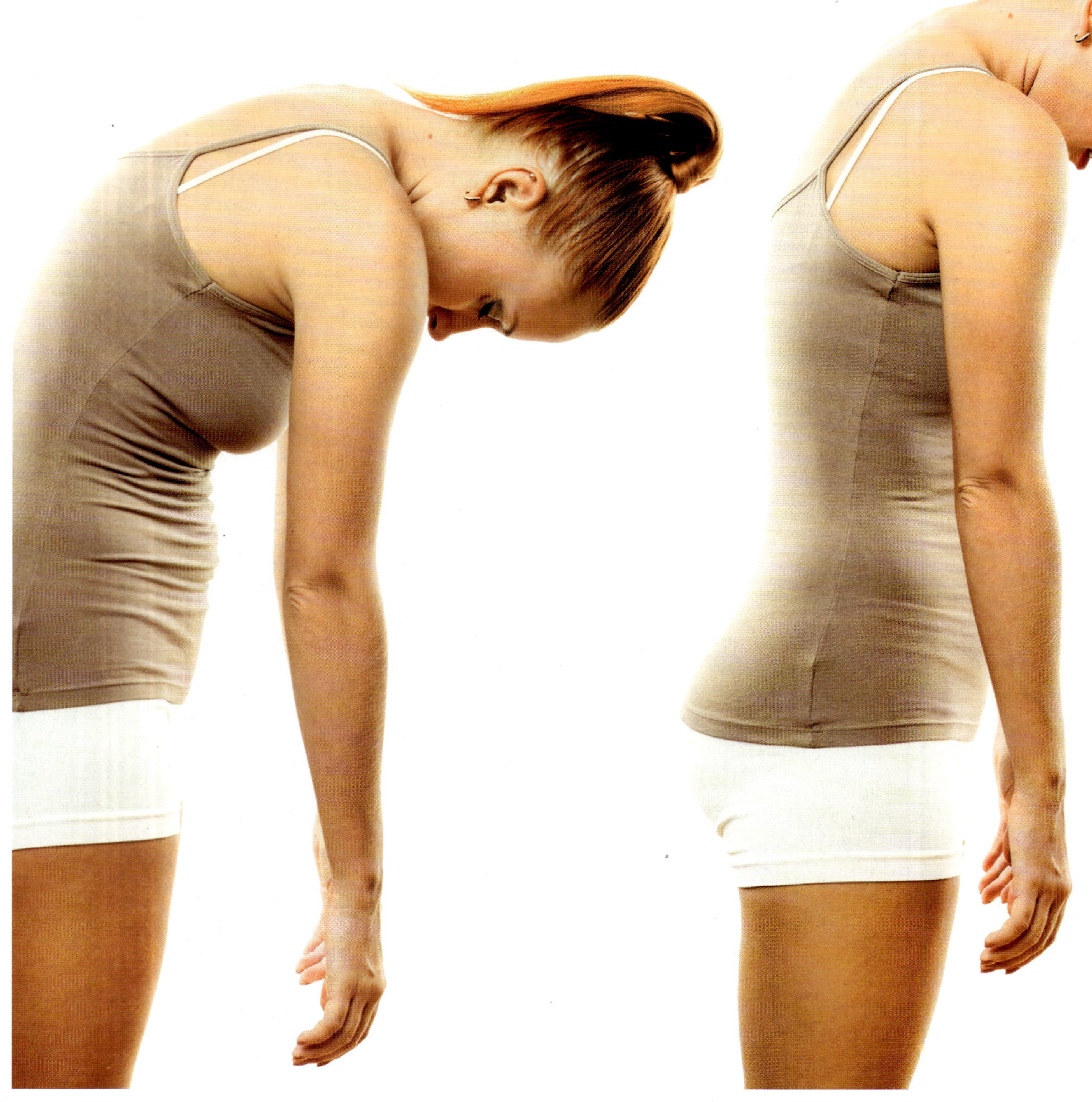

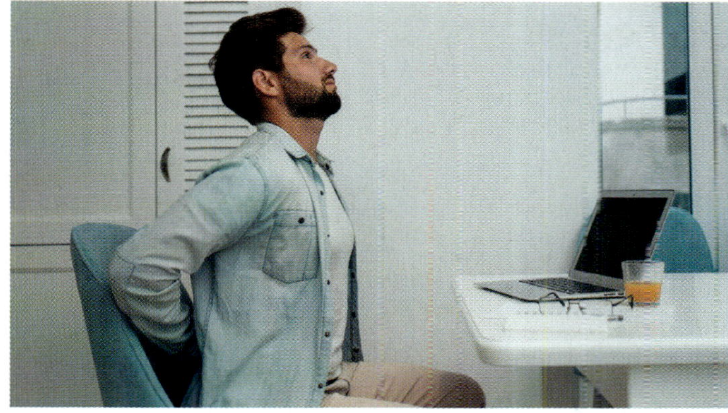

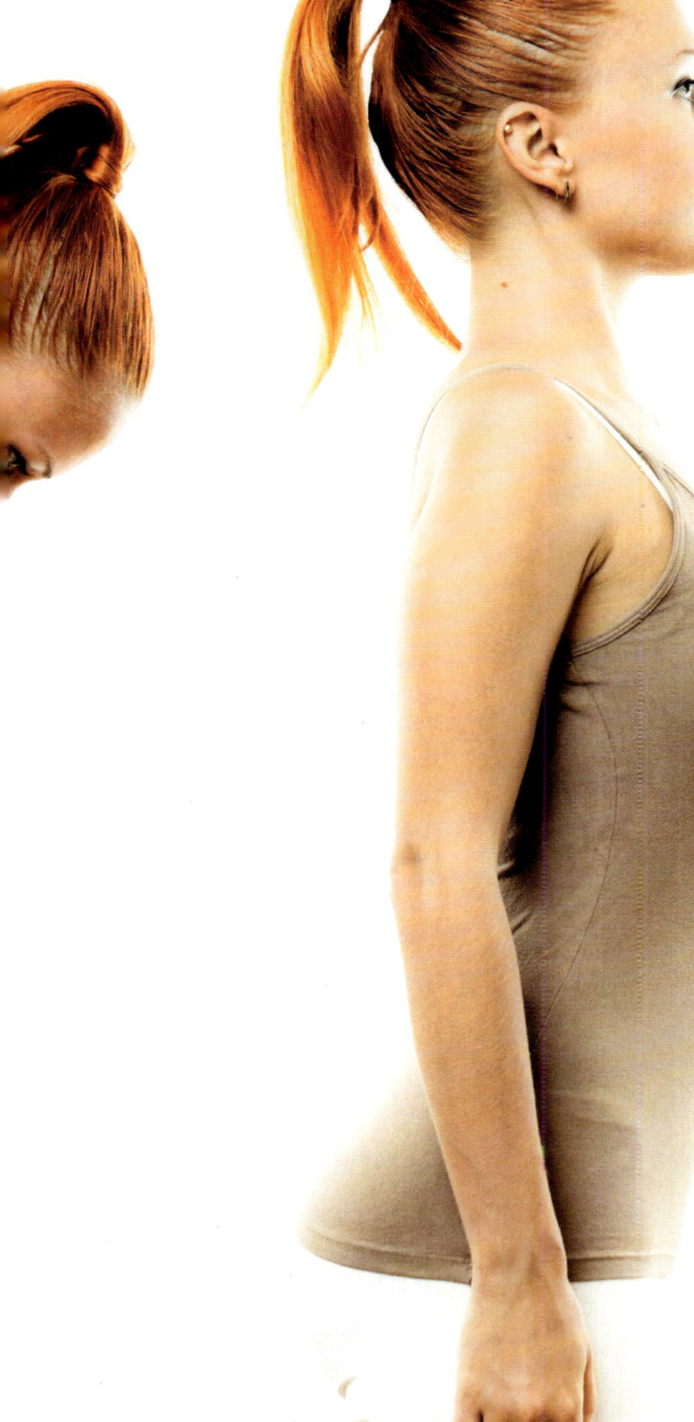

Many of us have been told to "walk tall" or "sit up straight". Which is all well and good but easily forgotten the moment you must focus on something or get distracted with something else.

Every muscle in the body has a specific job that it was designed to do. If a group of muscles, for whatever reason, are not doing their designed jobs, others will have to step in to take up the slack and this will have ramifications throughout the body eventually leading to chronic aches and pains in these 'stand-in' muscle groups. This is becoming more evident as we lead increasingly static lives, compared to the activity our bodies were designed to do daily. How often do you do tasks on the floor while squatting for example?

Our bodies stiffen up because of this unnatural lack of movement. As a result, the muscles develop dysfunctions that then pull our joints out of alignment, and eventually pain and tension take hold.

Most people have always understood the advantages of having good posture in terms of how you function and present yourself to the world. However, strong posture is also important for your general health and well-being, providing significant psychological and physical advantages. Postural alignment therapy is also extremely valuable for pre, during and post pregnancy as our bodies have forgotten how to do this naturally.

It's not WHAT you do. It's the body you bring to what you do. That tennis elbow isn't necessarily the result of repetitive action but may well be caused by the muscles in the shoulder having forgotten their jobs. The shoulder then ends up rolling forward, so rotating the ulna (the upper arm bone) and putting strain on the muscles around the elbow.

Chronic pain can make it seem as though you'll never be able to feel better or live the life you desire, especially if you've tried alternative therapies and drug-based treatments that either didn't work at all or only worked briefly. The ability to move with ease and with minimal effort on the joints, muscles, and ligaments, depends on those muscles doing their job as they were designed to affect the body's alignment. When working well this reduces neck and back discomfort, boosts vitality, boosts self-esteem and expands lung capacity.

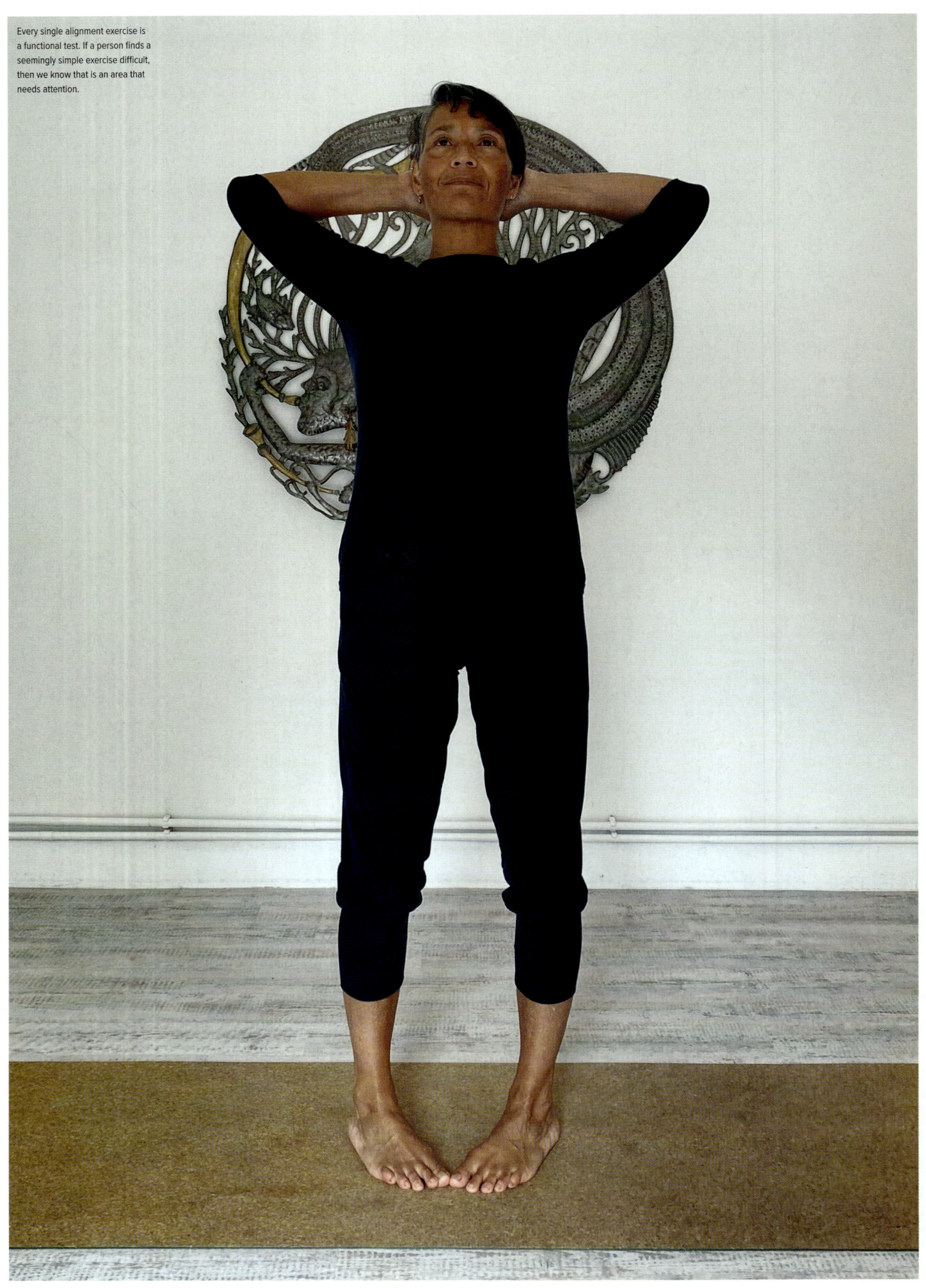

Every single alignment exercise is a functional test. If a person finds a seemingly simple exercise difficult, then we know that is an area that needs attention.

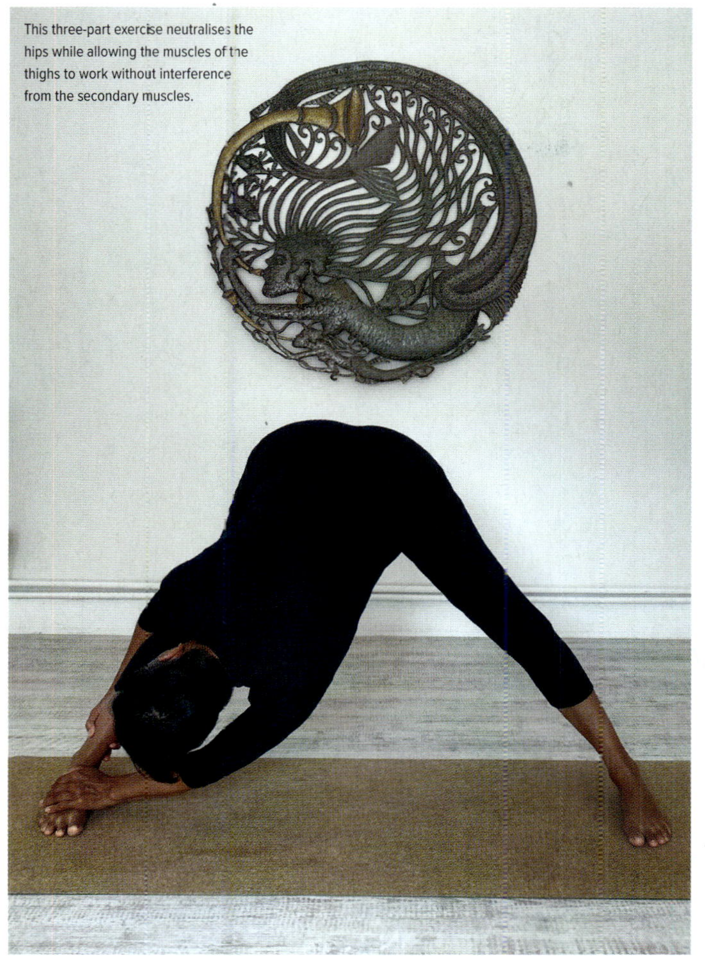

This three-part exercise neutralises the hips while allowing the muscles of the thighs to work without interference from the secondary muscles.

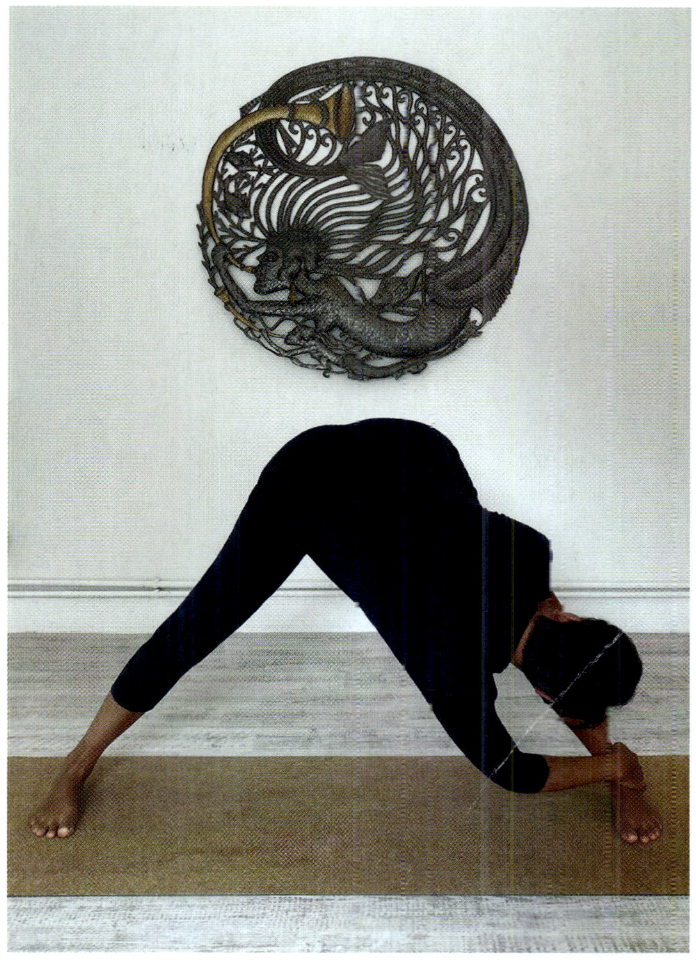

Keeping the second toes pointing forward maintains forward alignment of the femur, tibia and fibula to get the thigh muscles understanding how to hold the leg properly under pressure.

Muscles working well reduce neck and back discomfort, boosting vitality and self-esteem.

Postural alignment therapy corrects the postural problems that are the cause of your continuing discomfort, in contrast to most chronic pain treatment techniques that just target symptoms or attempt to deaden or conceal pain. This questions WHY you have chronic pain and instead of looking at the area that hurts, zooms out, and looks at how the whole body is being held to find the answer to this question.

According to the findings of a 2016 study, hip discomfort in a person with chronic hip issues may be immediately affected by changes in posture and movement patterns. The research shows that correcting incorrect posture can be a successful and non-invasive solution in situations of mild structural hip problems and suggests a role for non-surgical intervention in such circumstances. *(Lewis, PT, PhD, Khuu and Marinko PT, ScD, OCS)*

We are designed to be flexible and well-balanced, with good muscle function, lasting a lot longer than most of us are right now.

But all is not lost. Just as muscles adapt to a lack of movement, they can relearn how to move to their

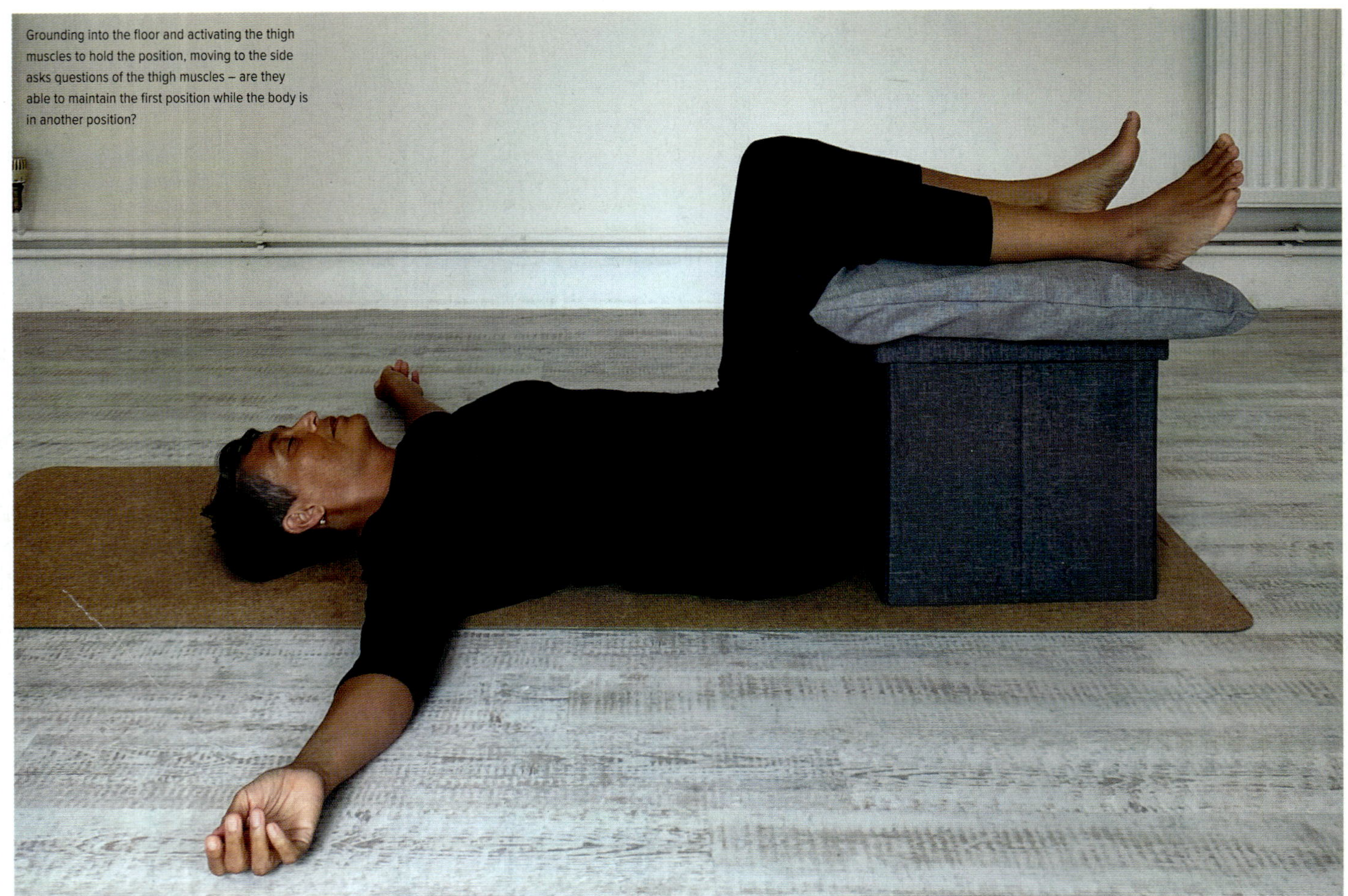

designed function, helping you to feel taller, lighter, to move more freely and enjoy everything – and that means EVERTHING – that you do. The exercises developed for postural alignment therapy teach the muscles throughout the whole of the body to work as a team together. All muscles must play their part for the body to be held in proper vertical alignment when standing and moving in any activity.

And it is not just on a physical level that postural alignment therapy is useful. Given how intricately interwoven the body is, it makes sense to think that both our posture and mental health are influenced by one another.

An example comes from a recent study, published in the journal NeuroRegulation, that examined how posture affected students' views of their performance. San Francisco State University health education Professor Erik Peper's study examined how students' posture affected their performance on basic maths problems. According to students who felt nervous about maths, slouching made them feel less capable of calculating. However, when they sat up straight, they felt more confident. When we

> ## It makes sense to think that both our posture and mental health are influenced by one another.

slouch, we crush the lungs and prevent the diaphragm from moving freely. The amount of oxygen accessing the body's tissues, including the brain, can be reduced by 30% leading to slower thinking times, even anxiety and depression. This means that having stronger posture will enable your body to utilise your lungs in the best possible way, optimising intake of oxygen and vital nutrients and resulting in a feeling of calm, improving decision making and imparting a sense of confidence.

Postural alignment therapy can be done in a class session or on a one-to-one basis which provides a more specific programme for your needs. Either way, you can learn so much about your body, how it works, and regain control of it instead of allowing aches and pains to control you. **M&B**

Postural alignment therapy can help with...
Headaches, tinnitus, vertigo, TMJ (Tight Jaw Syndrome), sinusitis, balance issues, neck pain, shoulder issues, rotator cuff syndrome, elbow pain, wrist pain, mid and lower back pain, sciatica, bulging/herniated/slipped discs, scoliosis, hip pain, groin pain, knee pain/issues, ankle pain, foot pain, plantar fasciitis, Ehlers-Danlos syndrome, chronic pain syndrome, hyper-mobility, irritable bowel syndrome (IBS), digestive issues, breathing issues, insomnia, pelvic floor issues.

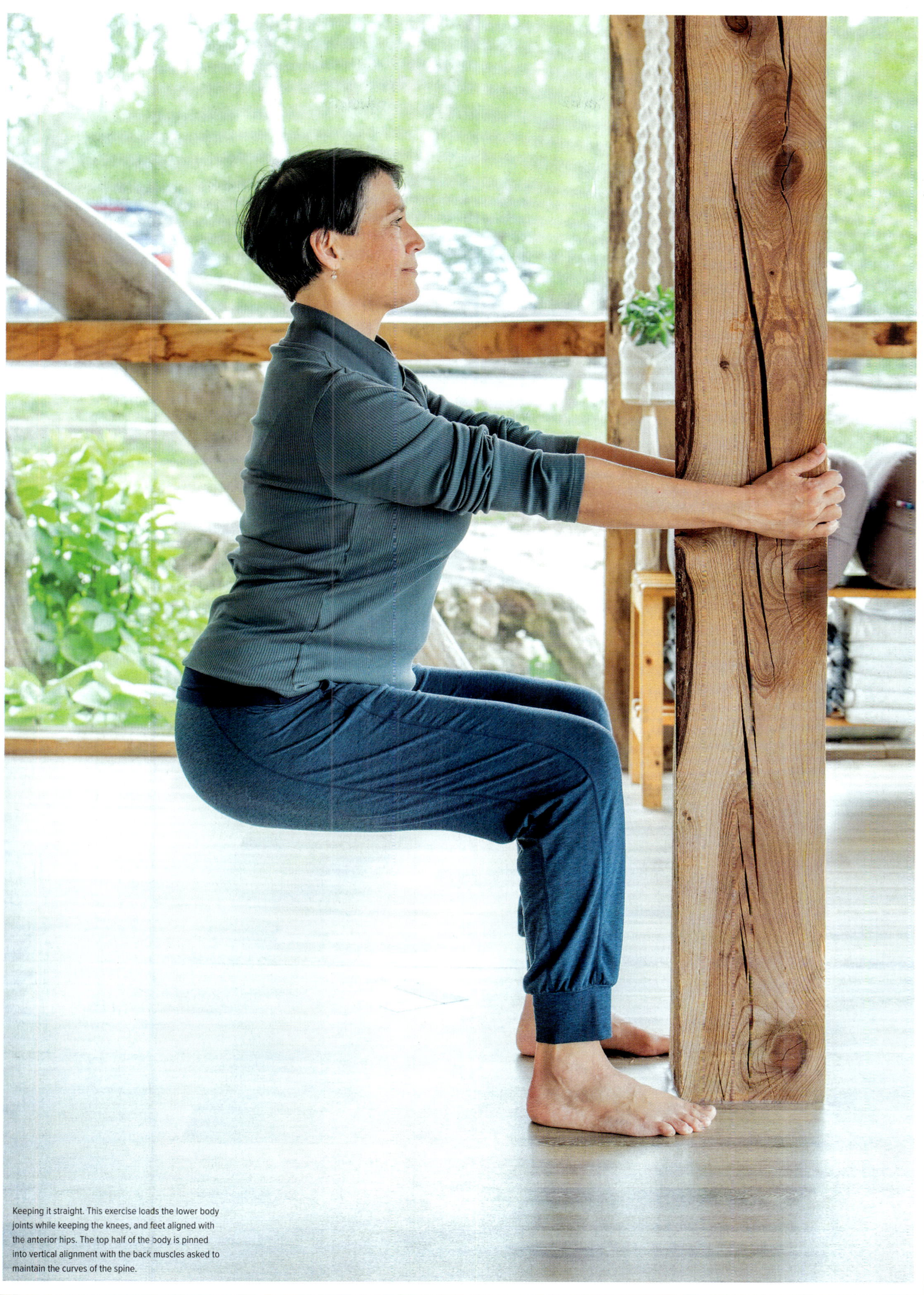

Keeping it straight. This exercise loads the lower body joints while keeping the knees, and feet aligned with the anterior hips. The top half of the body is pinned into vertical alignment with the back muscles asked to maintain the curves of the spine.

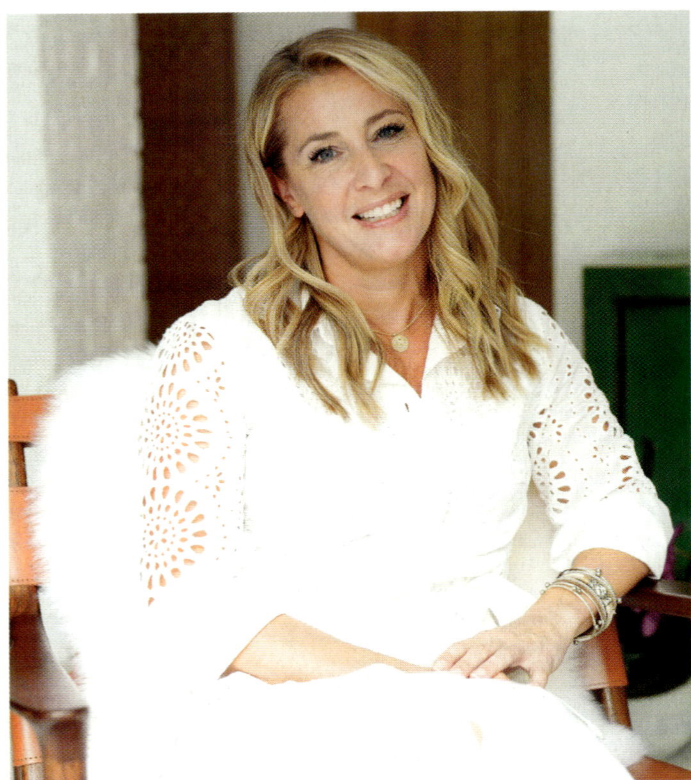

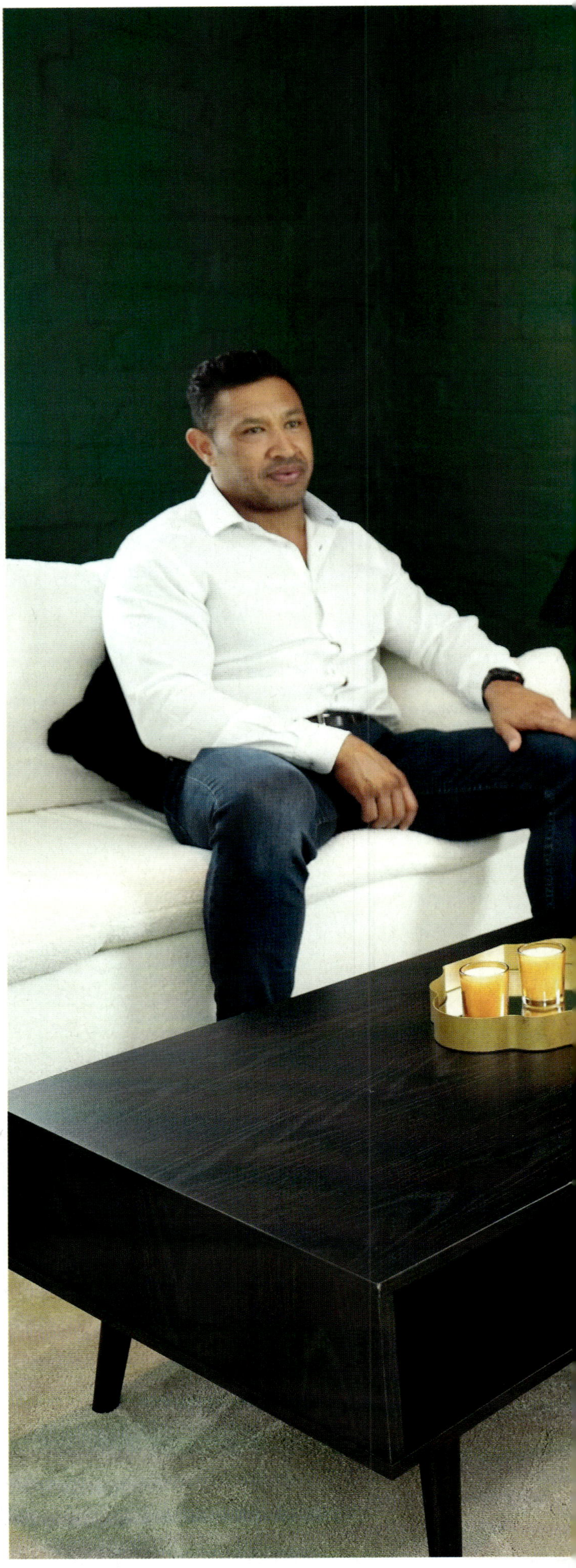

Think you know about hypnosis?

Leonie Garner is a practicing Clinical Hypnotherapist, with 10 years of experience in this ancient therapy. In this chapter let her guide you through the history, the methods used and the results that can be achieved.

can never decide whether those myths and misconceptions around hypnosis are useful or not when I say what I do for a living.

As a Clinical Hypnotherapist, it often feels akin to stating I am the neighbourhood witch, given the responses at social events or school gate. I see many a raised eyebrow, followed by the inevitable question, 'but does it actually work?'

Stage hypnotists are fun and memorable, and I suppose they do show that hypnosis is a powerful psychological tool. I cringe inwardly to watch it being used for spectacle though, as a rather unwieldy and blunt instrument, rather than with the beautiful precision I know it to possess.

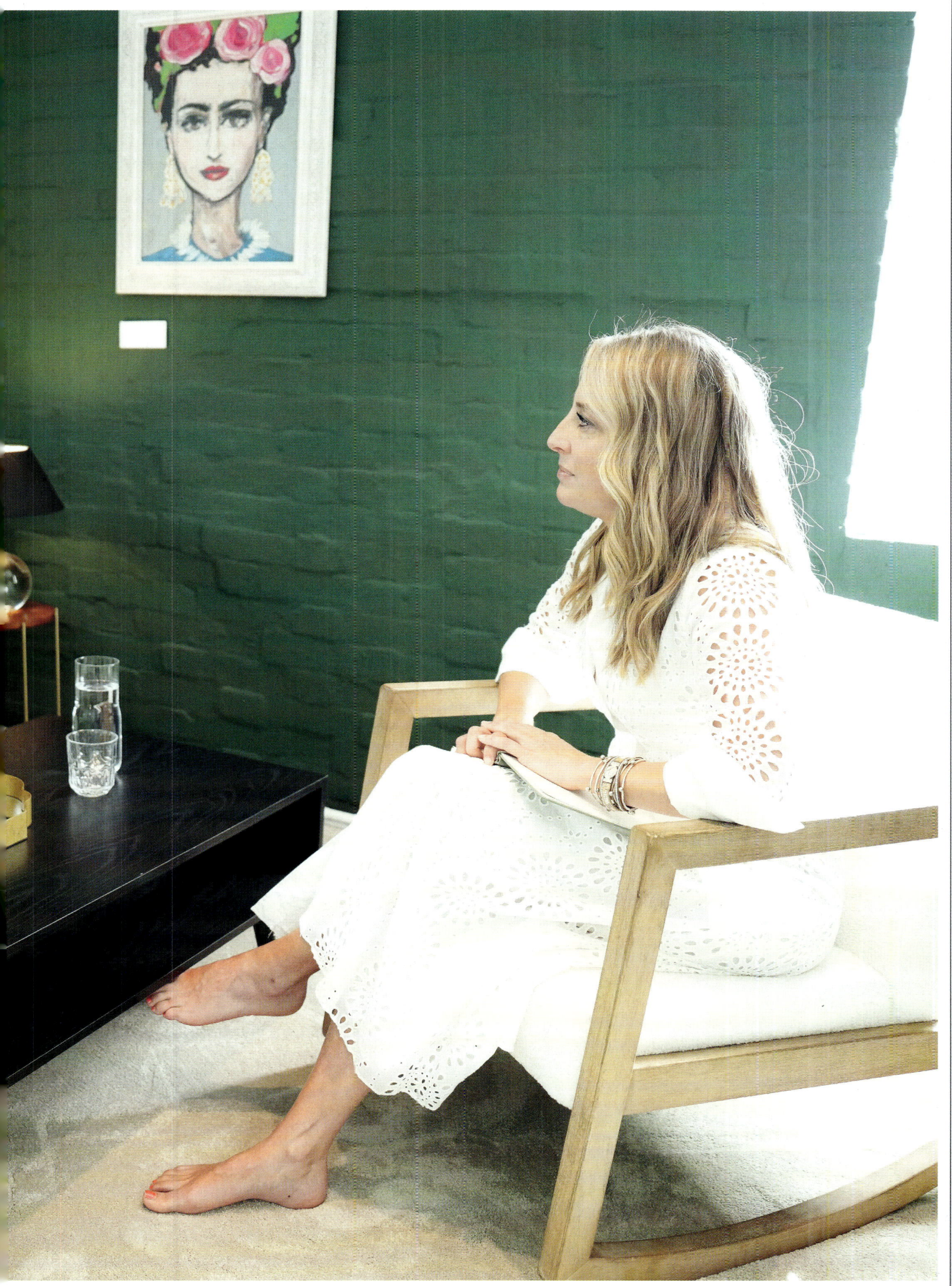

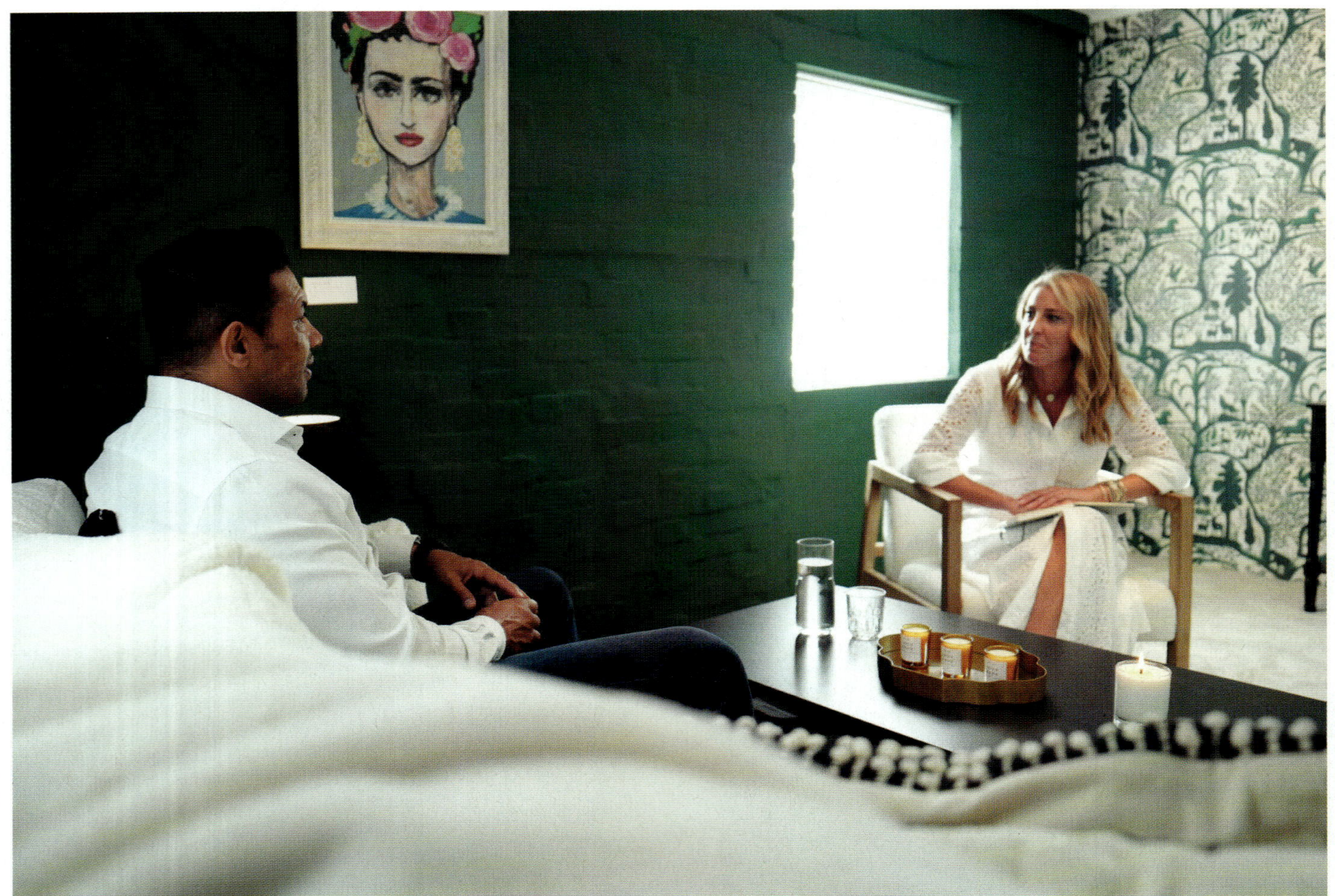

It's also, if you'll pardon the pun, really missing a trick. There is so much more to hypnosis and hypnotherapy to benefit from, and potential to be further embraced within the world of health and wellbeing.

So, I will say now that I have never made anyone fall in love with an inanimate object, bark like a dog, or eat a raw onion like an apple. I don't own a pocket watch or use the phrase 'look into my eyes'.

But does it actually work? And how?
Therapeutic hypnosis is as far from those strange cultural images and ideas around mind control as it is possible to be. In practical terms, it can bring you much more control over your own mind and instinctive behaviours, with the potential to make positive, measurable changes to your mental, emotional, and physical wellbeing.

Most of my clinical time is spent treating anxiety, which has many emotional forms but manifests physically in the body. I help clients combat physical ailments such as irritable bowel syndrome (IBS) and migraine far more often than just helping people stop smoking or dismantling phobias about flying (although I can do that too).

Hypnosis in the hands of a skilled and qualified Hypnotherapist is applied with precision and care, and importantly, with integrity. We make good use of the body's own natural deep relaxation response in a calm therapeutic space, but it can at times feel transformative

Hypnosis in the hands of a skilled and qualified Hypnotherapist is applied with precision and care.

and 'like magic.' Hypnotherapy can create profound change in how people think about and respond to their world, and crucially, how they feel.

The ability to respond to suggestion varies from person to person, but even those who believe themselves very much non-suggestible can still find it very helpful.

Ironically, and much to their surprise, these are often the people who benefit most. As a clinician, this makes them wonderfully rewarding to treat.

If you are sceptical but curious about how hypnosis feels, just think about the novel you became engrossed in on holiday, that film, or TV series you're currently watching that feels really absorbing.

As involved as you are in the character narratives and stories, they're not controlling you. You engage with these things willingly because it feels so good to be fully immersed, enjoying paying close attention to them. They ignite imagination and we suspend disbelief, naturally becoming invested and engrossed.

This is at the core of how hypnosis feels, and works, by switching on a focused state of attention, reducing critical barriers and in doing so, finding the mind's learning mode'.

You may be surprised to know that even now, as you read, you are likely to be in a form of light hypnosis, because your awareness is slightly narrowed, and your attention is focused. Your brain is screening out many distracting background sounds, movements, and details from the environment but you're still awake enough to respond to others' words or cues. You may even be hearing the internal 'voice' you use to think with, to yourself, as you read. Can you hear that?

In observing that inner voice, even just for a moment, you've been able to step outside of and notice your thoughts from a slightly different angle. You have practised a little bit of mindfulness, too.

Historic references to hypnosis go as far back as ancient Greece and Egypt. Hypnosis is even documented within the beginnings and teachings of Chinese medicine, as early as 2,600BC.

Trance-like focused states of awareness have been used throughout human existence to still the mind, seek wisdom and insight, and improve health and wellbeing through meditation, yoga, breathwork and chanting.

They are woven through the fabric of most religious ceremonies and practices in some form or another, particularly prayer.

The concept of neuroplasticity, the brain's ability to adapt its circuitry, is even beautifully captured within a Buddhist saying, "The mind takes the shape of what it rests upon".

This glittering little gem of ancient wisdom is now supported by modern research.

Neuroscience shows us that the human brain has amazing plasticity. It is flexible, continuously learning, forming, changing, and reinforcing networks of connections between billions of nerve cells, called neurons, in response to your thoughts, feelings and experiences. You may have heard the phrase 'what fires together, wires together' in relation to neuroplasticity.

This, and a phenomenon known as the Quantum Zeno effect within neuroscience shows that whatever you continuously focus on creates observable changes within the brain. That which you place your awareness on, becomes amplified.

In other words, how you use your brain changes what your brain can do – and the body responds accordingly.

So, what does hypnosis have to do with this?
Hypnotic states slow our brainwaves down (through Alpha to Theta) mimicking the brain's dreaming and learning modes, allowing it to rehearse and accept change more readily, becoming more open to suggestion.

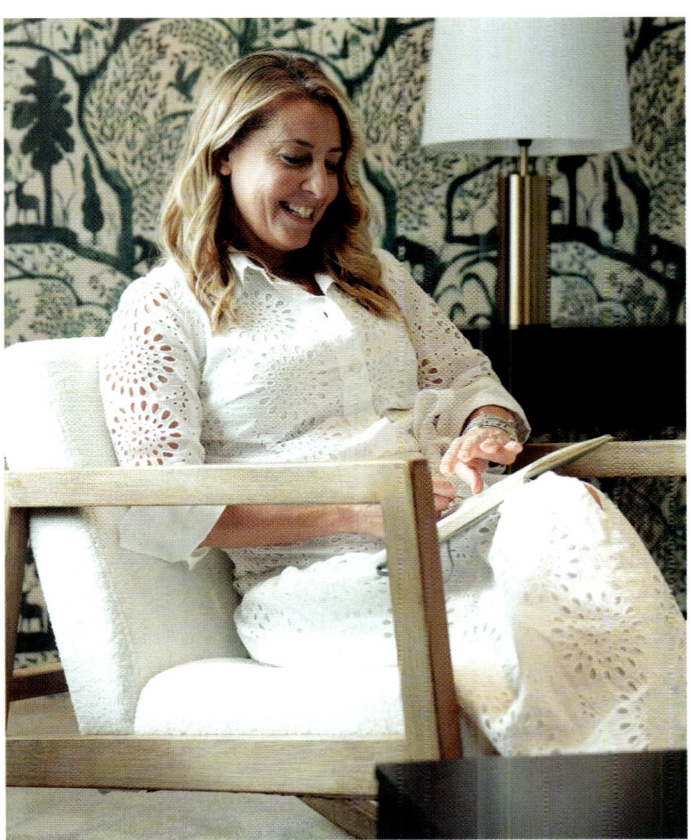

When we understand this, we can find the true potential and purpose of this amazing psychological tool: to improve health and wellbeing and overcome thoughts, fears and habits that can limit us. We gain the ability to better manage our own mental health and resilience, consciously defend against stress-related illness and build healthier brain practices into daily life that can help us thrive.

For example, imagine the changes you could make in how you approach food and exercise if you truly felt differently about them? What if you focused mental energy on nourishing your body really well, rather than criticising it? What if it felt weird to have a week go by without a hike, swim, or yoga class?

How could fundamentally changing those old, well-worn, autopilot thoughts change your life for the better? What could you do that you weren't able to before? Who would that allow you to be?

I recently completed some additional training in Positive Psychology. Evidence for this approach in treating all kinds of emotional and mental illness, is substantial. I was struck by how Hypnotherapy is, in so many ways, a way of applying positive psychology in its purest form i.e. when we habitually place our focus on

This glittering little gem of ancient wisdom is now supported by modern research.

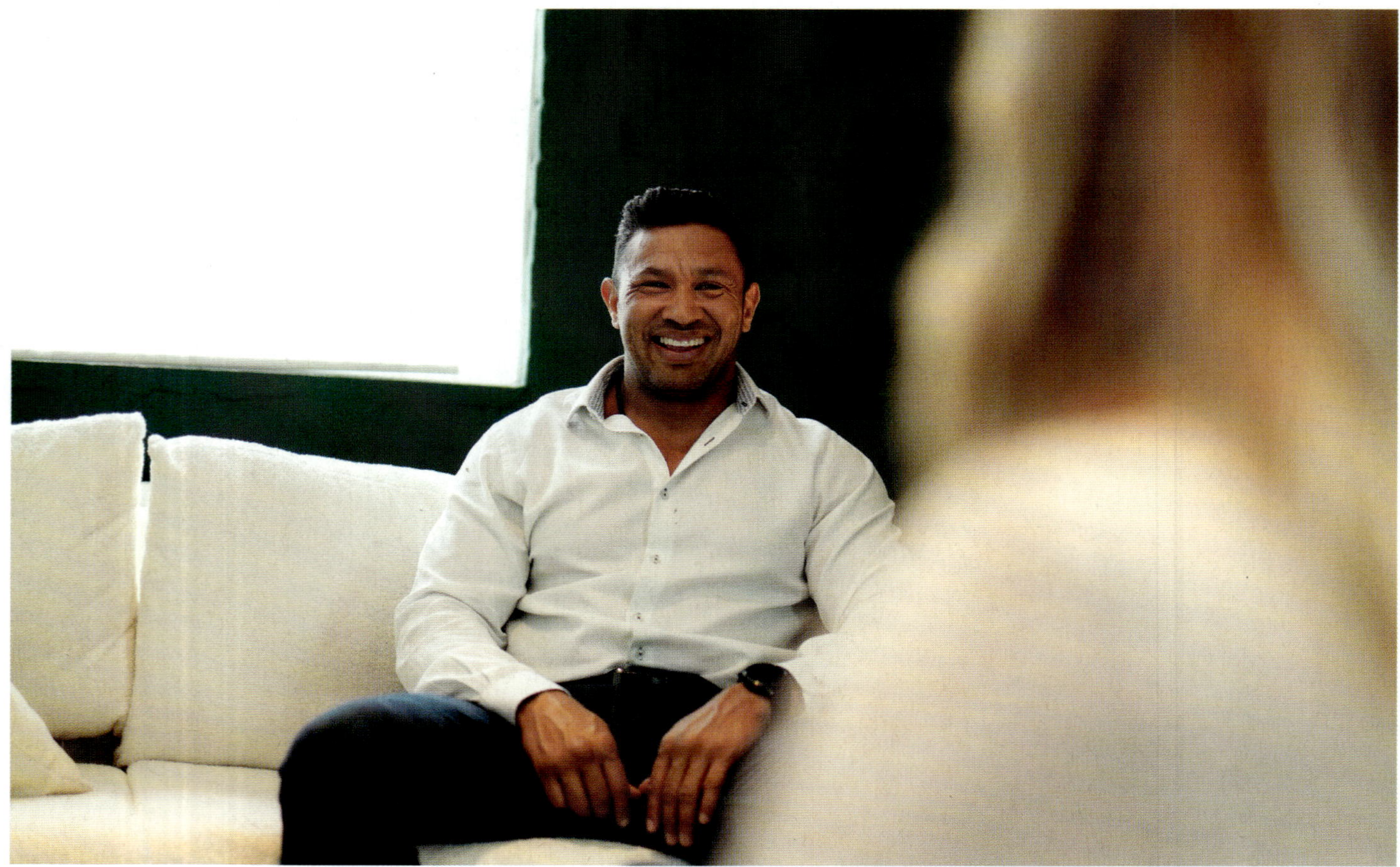

solutions, strengths and resources, positive emotions, perspectives and strategies, rather than fear and weakness, it builds physically healthier, more resilient minds. In turn we build healthier, more resilient bodies.

Put simply, Hypnotherapy asks the question, 'how would you like to be instead?' then plots the route for your mind and body to follow.

But we can go beyond this too, and into the mental and physical benefits of lifting trauma. Research into the nature of memory, and memory consolidation shows that within hypnosis we can shift and update emotional states encoded within our memories, calming down emotions attached to the past, so that the brain consolidates – rewrites and re-saves – those memories with much healthier emotional content and a more regulated, updated emotional blueprint.

Because of the mind-body connection we all have, it makes sense that, alongside psychological change, hypnotic techniques can play a role in recovery from physical illness too.

Whilst hypnosis should never be claimed to be the miracle cure-all, the use of profoundly deep levels of

Stress has an impact on some of the mechanisms underlying dementia.

relaxation involved, and careful guidance of certain body processes means we can positively influence how we experience our minds and bodies.

Hypnosis has been shown to reduce stress hormones like adrenalin and cortisol, to lower blood pressure, speed up healing, boost immunity, to shift the perception of and management of pain, and reduce inflammation – often most helpfully within the digestive system (amongst others).

Inflammatory and auto-immune conditions are known to be exacerbated by stress (often triggered by illness itself, creating a feedback loop). Memory difficulties can be linked to high levels of cortisol. Stress has an impact on some of the mechanisms underlying dementia.

So, in the interests of longevity, and proactively staying well into old age, learning to use tools such as self-hypnosis to reduce physical effects of stress can be of tangible benefit.

As humans we evolved to live in hunter-gatherer groups, to act on gut instincts, be supported by and connected with one another, and intrinsically synced with our natural environment. Our brains and bodies haven't evolved much beyond this. Our internal operating systems haven't had an upgrade in the last few thousand years, to keep up with the pace and demands of modern life.

Our primal stress response to this clash of worlds, when triggered by a traffic jam, email, or even daily family life, powers down non-essential bodily functions like

Hypnosis can also help with conditions such as insomnia, not only speeding the transition into falling asleep, but by adjusting the frequency of REM dream-sleep cycles.

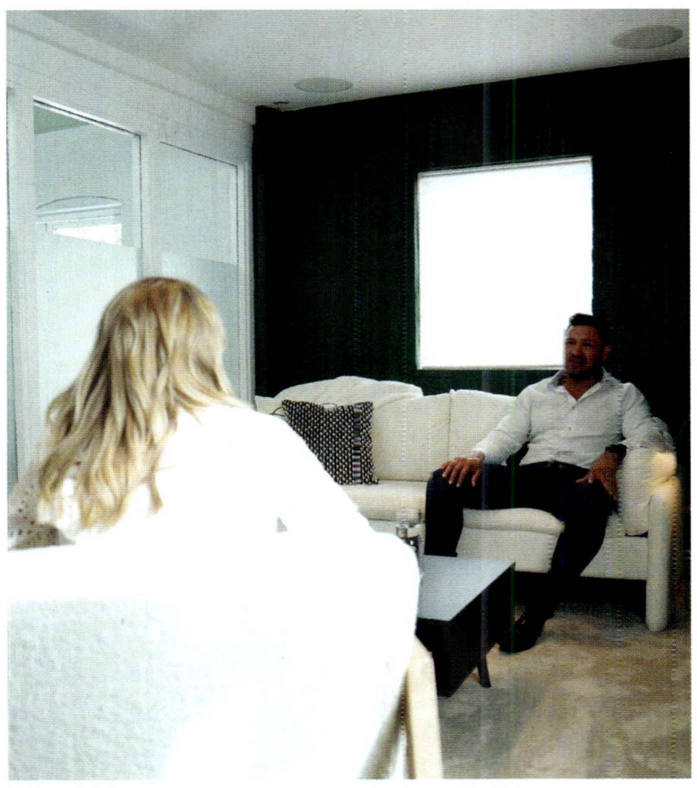

digestion and immunity, diverting energy and resources to systems that will enable us to survive short term danger. Not a problem when occasionally outrunning a predator, or enemy tribe, but if this becomes regular or prolonged, these systems understandably suffer. Modern life can mean our minds and bodies become switched into this Fight or Flight mode for hours, days, sometimes weeks and months at a time.

So while we can't really change modern lifestyles that normalise the opposite of our instincts and earlier existence, i.e., stress, materialism, individualism, and constantly bombarded with information, we can adapt all the better by regulating our minds and bodies.

Hypnotic guided deep relaxation can be the antidote to states of stress, returning the nervous system back into Rest and Digest mode, reducing inflammation, and supporting and activating natural healing mechanisms within the body.

For this reason many studies have shown hypnosis to be an effective tool in treating irritable bowel syndrome, reducing debilitating symptoms and preventing relapse long term, without potential side effects of medication. Gut-directed hypnotherapy is recommended by the NHS for people who don't respond to traditional pharmacological treatment.

Hypnosis can also help with conditions such as insomnia, not only speeding the transition into falling asleep, but by adjusting the frequency of REM dream-sleep cycles, to improve quality of that sleep. If you're familiar with the feeling of waking up, even after a full 8 or 10 hours of sleep, still feeling exhausted, you might be over-dreaming. A stressed brain dreams up to three times more than normal and can feel almost as busy as being awake all right. Stress and worry disrupt sleep cycles, skewing the balance towards more busy dreaming (REM) cycles, and less restful, slow-wave, deeper sleep

Using guided hypnotic imagery or self-hypnosis at bedtime can help de-stress the mind and body before going to sleep, engaging the imagination better, halting worry, and creating calm distance from those things that might not be so easily resolved in real life

(but certainly cannot be solved at 3am). This helps the brain relinquish those emotional puzzles and thought loops from going round and round overnight and sleep becomes restorative.

So, hypnosis is so much more useful to wellbeing than those 'mind control' parlour tricks and illusions.

In the right hands it can help reduce post-traumatic stress disorder (PTSD), depressive illness, anxiety, insomnia, IBS and chronic pain. Hypnotherapy is recognised by the NHS as an established complementary therapy that can supplement conventional medical treatments, it can be used alongside antidepressants, and make other therapies such as cognitive behavioural therapy (CBT) even more effective.

But is it helpful for everyone?

Hypnosis is not suitable for people experiencing psychosis, and The Royal College of Psychiatrists recommends sharing any existing health conditions with your therapist or GP before receiving hypnotherapy.

It is of course important to find a practitioner from an accredited register, such as the Professional Standards Authority or National Hypnotherapy Society.

Other than that, studies have shown that hypnosis within hypnotherapy can provide help for many people experiencing a whole range of difficulties, and it can super-charge the wellbeing of most. Without a swinging watch in sight. *M&B*

Leonie Garner
leoniegarner.com

The Route to an Even More Fulfilled YOU

The coaching world has exploded in the last decade, becoming one of the fastest growing industries on the planet. International coaching expert **Amanda Slade** tells us more…

More and more people are seeing the power of investing in a personal coach to elevate and educate themselves to new possibilities.

But what exactly is a coach and how would getting one assist you, a friend, or family member to become a more self-aware and therefore an even more empowered human?

Ultimately, success can be measured by performance. So, whether a business invests in corporate coaching, the success of any corporate, depends on humans and their personal performance. When we can hone any skill and become effective, it delivers change and growth. Everyone benefits.

Good quality coaching is designed to instantly bring out the best in people, it sharpens and perfects skills and talents, that may be evident or may be yet undiscovered to the client.

Coaching contributes to fulfilment by using skilled questions and relationship building with the client, so they can understand where they are and identify where or who they would like to become.

As people, we are shaped and sculpted by everything in our lives from the moment we enter this world, the family and environment we are born and raised in, influences and connects us to certain beliefs and systems of what becomes 'normal' for us. Overtime and practice these beliefs get hardwired into our personality and creates the YOU that you recognise.

There is often a perception that change is hard or difficult and yet in my experience change is much easier when we have someone who can see and lead people into a deeper understanding of the why's behind the how. We may like to stretch and grow in new and more exciting ways.

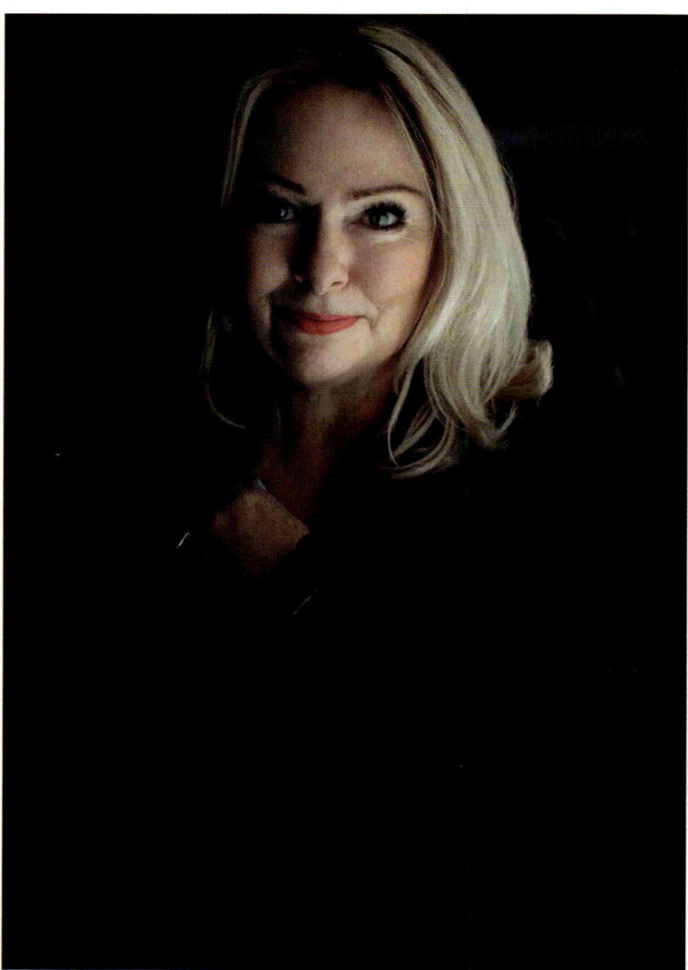

Effective coaching will enable a client to feel empowered, which means they will be able to accept responsibility for the results.

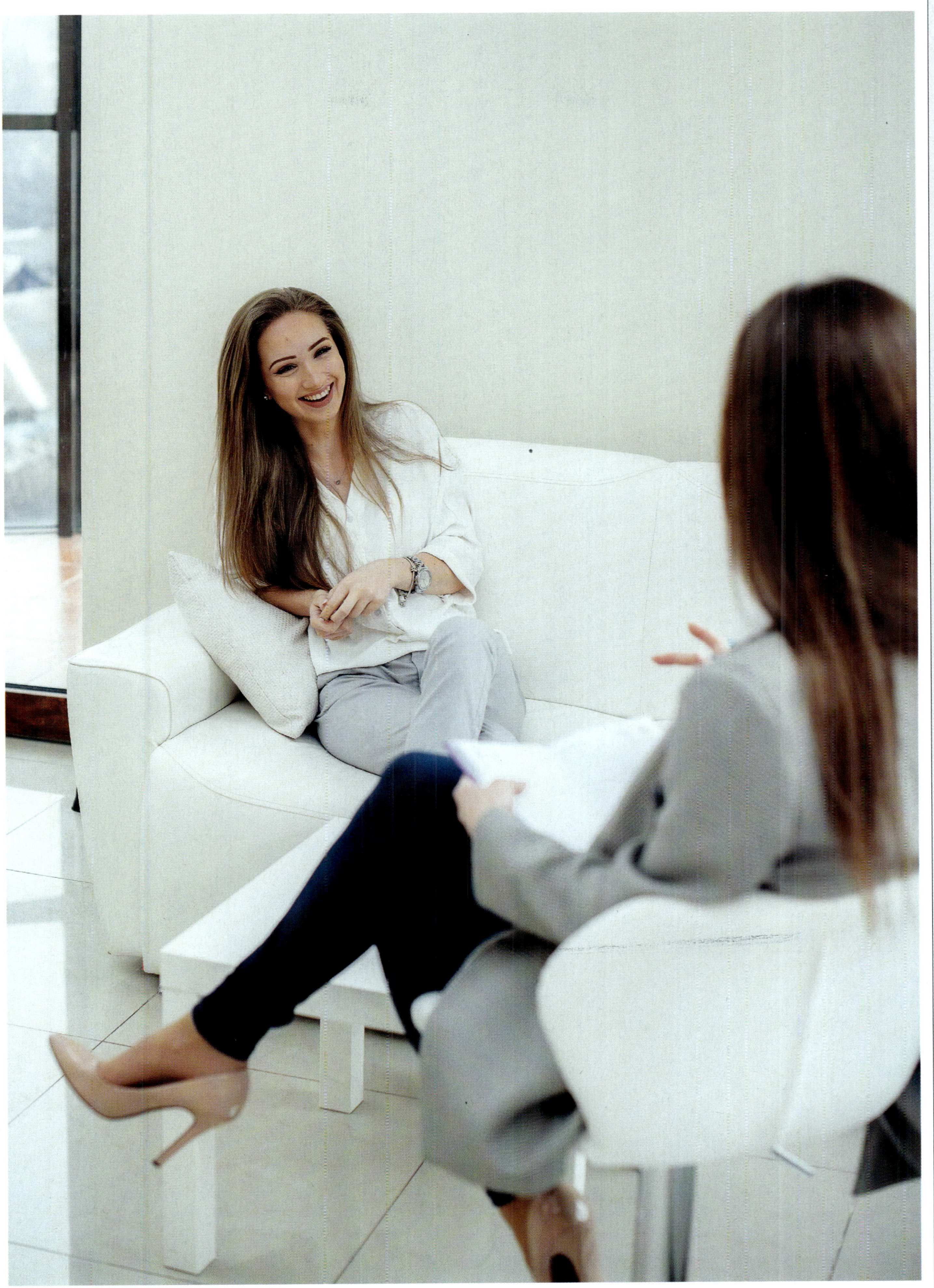

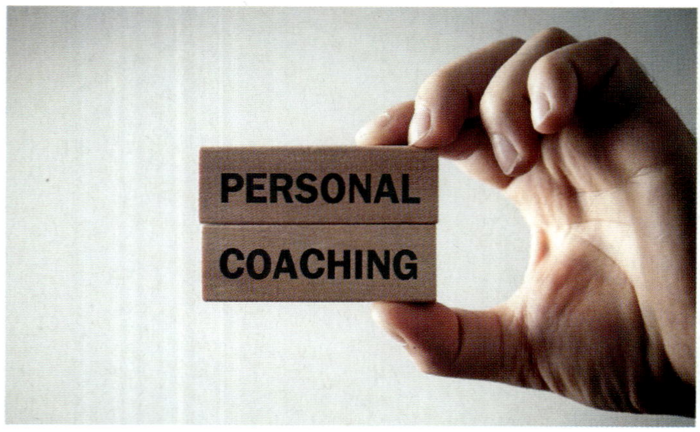

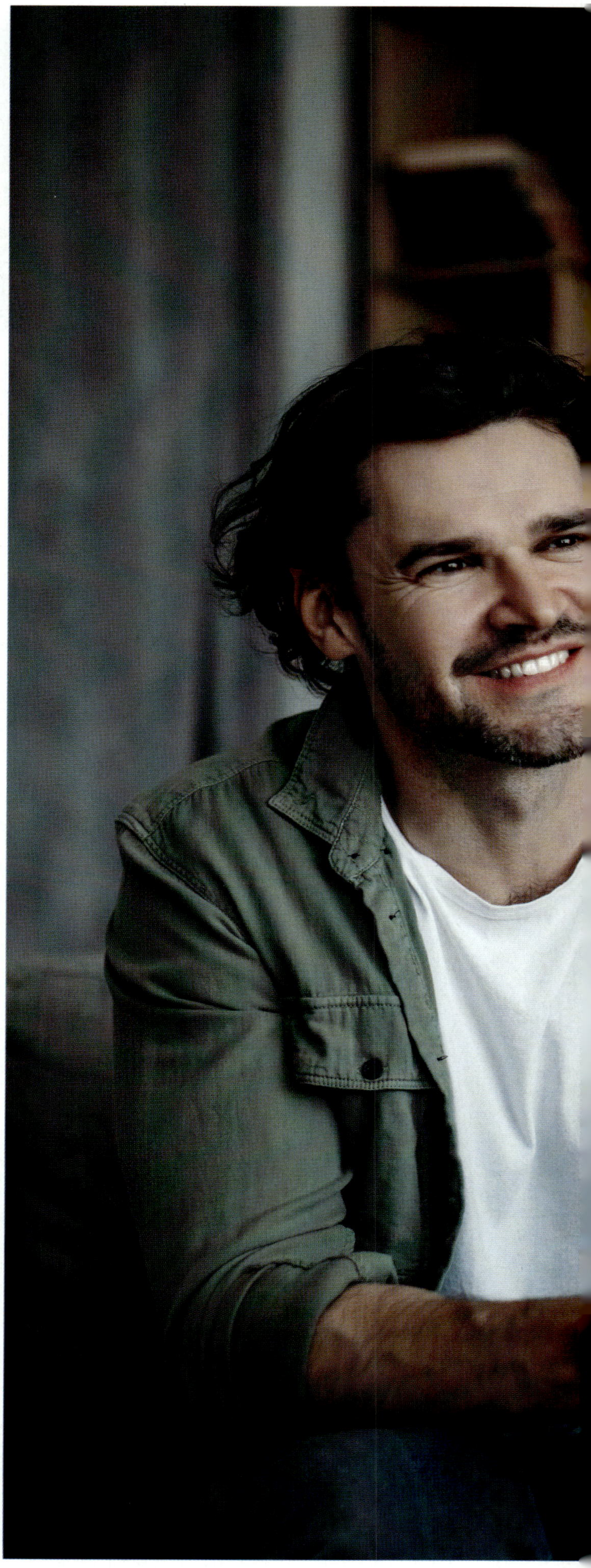

Tony Robbins who is one of the world's leading coaches says "The quality of your life will vastly depend from the quality of the questions that you constantly ask yourself." A coach will ask you some different questions which by there very nature will cause you to have some new and revealing answers.

Coaching stimulates individuality. It allows the client to decide for themselves the best approach to finding resolution in whatever segment of their life they perceive there is maybe an issue or challenge. Sometimes called a problem!

Effective coaching will enable a client to feel empowered, which means they will be able to accept responsibility for the results, rather than hiding or blaming others or situations for their results or lack of results.

Coaching can enable a client to identify any blocks or misunderstandings of themselves and others, so they can find the route to new levels of contentment and harmony.

Coaching is the on-going developmental interaction between two people, client and coach.

I have seen businesses created and built; huge deals made that seemed impossible before.

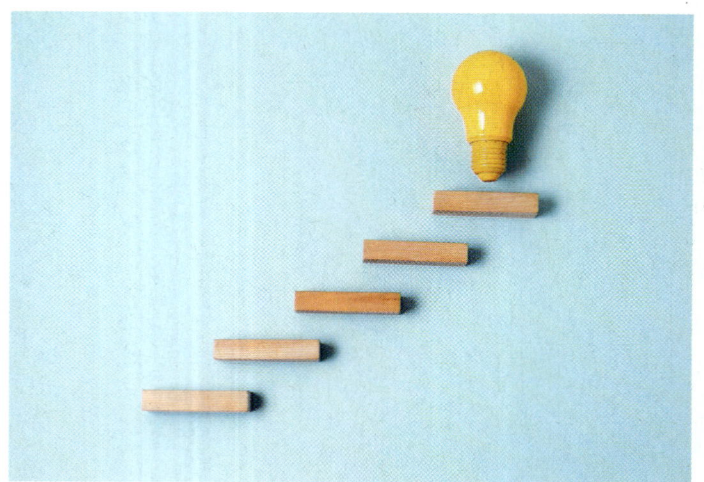

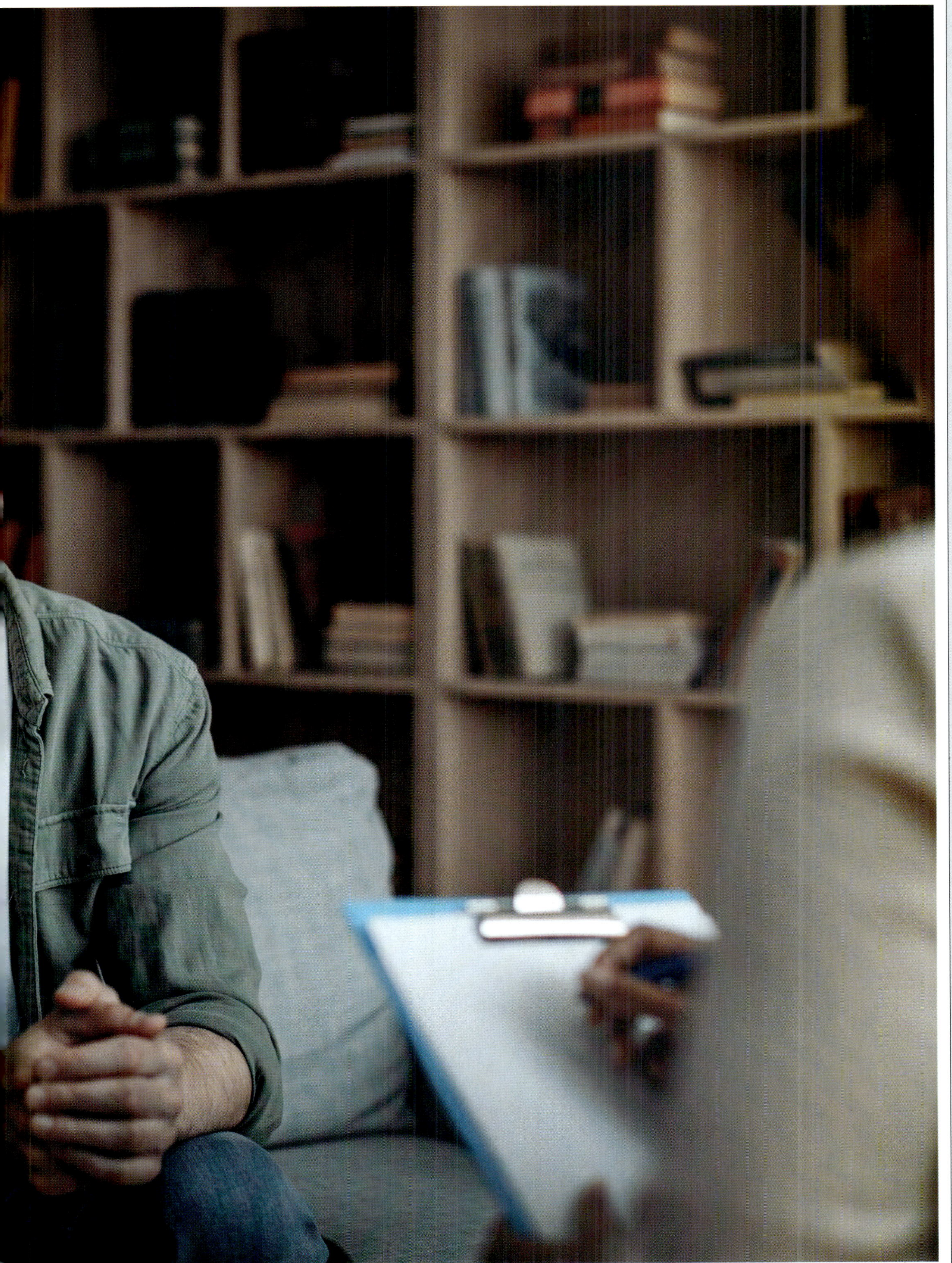

The coach will use questions, reflections, explanations, metaphors, tasks and guiding focus exercises to literally re wire the beliefs of what is possible. Creating potential that was previously unseen or seemingly unavailable before.

Here are a few things coaching can and is used for: Accepting change or transition in life, commitment issues, anger management, career development, communication and conflict resolution. Decision making, leadership, negotiation. Feeling left out, or left behind, presentation and motivation. Procrastination, self esteem issues, confidence, and organisations skills. Work life harmony, relationship challenges. Stress and disorderly thinking, the list goes on and on. Really any life situation or experience will benefit from a skilled coaching session or programme.

Life these days seems often to have become complex and busy beyond what seems previously experienced in history. Stress and anxiety are on the increase, mental wellness seems to be challenged and people are medicating pain and loneliness with many unhelpful long-term methods, such as drugs and alcohol to name just a couple.

Looking for the why behind these behaviours is key to resolution. All of the above is some education and information about coaching and the benefits.

I would now like to make it a little more personal and tell you a little more of my experience of seeing the benefits of coaching in my own life.

Having a coach enabled me to see things from a whole new perspective, break some unhealthy and unhelpful patterns in my own life that I had previously believed were personality flaws or quirks. Helped me to let go of some very firmly held self-beliefs, release some unexpressed and unresolved negative emotions. Change negative self-image and my self-talk about my body. Enable me to be more honest about my feelings and thoughts without fear of judgement from others or myself.

These are just a few of the huge changes that have happened. Since qualifying as a Master Coach, Master NLP practitioner and relationship coach, I have seen these skills work time and time again freeing and releasing people from their own limitations and self-beliefs.

I have seen businesses created and built; huge deals made that seemed impossible before. Relationships rebuilt or created. People speaking up and reinventing themselves entirely. Overcoming of seemingly huge obstacles and challenging situations to finding a new route completely.

There are many styles and methods of coaching and the important thing when choosing one for you is rapport and relationship.

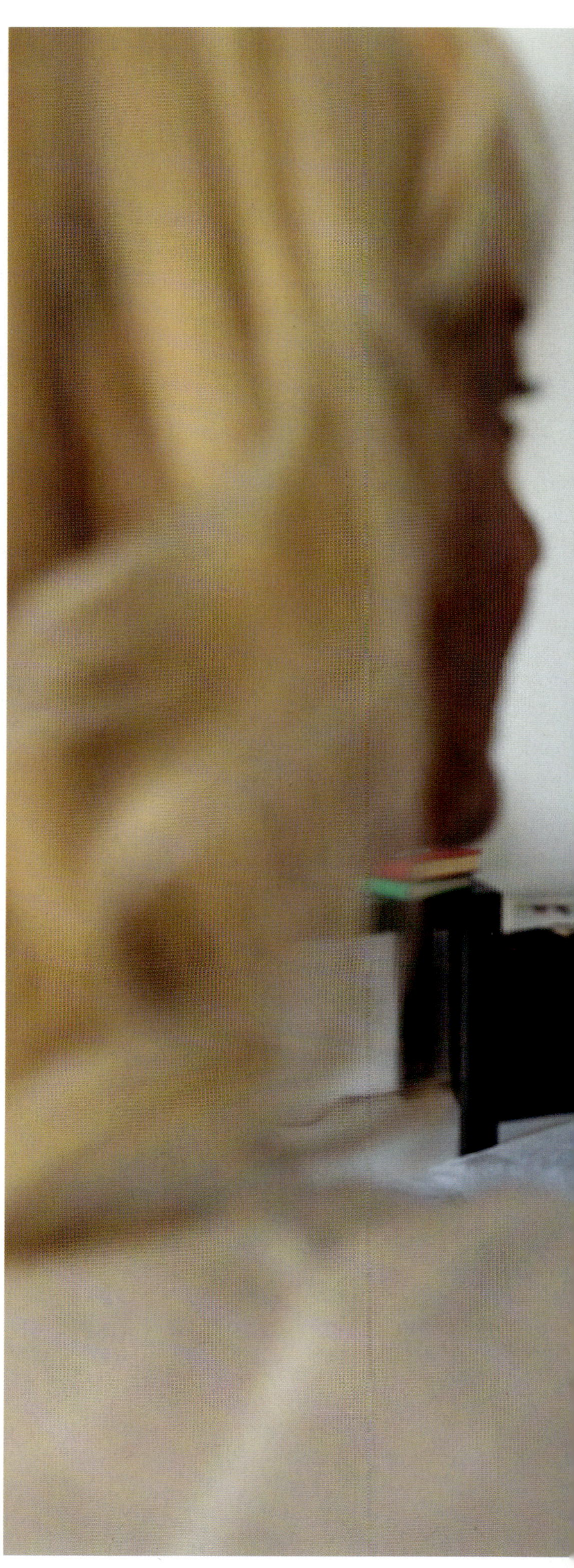

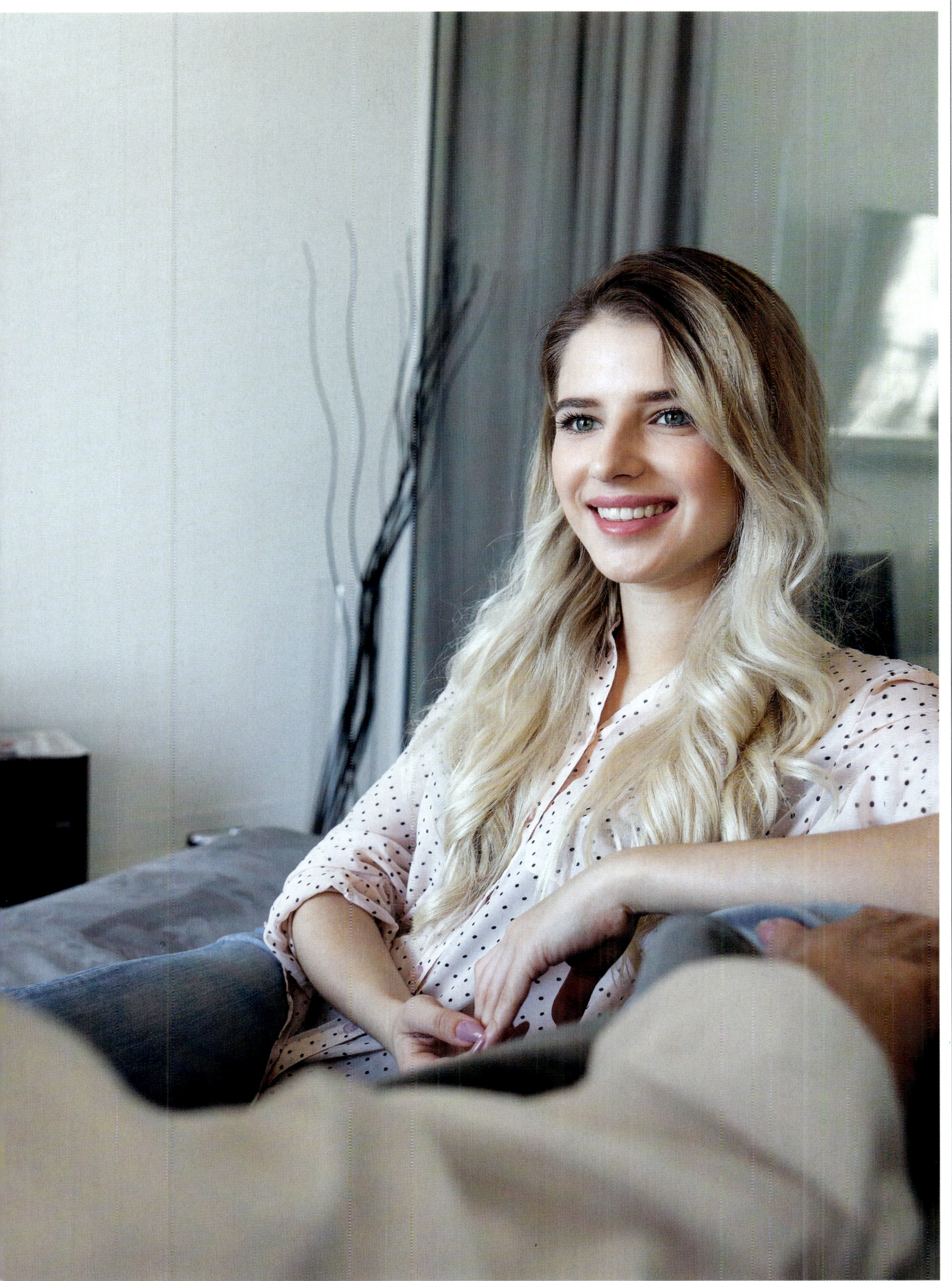

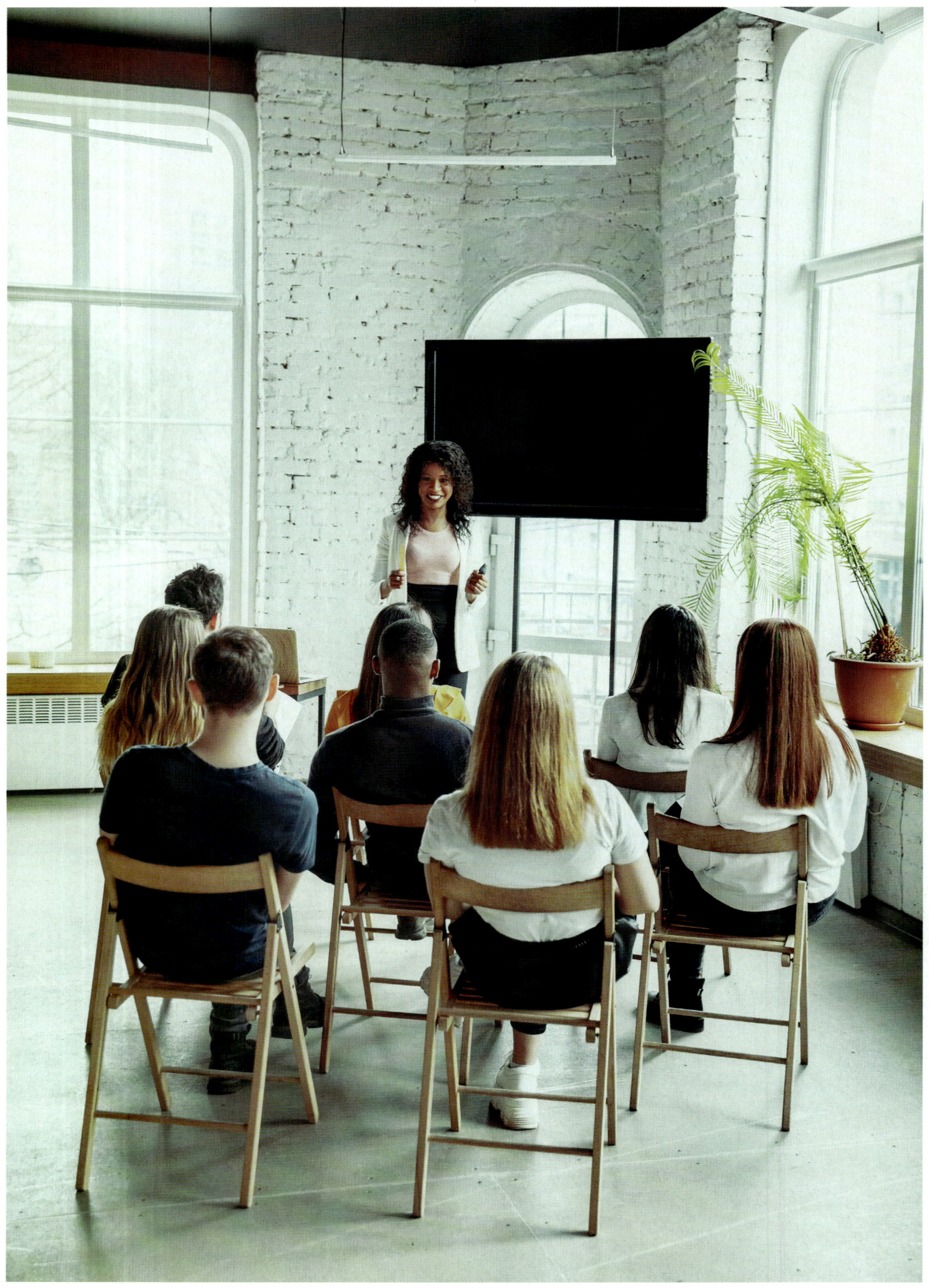

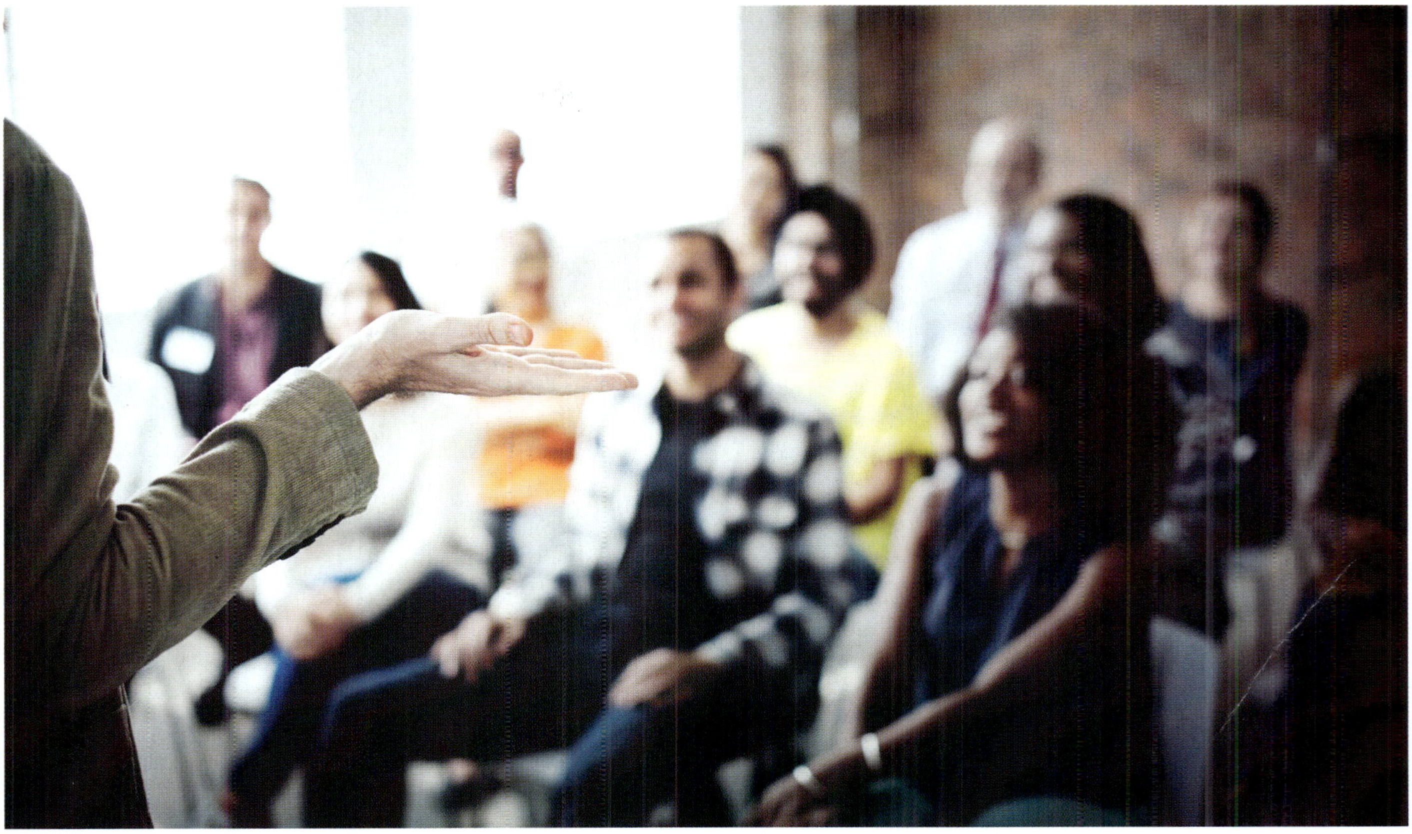

This will enable you to get the results you desire faster and effectively. My personal practice is very results based and is full of patience and honest story telling seasoned with fun and laughter, as we all learn better in a relaxed environment of encouragement and joy. Especially when often our previous experience of learning (school) can bring up painful memories of something else completely.

I am aware that you may or may not have thoughts about coaching. So I want YOU to get something for YOU from even reading this little piece or slice of input.

I work with this guaranteed method, stay with me here.
- You have some results in your life that you are not particularly enjoying. Let's call this (R).
- These results have come from some behaviours or lack of actions. We will call this (B).
- These behaviours (or no behaviour) come from a certain mood or state at any given moment in time (S).
- The state comes from your conditioning, (meaning all your learnings and beliefs that you have taken on up until now) (C).

I use this framework to trace back the route of the belief that is attached to the result, **C > S >B >R.**

The idea though is not to analyse everything with a fine-tooth comb, the benefit is to learn self-mastery of any state at any given moment in time which is a tool a client can use in between sessions because the aim is not to be dependent on the coach, but to have freedom to master experiences that previously mastered them.

Freedom is the name of the game! The ultimate result. Freedom to choose how you feel about life itself.

Looking for the why behind these behaviours is key to resolution.

Most people have three things that they think about.
- I am NOT enough.
- I am disconnected from myself or those around me.
- Something is unavailable to me.

Learning that these three things can be seen so differently, which by itself can alter the limiting belief to a much more empowered state.

As I started to write this I was very struck by how many things and situations could be altered so much more easily when we are just able to make a connection with another person at a deeper level than "I am fine" or small polite conversation.

I hope that this little introduction has enabled you to get a little curious and maybe dip a toe into new possibilities. You could just find that it's just what you have been looking for.

Remember, the first step is often the biggest, after that every step feels easier.

I wish you all the best on your continued journey, may you find what you are looking for inside. **M&B**

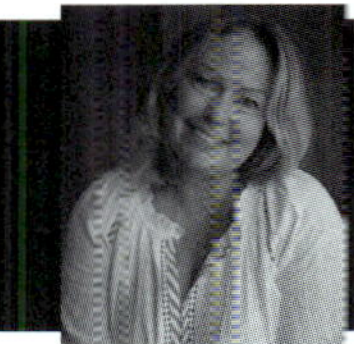

Amanda Slade is a fully qualified Master NLP practitioner, Master relationship coach and Timeline practitioner. She lives and works in Dubai, is married with three children and six grandchildren.

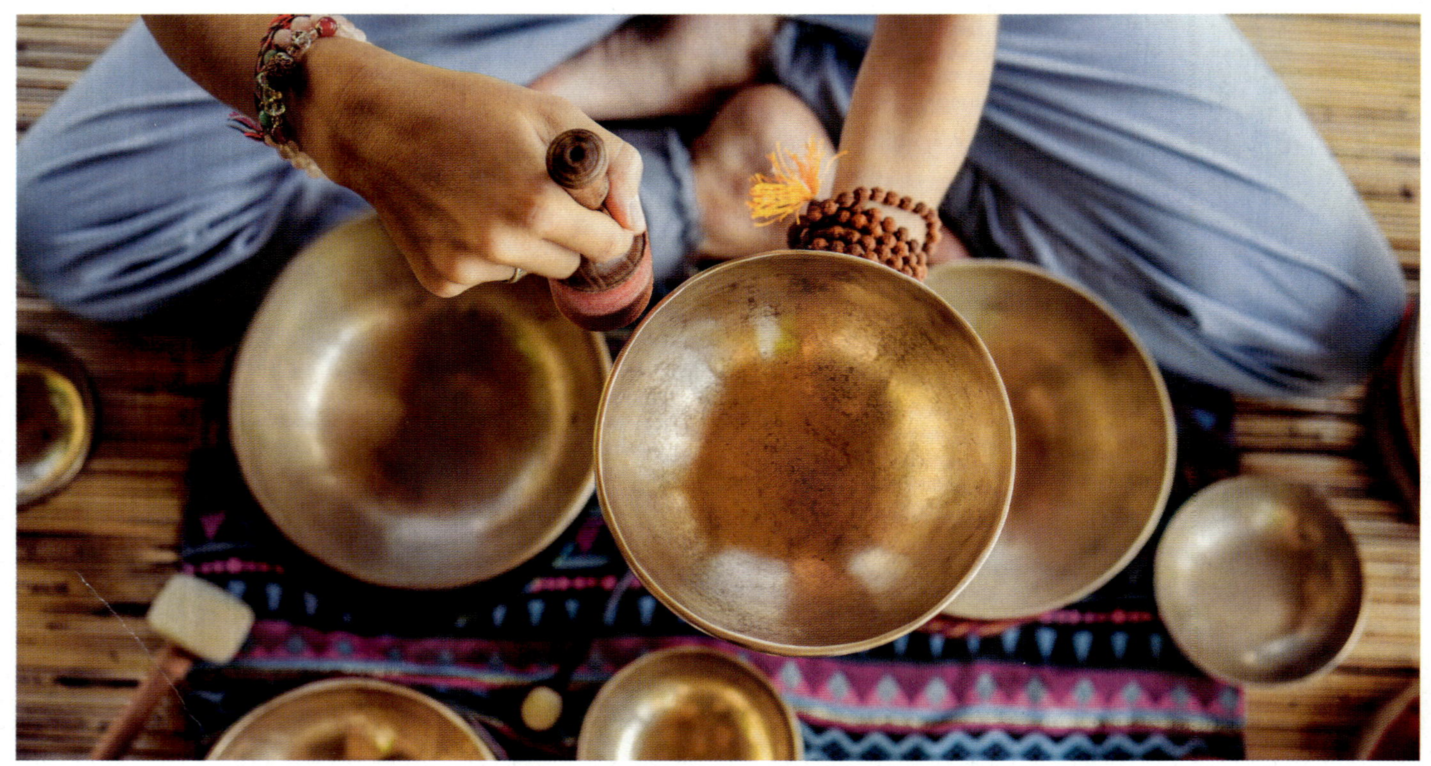

Good Vibrations

Hayley Bee explains that the body is a complex vibratory system and everything within it, our organs, bones, tissue etc, has a signature resonant frequency - the rate at which it vibrates. When we're in a state of wellness, we aptly refer to this condition as being 'of sound health'. When a part of the body is out of balance, harmony within the body is disrupted and we call this condition 'dis-ease'. Many things such as environmental pollutants, processed foods and other ingested toxins, radiation, negative thoughts and heavy emotions all disrupt our health. Sound healing, whether it's from an electronic device, acoustic instrument or the human voice, works on the basis of helping natural resonant frequencies to be restored to the physical, emotional, mental and spiritual body.

It feels appropriate to tell you a bit about my personal journey with sound, so that you can understand what's lead me to be in a position to talk about it.

I've been very sensitive to everyday sounds and deeply moved by music for as long as I can remember.

I started playing my first instrument at five years old and throughout my childhood years music was, and always will be, one of my biggest sources of joy and comfort.

One of my earliest memories is when I was around seven years old. I used to lay in front of the hi-fi with speakers either side and listen to music for what felt like hours, eyes closed, absorbing the sound. My mum loved classical music and had a huge collection of tapes featuring all of the well-known composers and their musical works of art. I had my favourites of course and when I look back, I can still so clearly remember how good they made me feel. Music held me in a bubble of safety and calming energy in an otherwise very tumultuous time.

Within the space of a year or so, my grandad passed away, which was my first experience of someone dying, we moved house, I started at a new school midway through the academic year and then shortly after that, my parents separated. Without realising it, 'd found one of the healthiest ways possible to self-soothe, with sound.

Travelling the world in my 20s and 30s, I was fascinated by the native music of each country as t seemed to communicate so much about a place and the heart of its people. From the Spanish guitar to the panpipes of Peru and Australian didgeridoo, to name just a few, these soulful sounds still never fail to give me goosebumps.

In the last few decades, science has proven beyond all doubt, that music has a measurable impact on our mental, emotional and physical health.

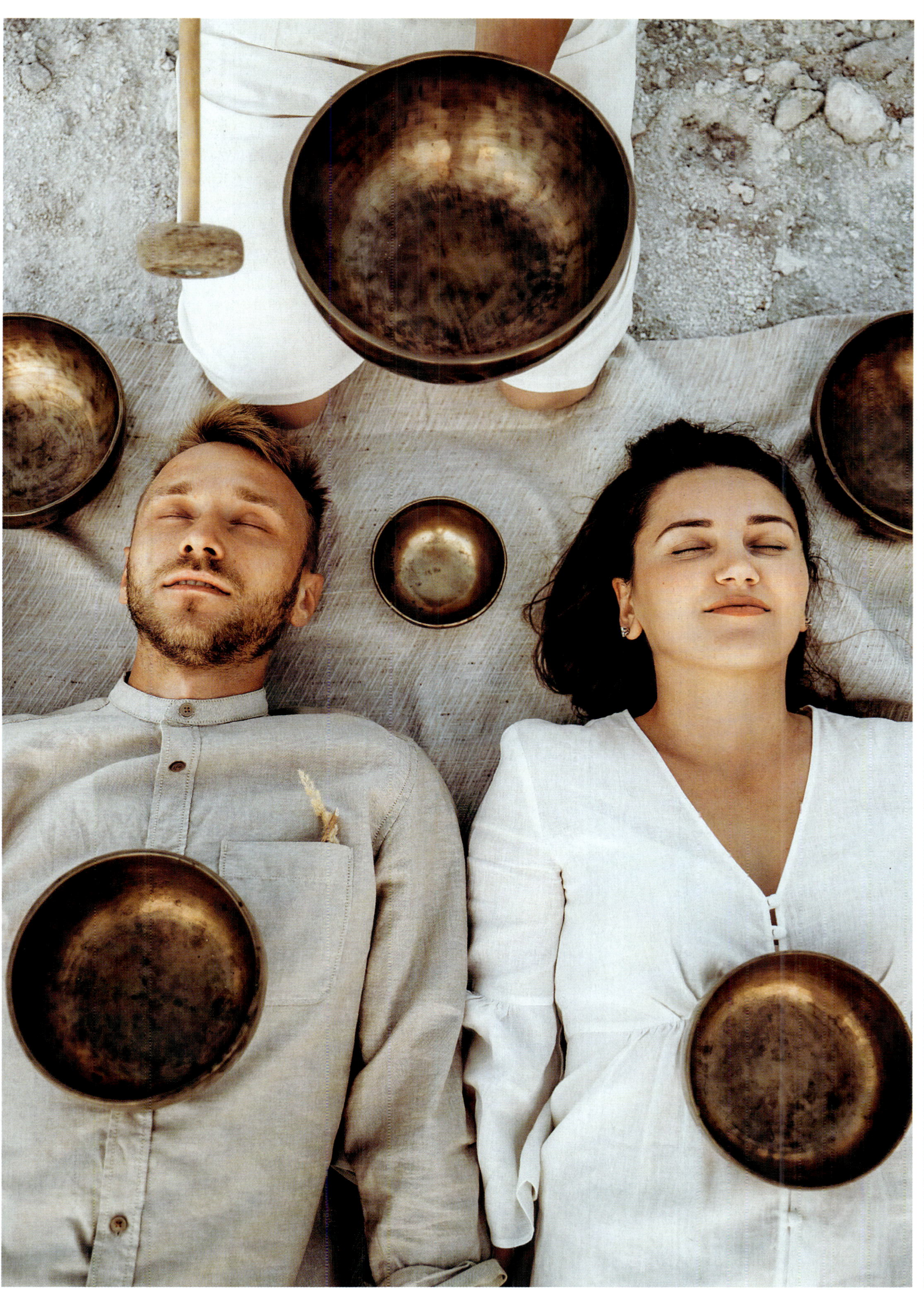

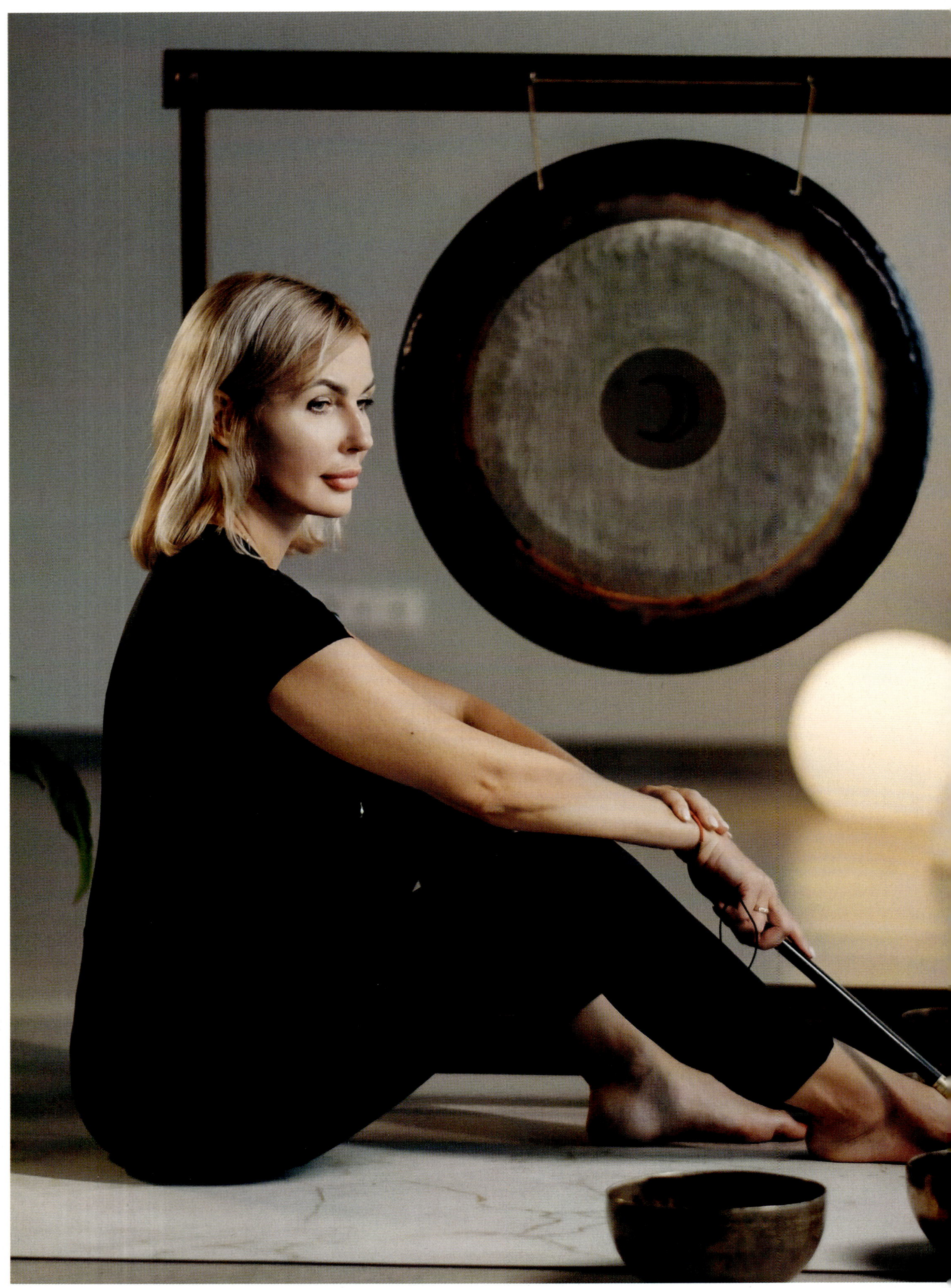

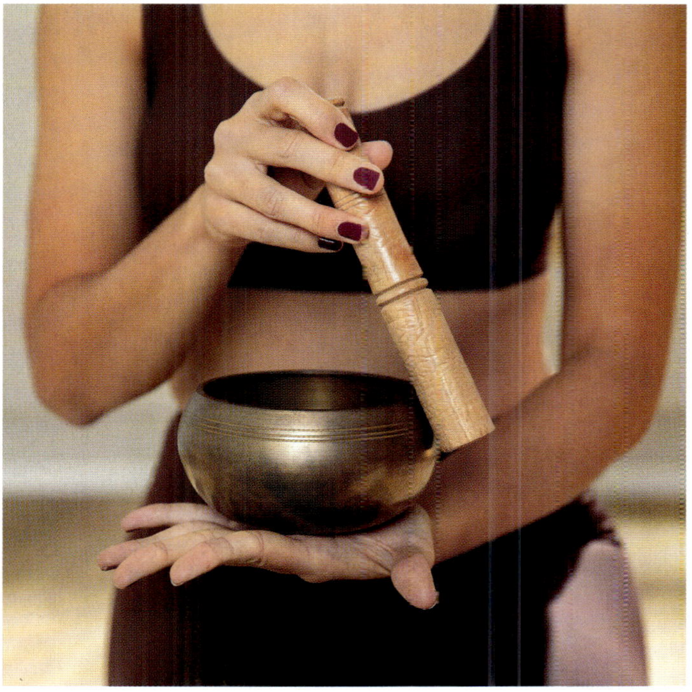

This knowledge spurred me on to dive even deeper into the world of sound and learn how to use it as a healing tool for myself, particularly in times of personal crisis.

In 2021, when I was home alone with Covid for a couple of weeks, I had lots of time on my hands and I listened to music almost every waking hour of the day. During this time I found myself assessing life and asking myself some deep existential questions. It was here that I felt a calling.

I signed up to do a course when I was well again and after 10 months of studying and practical training, became a certified Sound Healing practitioner.

Now I can help not only myself, but others too.

The science

Although sound healing has grown in popularity in recent years and to many still seems a bit 'new-agey', it's actually been around for a very long time.

Ancient civilisations have left evidence to suggest that vocal chanting and shamanic drumming were used in healing ceremonies and to achieve higher states of consciousness. Tribes in certain parts of the world still practise this in fact.

It's important to make clear at this point that the word 'healing' doesn't imply receiving a 'cure' for your ailments. Its power is in being able to induce a deeply relaxed state, which in turn stimulates the body's own innate healing abilities.

Music has a measurable impact on our mental, emotional and physical health.

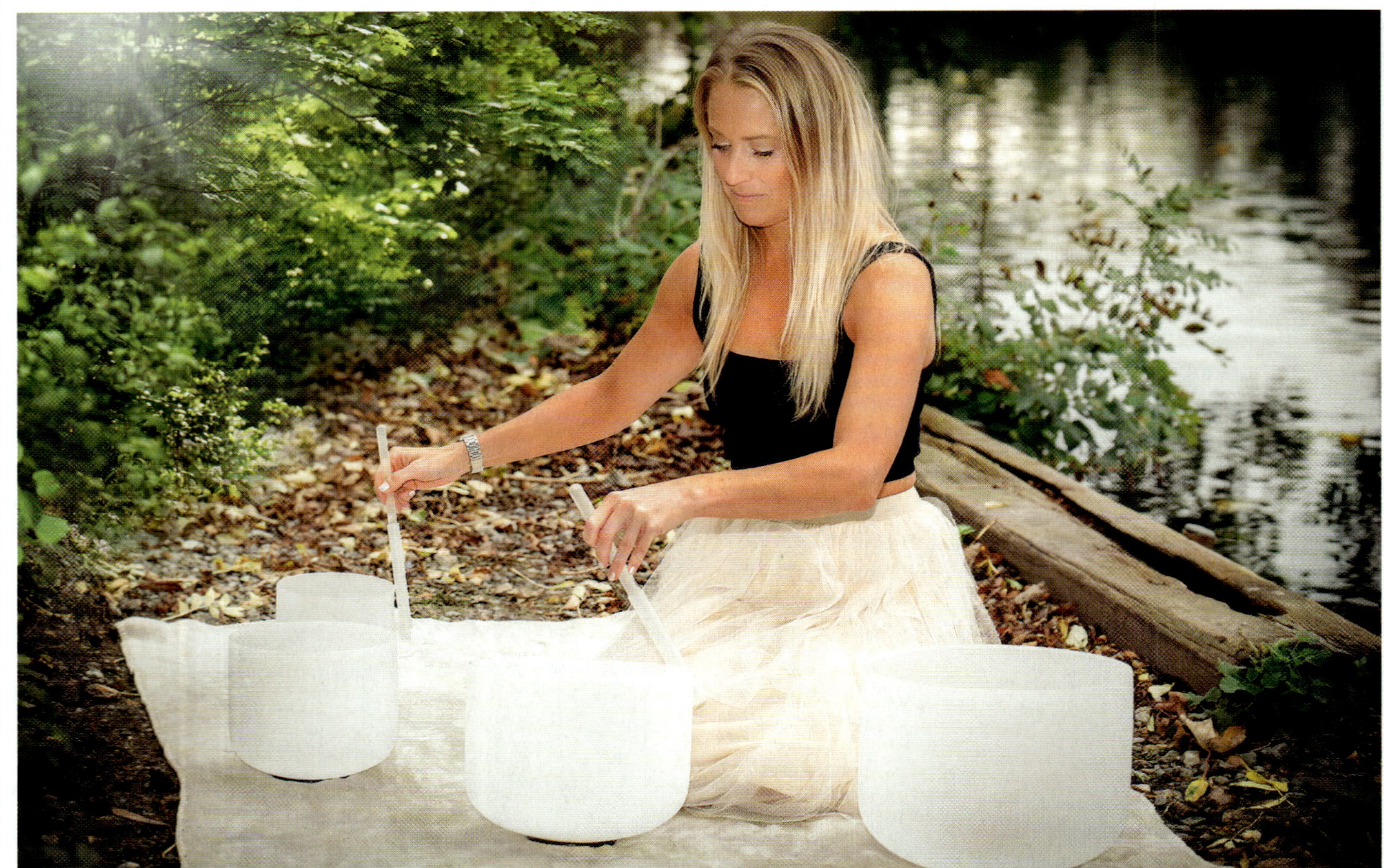

The power of sound as a tool for 'healing' is gaining momentum in the medical world which is exciting to see. A few examples include its use to break down kidney stones, calm the tics of Tourette's Syndrome and to stimulate movement in paralysed Parkinson's patients. Drumming has shown measurable improvement in brain and social function for those with Autism and ADHD and a new study at America's prestigious MIT that carried out studies on a small cohort of people with Alzeimer's displayed.

So what does a sound bath involve?
You'll typically lie down on a yoga mat and cosy up with a blanket, cushions and an eye pillow, whilst wearing your comfiest of clothes.

Other multi-sensory elements are often incorporated as well, such as ambient lighting and and the fragrance of essential oils/incense that further encourage the mind and body to relax.

Then you simply close your eyes and listen, as a variety of instruments are played that "bathe" you in soothing sound vibration and lull you into a meditative state.

We're hearing a lot these days how meditation is good for us, but the reality is that for most people, quietening the mind can be a real challenge. Using sound as a tool however, makes this ultimate well-being practise much more attainable.

Various types of singing bowls, gongs and chimes are the instruments that practitioners use most often in sound baths.

A variety of instruments are played that "bathe" you in soothing sound vibration.

Some practitioners use just the singing bowls, or a selection of gongs. Each have their merits and I personally enjoy offering a wide range of sounds that takes people on something of an acoustic journey.

I hold private sound baths for one or two people at home. These are really well-suited to first-timers or those who are currently experiencing or wanting to deal with some past, emotional trauma. Sound is very effective in bringing stuck emotions to the surface in order to be released. For those in the process of doing some deep, inner work, sound is a powerful tool for healing the heart and restoring balance to the emotional body.

I also host events for larger groups and enjoy collaborating with others who bring their own unique energy and skills, such as those who can play the didgeridoo, breathwork, or can offer Reiki healing, for example.

As a lover of music, as much as sound, and in understanding its powerful therapeutic effects, I always include one piece of carefully selected music at the end of my events as well, which often features the voice, as this is one of the most beautiful instruments of all in my opinion. Plus, it's one we all have and can relate to.

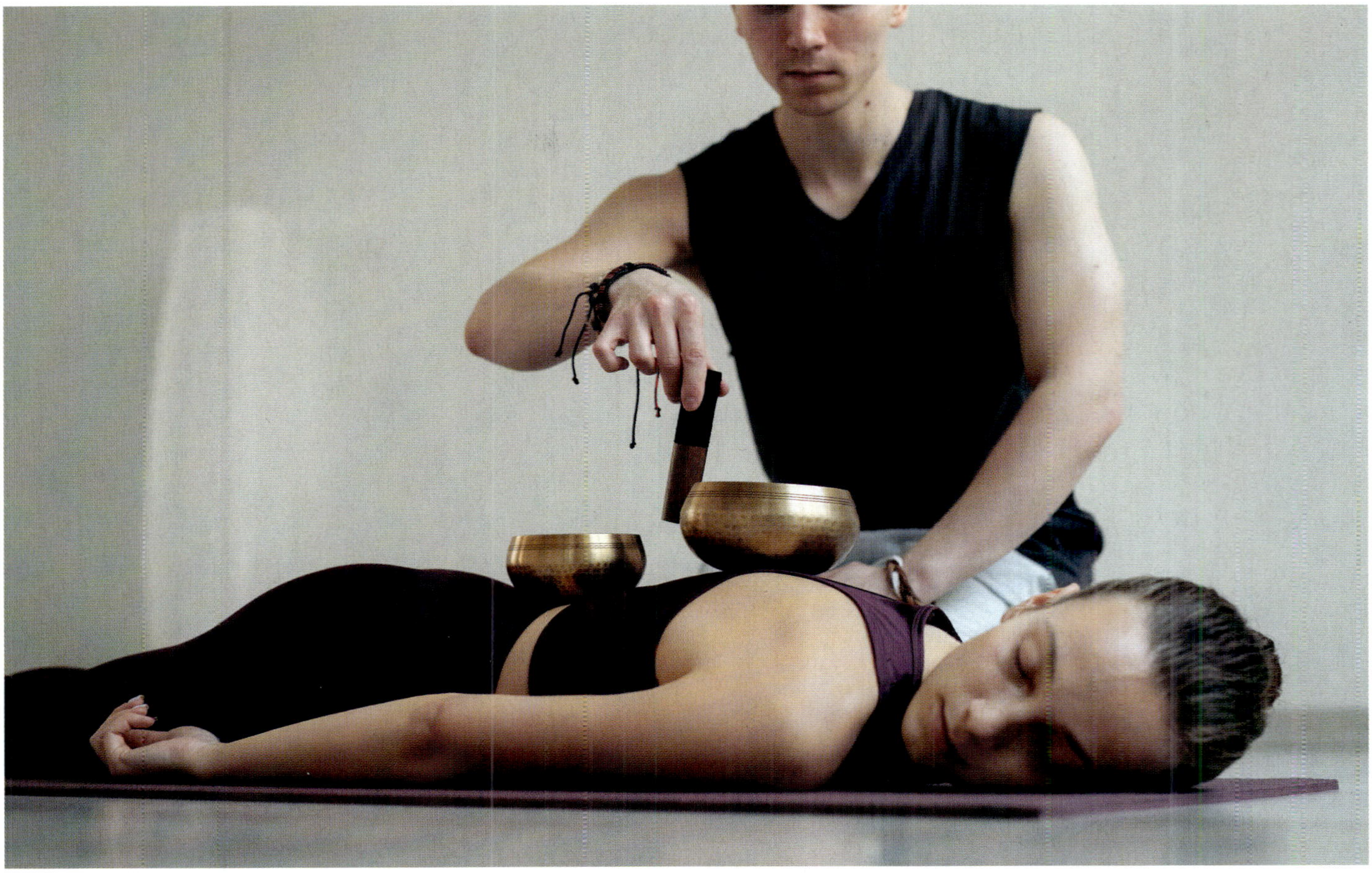

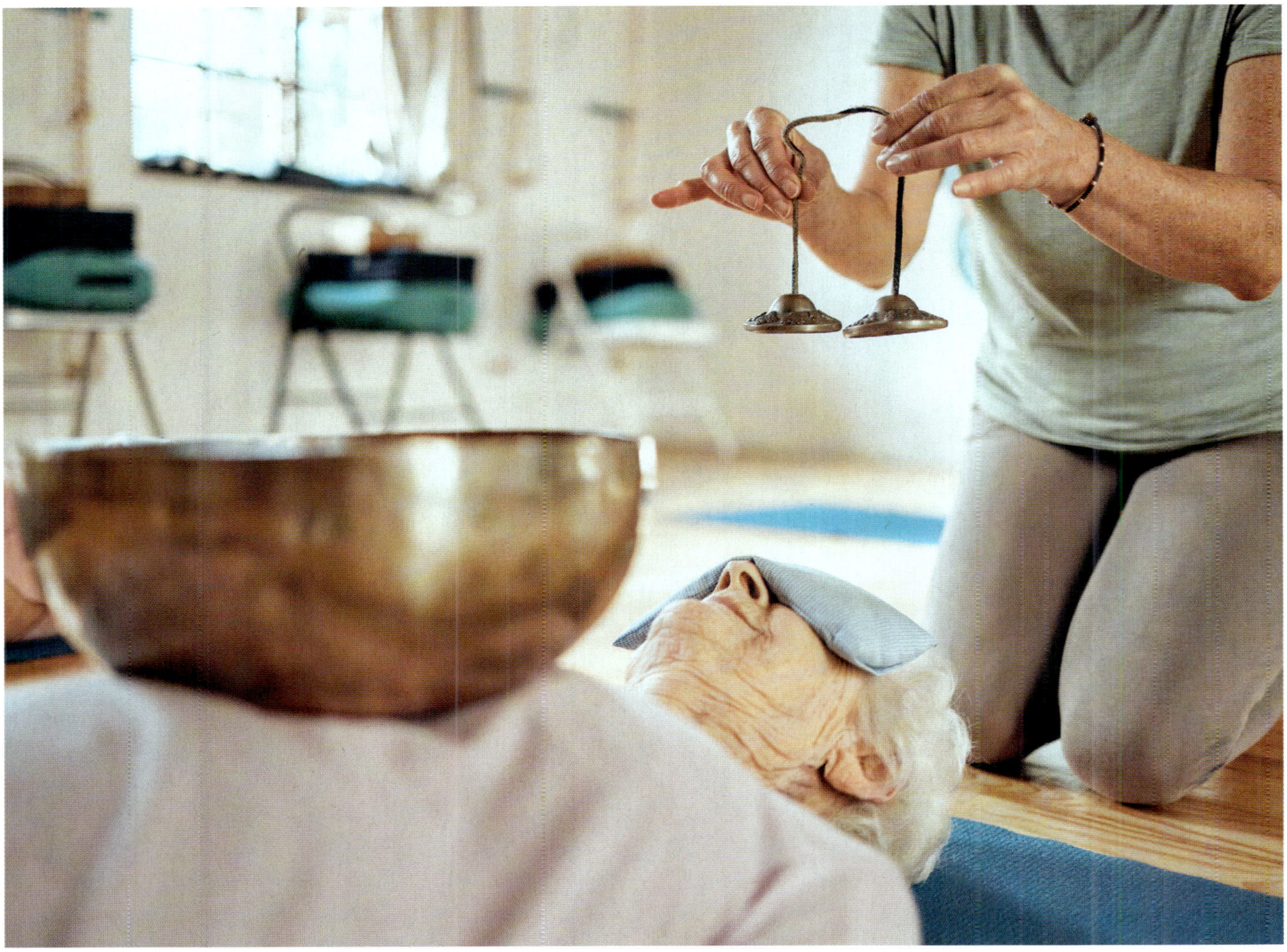

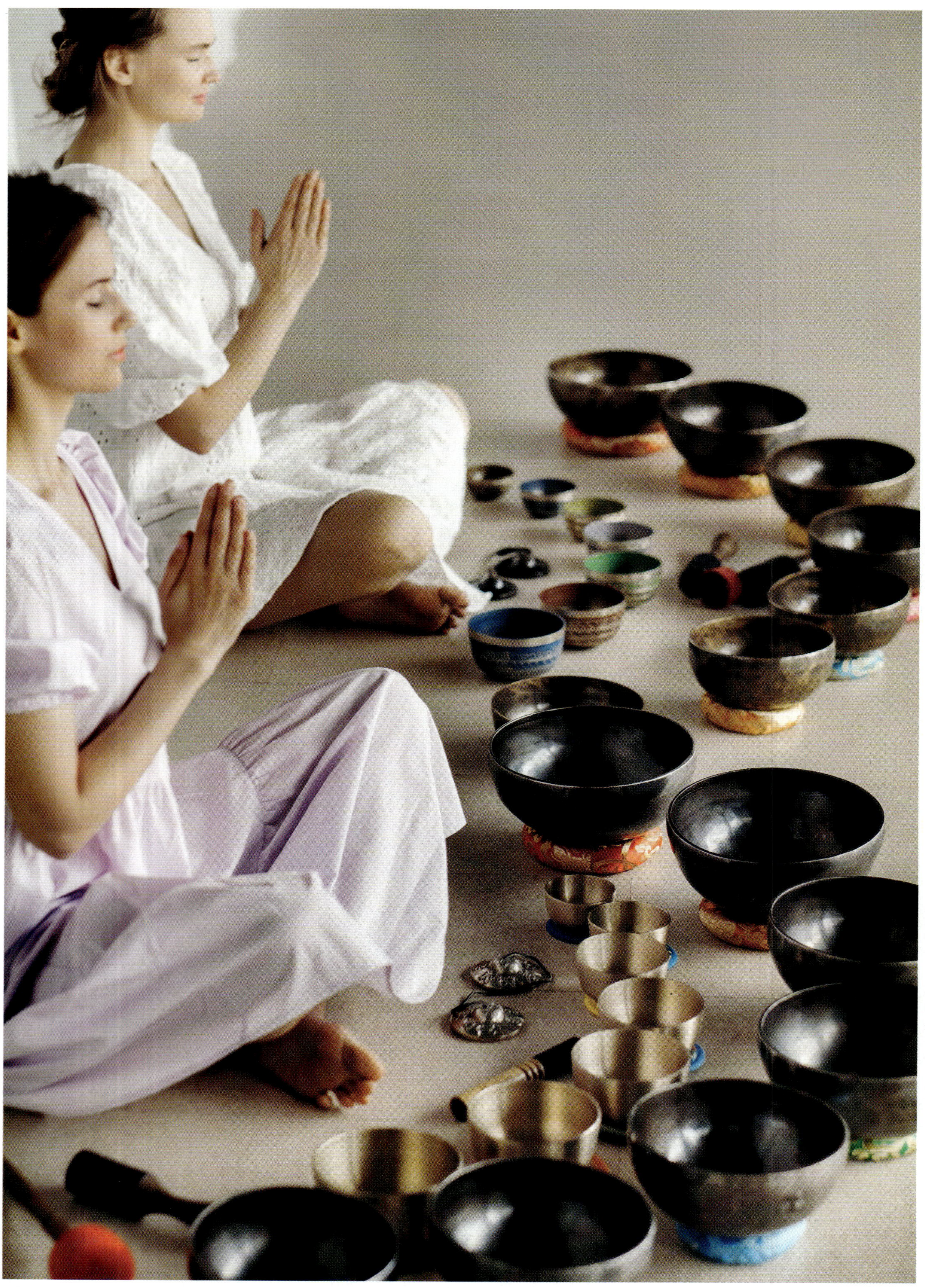

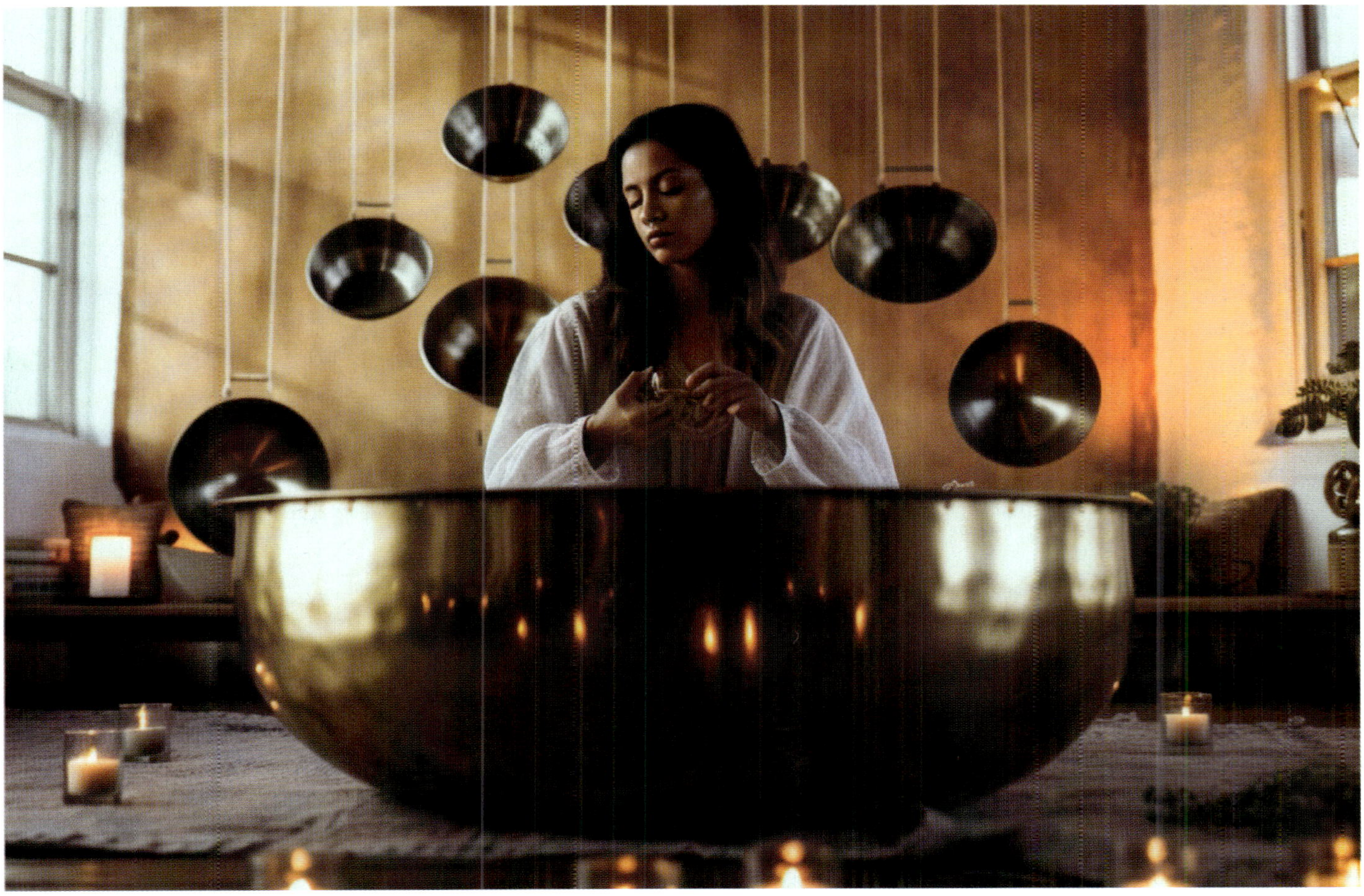

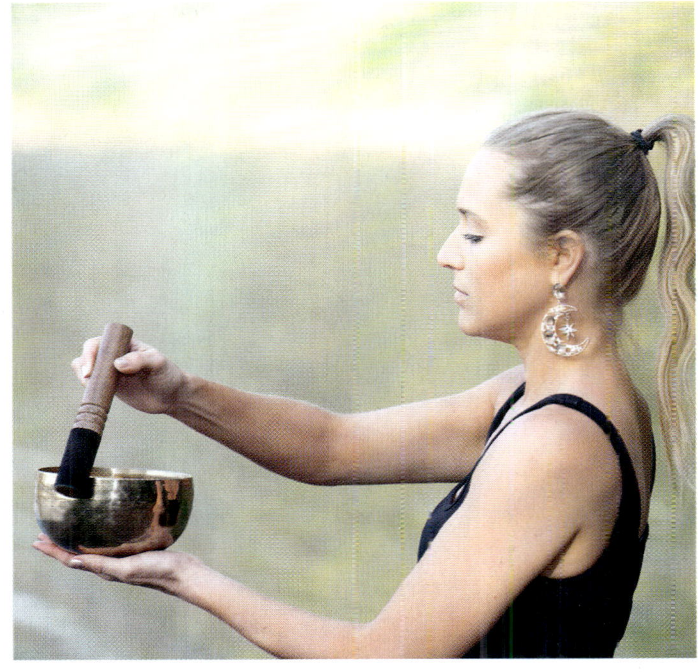

Sound baths typically last an hour. That provides an ample amount of time for people to reach a deeply relaxed state and be in healing-mode, before getting too uncomfortable.

One of these events includes the luxury option of laying in a silk hammock instead of on the floor, which takes the experience to a whole new level of enjoyment, as there's no pressure on the body. Cocooned in your own private, womb-like space, the ability to rock gently back and forth releases endorphins in the brain, one of our natural feel good chemicals.

Sound baths can be shorter and still effective, such as a 20-30 minutes after a yoga/Pilates class. Essentially, any time with sound, that allows our brainwaves to slow down, inducing a deeply relaxed state is extremely worthwhile.

The after-effects include and extend way beyond:
• improved quality of sleep
• reduced anxiety
• alleviation of physical and/or emotional discomfort
• clearer thinking
• a deep sense of peace
• feeling much more in balance

Best practice
There are few people that sound baths aren't recommended for, which include those who:
• are in their first trimester of pregnancy
• have any type of brain injury
• have epilepsy
• have severe heart conditions
• have any bio-medical devices such as a pacemaker, metal plates or cochlear implants
• have any severe, clinical mental health conditions *M&B*

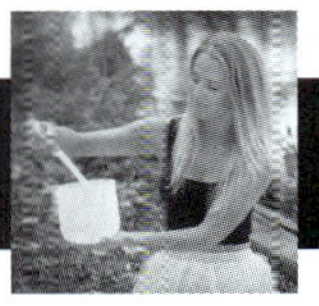

Hayley Bee
hayley@heartfull.soundhealing

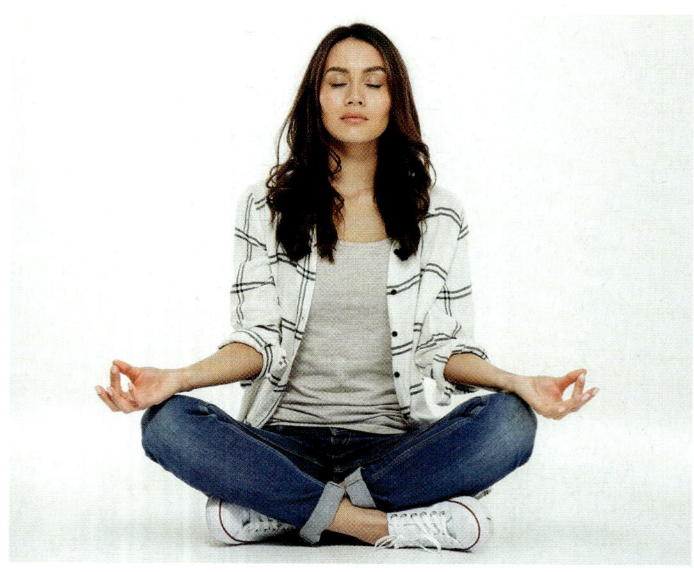

Calming Clarity

Meditation evokes a state of greater calmness, physical relaxation and physiological balance, a moment to become mindful of thoughts, sensations and feelings.

Meditation, from the outside, can be a pretty daunting topic. I think for many of us it draws visuals of sitting on top of a mountain, cross legged, eyes closed and looking peaceful. I think this is the first misconception to squash. Anyone can meditate and you can meditate anywhere and anytime of day. Meditation is a craft and becomes easier with practice like everything. The intention is to find a pocket of your day to be still and completely present in the moment. Now when we start its normal to be thinking about work, children, whats for dinner. The knack is to let these thoughts pass, float by, don't ignore them just notice them, accept them and be.

History of meditation

The word meditation comes from the word meditatum a Latin term "to ponder" experts know that the practice of meditation has been practiced for thousands of years. Widely used in many denominations of faith and cultures. Meditations can be described as mindfulness, communion, a trance state, chanting, contemplation and many more. Or, to simplify, to become familiar with one self.

In the 70s Dr Jon Kabat -zinn opened a centre of mindfulness at the university of Massachusetts medical

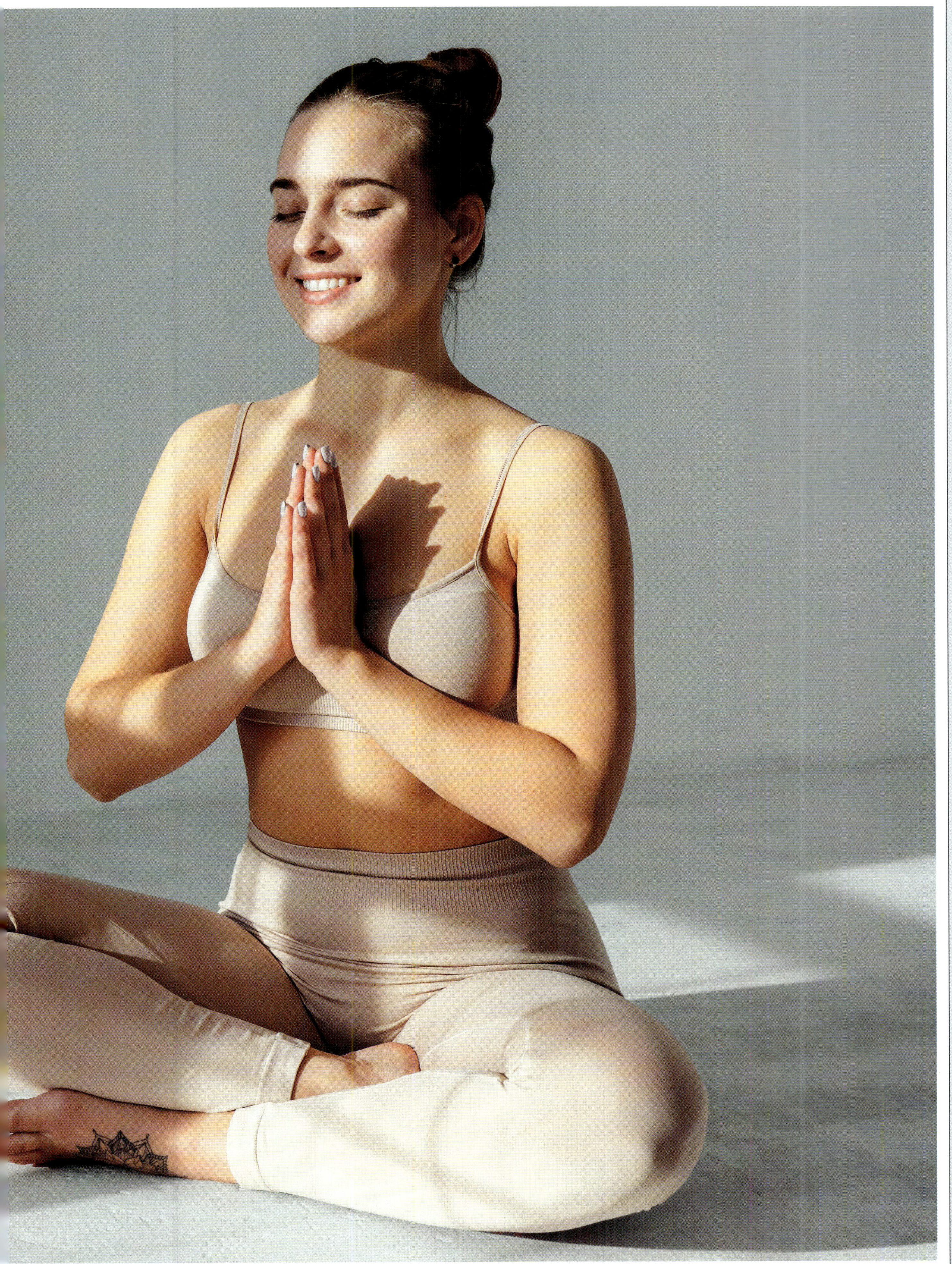

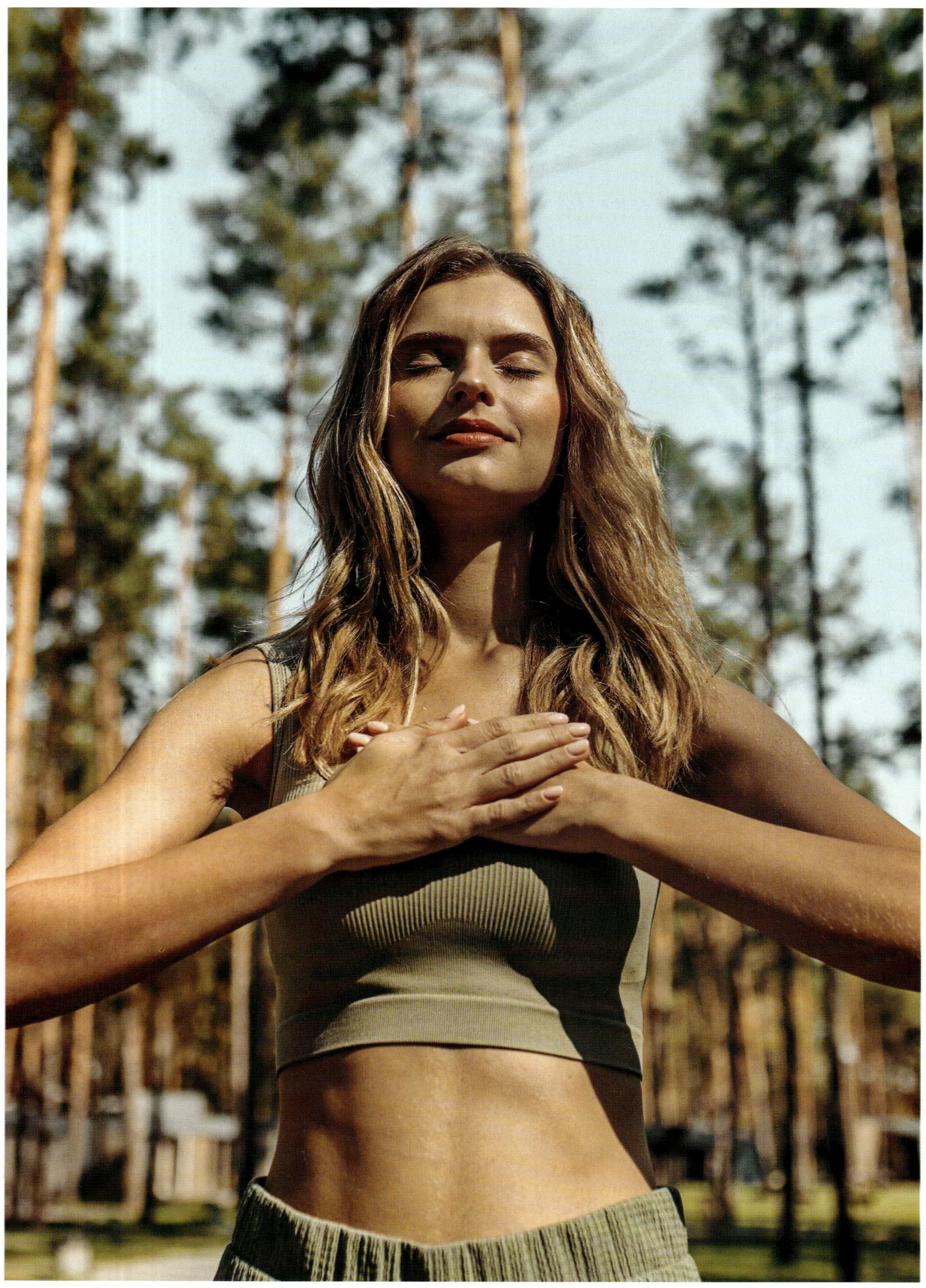

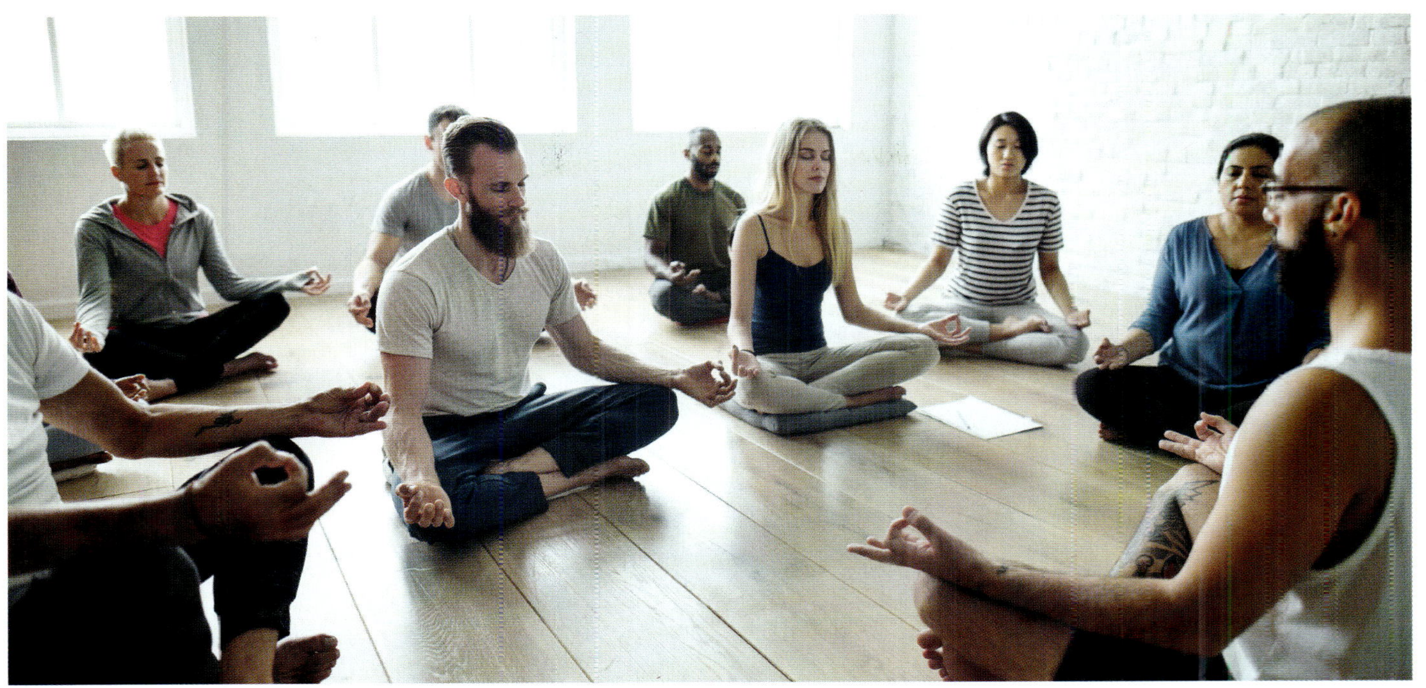

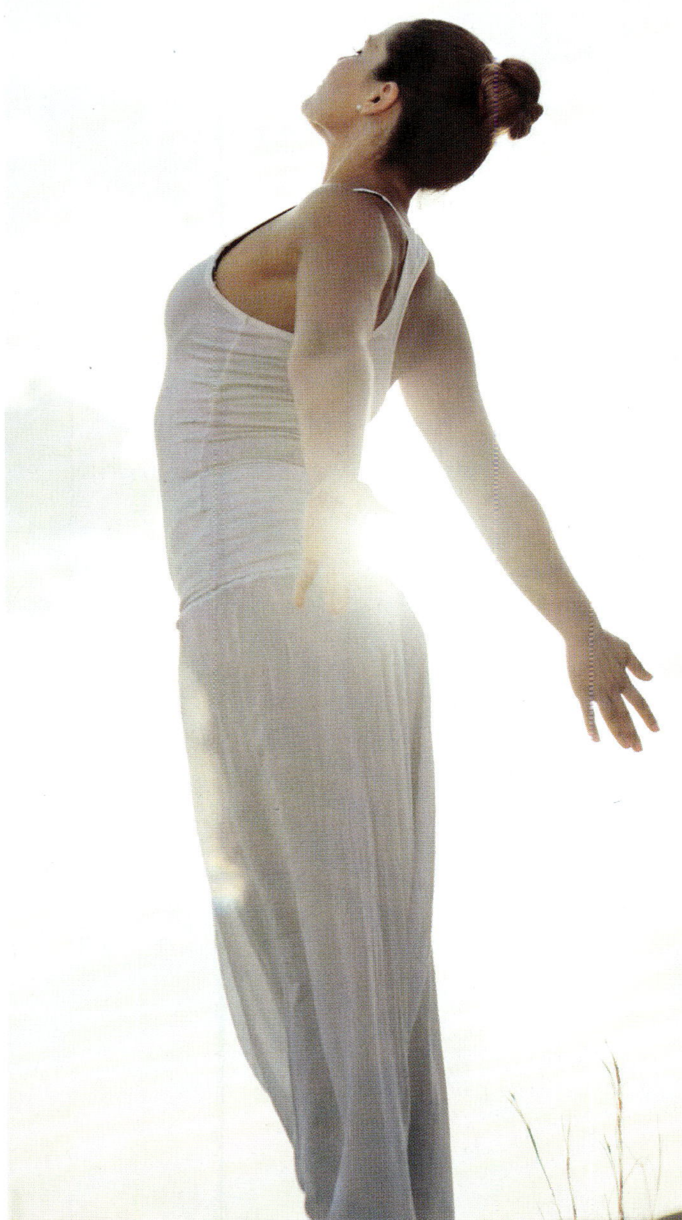

school and focused on the science behind meditation, moving away from the religious connotation and teaching men and woman that the benefits of meditation aside from clarity and peace that it had other huge physical and mental benefits not linked to any religious practice.

The science

Practicing meditation is really the art of being fully present in that moment but there are so many health benefits with long lasting effects. Meditation is proven to sharpen the mind and help you have the ability to focus on tasks and problem solve with a clearer mind. Practicing meditation has proven to reduce day-to-day stress with fewer effects to the body. People who meditate consistently are shown to have healthier relationships, better sleep and good food habits.

The positive impact on people practicing meditation hugely reduces their anxiety and depression. Studies have also shown that those who meditate increase their telomeres (cell length) which in turn can boost immunity and help slow the effect of the ageing in the brain. Meditation is proven to reduce inflammation in the body, by allowing yourself to 'let go' of external stress and expectations.

How do I meditate

The first thing to remember when meditating is, what works for you? You can mediate on your own, in a class setting, you can follow guided meditations, the list is endless. Similarly, you can mediatate at end the of day, there really is no set rule.

Making meditation a daily habit, like anything, takes practice.

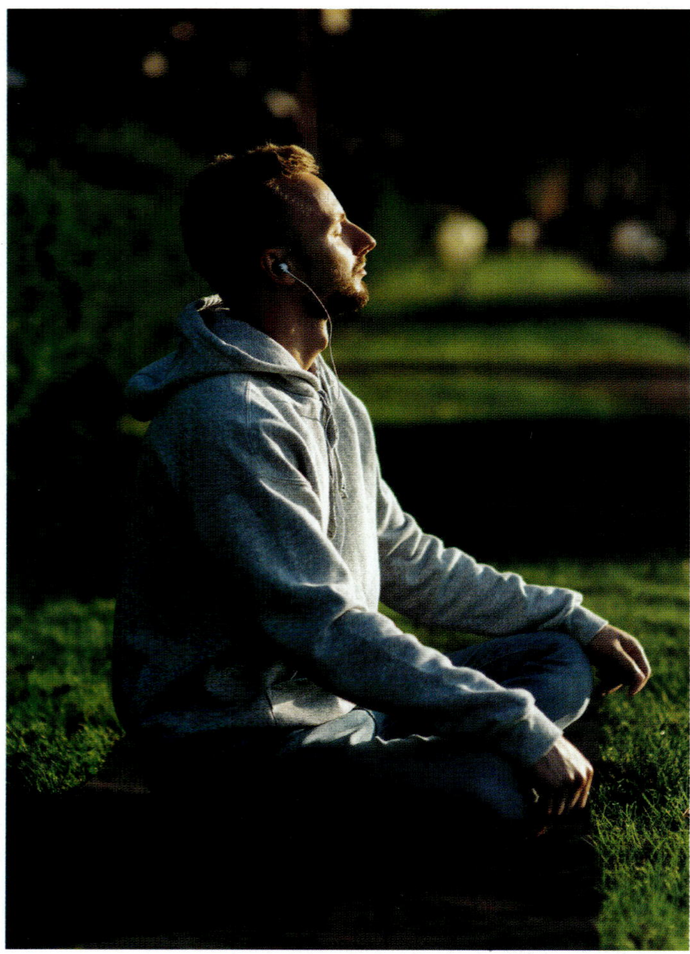

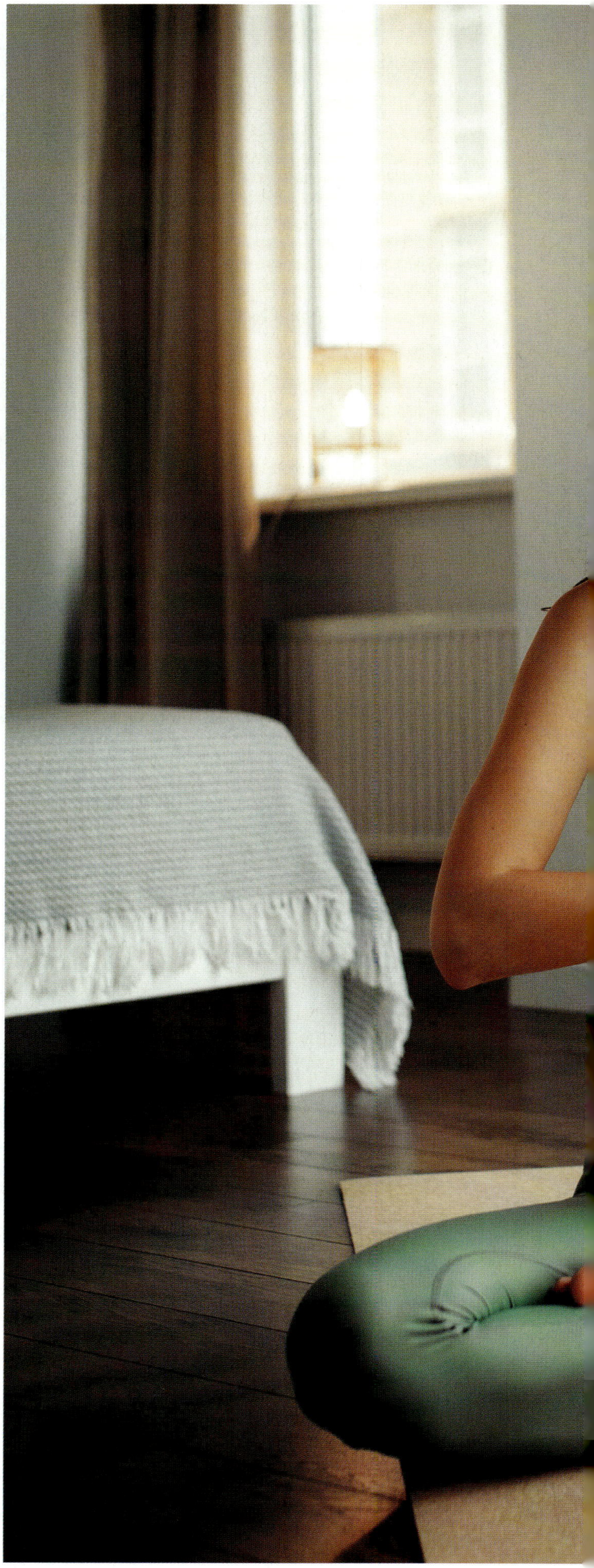

Lets get started you started

1. Let go of all expectations

2. Sit down. This can be on a chair or on the floor

3. Set a time limit (five minutes max to start with)

4. Close your eyes

5. Become aware of your body

6. Notice your breath

7. If your mind wanders bring your awareness back to your breathing

8. Notice your thoughts, allowing them to pass like clouds.

9. Take one long extra deep breath

10. That's it. Start to notice how now you feel ?

Making the exercise above a daily habit, like anything, takes practice. Eventually, overtime, it will become easier as you see and feel the benefits.

When starting a new habit, the key is to really make an informed decision, focusing on the benefits and making a commitment to yourself. Inevitably we sometimes, for all sorts of reasons, don't always live up to our own expectations. So, be easy with yourself if you miss a day or two. Recommit to starting again and reward your progress. Forget about perfection and rename it as progress. This will enable you to feel

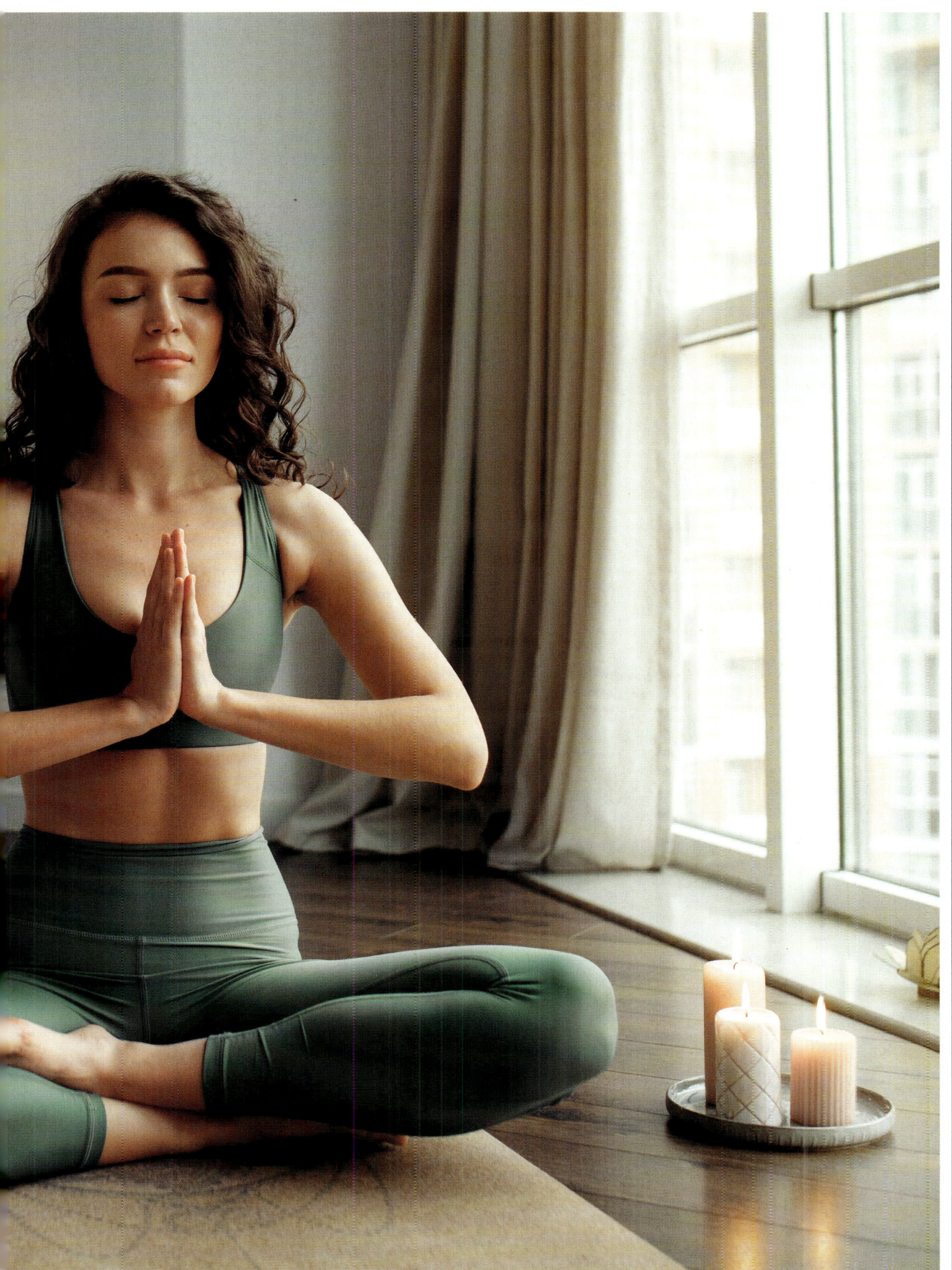

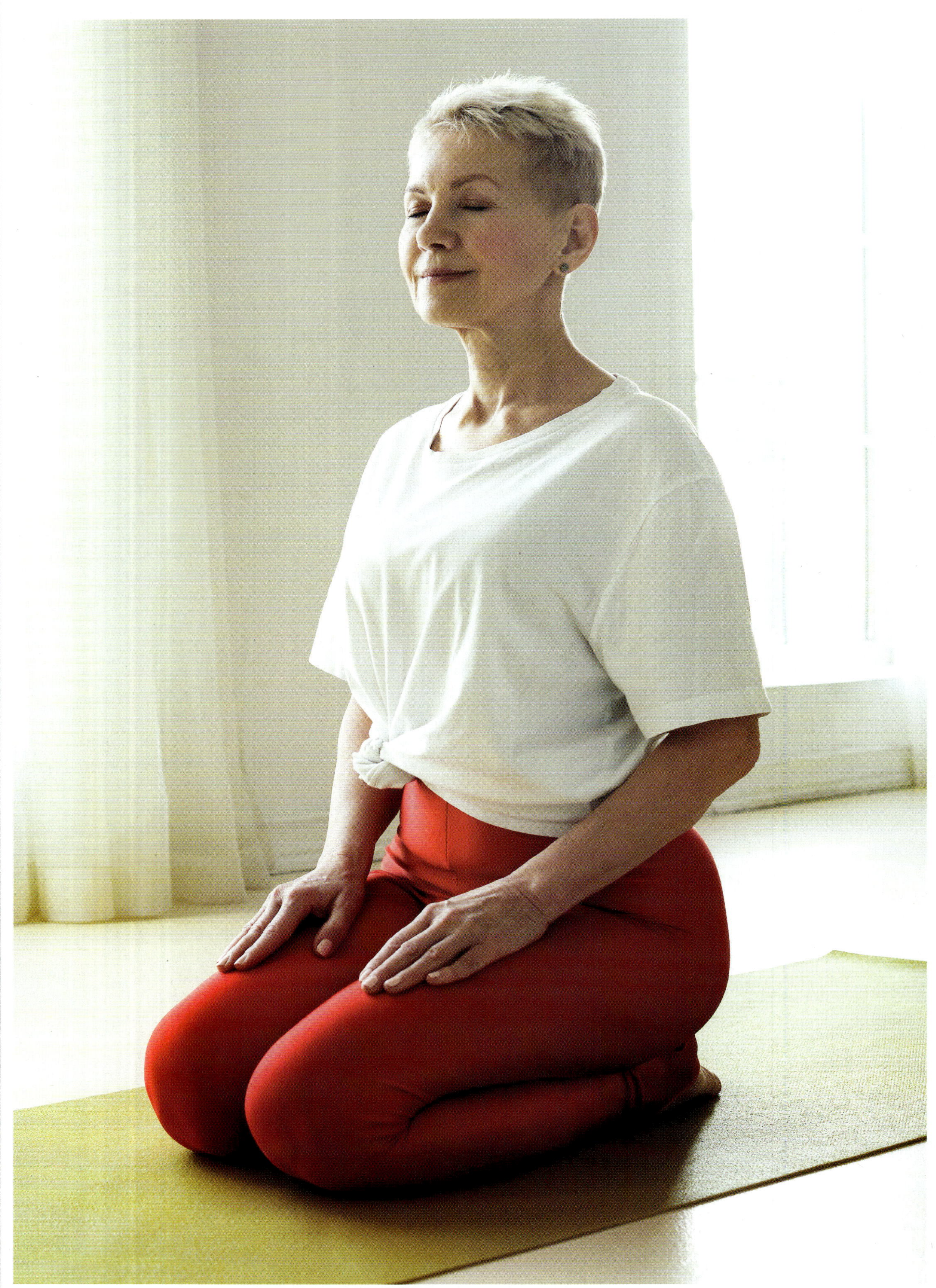

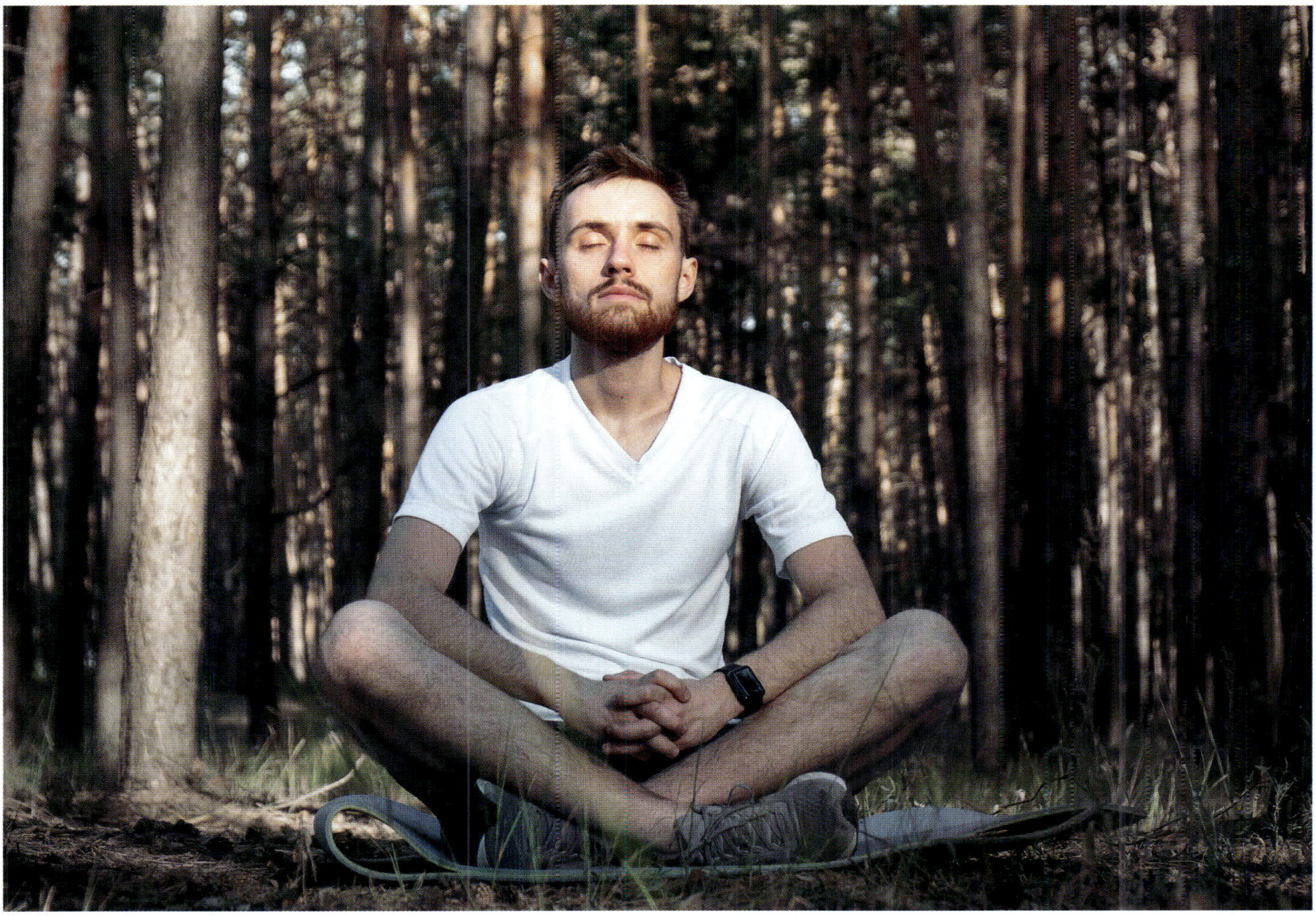

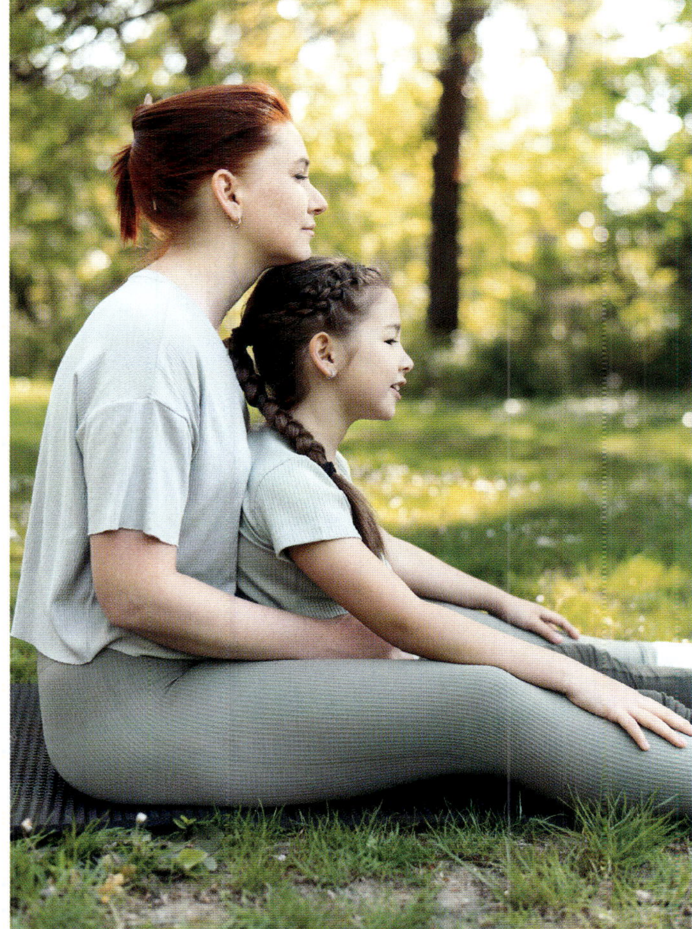

Remember, you are a 'human being' rather than a 'human doing'.

better faster. Looking for online guidance or investing in an app based programme to assist you makes bringing meditation into your life a lot easier.

Think of every five minutes you choose to meditate as a little check in with how you are being today. Remember, you are a 'human being' rather than a 'human doing'. No wonder many of us are feeling tired and stressed when we are always doing.

This practice is being used by many business leaders and CEO's due to the proven results in higher performance and expertise. As a side note for those of you who have children this is also a lovely time to reconnect leaving language aside for a while just connecting in stillness and silence. If only we had learnt some of these things as children...

Let's all make a commitment to do what we can when we can to uplevel our experience of this wonderful gift of life. There are zero side effects to mediation only benefits. Take a leap into something new that is available to all of us. **M&B**

SUBSCRIBE

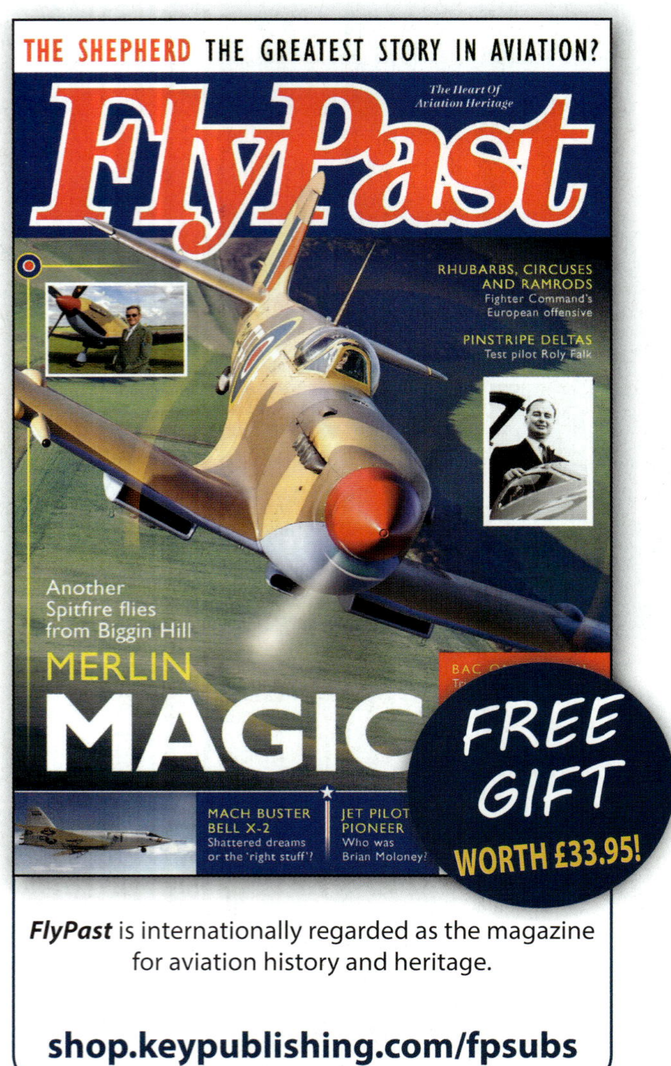

FlyPast is internationally regarded as the magazine for aviation history and heritage.

shop.keypublishing.com/fpsubs

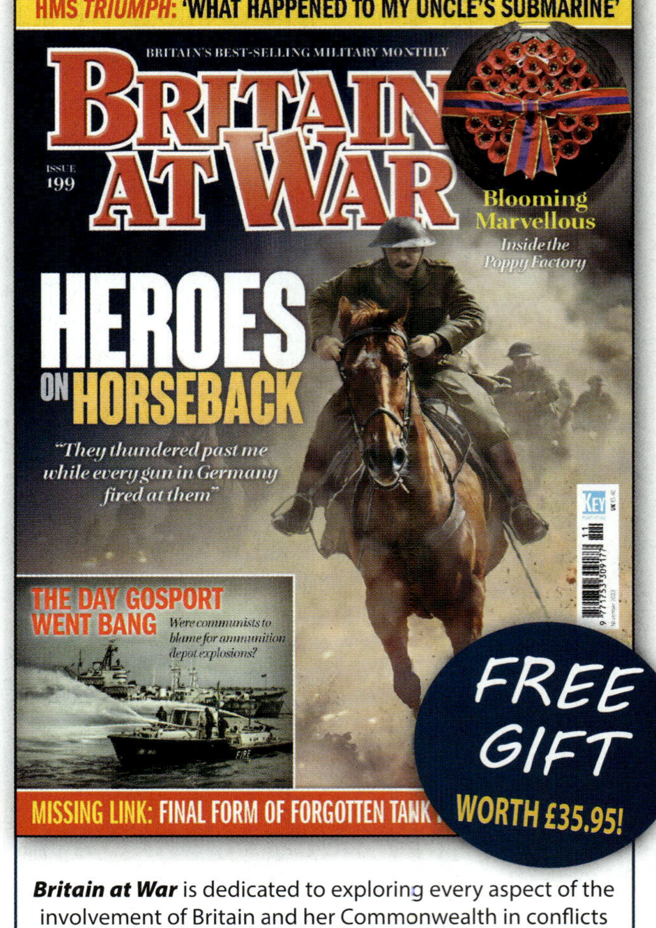

Britain at War is dedicated to exploring every aspect of the involvement of Britain and her Commonwealth in conflicts from the turn of the 20th century through to the present day.

shop.keypublishing.com/bawsubs

ORDER DIRECT FROM OUR SHOP...
shop.keypublishi

OR CALL +44 (0)1780 480404
(Lines open 9.00-5.30, Monday-Friday GMT)

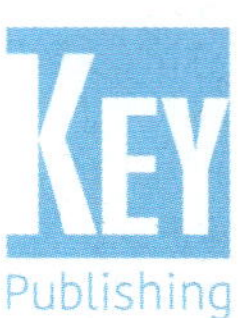

799/23